Agricultural Research at the Crossroads

Revisited Resource-poor Farmers and the Millennium Development Goals

Bo M I Bengtsson
Prof. Emeritus, Crop Production Science
Swedish University of Agricultural Sciences
Hoor, Sweden

Science Publishers

Enfield (NH)   Jersey   Plymouth

CIP Data will be provided on request.

SCIENCE PUBLISHERS
An Imprint of Edenbridge Ltd., British Isles
Post Office Box 699
Enfield, New Hampshire 03748
United States of America

Website: *http://www.scipub.net*

sales@scipub.net (marketing department)
editor@scipub.net (editorial department)
info@scipub.net (for all other enquiries)

© 2007, Copyright Reserved

ISBN 978-1-57808-514-9

Published by Science Publishers, NH, USA
An Imprint of Edenbridge Ltd.
Printed in India

Preface

Globally, current discussions of the Millennium Development Goals and the United Nations reform process aim for real progress by 2015, and beyond, with a reduction of poverty and malnutrition for millions of people. This is a necessary target. Considering the political rhetoric in light of past achievements in agricultural research and development, I still find reasons for concern. Over four decades, I have been involved in a variety of activities of agricultural research and development with an international focus, most recently as a professor in Crop Production Science at the Swedish University of Agricultural Sciences. I have also gained experience of development assistance to farmers in the field and from my responsibility for Swedish support to research in developing countries and international agricultural research institutions, serving as a member of international boards and attending numerous international conferences and meetings. Another point of departure is my childhood and youth on a small tenant farm in a parish of southern Sweden, a rural setting with a high degree of tenancy. All these experiences form an interesting background to actual developments in the field for resource-poor farmers in my own country, in Ethiopia and in Trinidad and Tobago. By revisiting the same farmers at several locations in these countries, in total more than 200 farmers, I have been fortunate to gain first-hand knowledge about their progress and technical changes in agriculture in the 1960s, 1980 and the early 2000s. This set of empirical data offers an interesting platform for policy analysis of agricultural research and development in the context of globalization and of ways in which global threats and challenges may affect future policy for actions to actually reach the poor. Finally, this leads to a discussion of research policy on future agriculture and land use.

It may be easier to concentrate on the global features and trends or to review or monitor effects of specific development projects over a few years. We seldom apply a long-term outlook before making the final assessment whether or not "development" has been positive and reached the most needy. Certainly, it is more difficult to conduct studies over a longer time and combine field data with a global perspective. Still, there is a need for combining empirical field data over longer time periods with proposals and ideas from political agreements in order to see what

development may look like for resource-poor farmers in reality. Initial positive changes, as part of a conventional concept of development, may sometimes turn into declining trends, if the time perspective is long enough. It is necessary to integrate field data relevant to policy with a global overview with up-to-date information for synthesis into scenarios and a vision of how future research and development can best help those who are most needy and have little access to productive resources. This is also relevant for science aid, in particular if it is to be increased over the next few years. The overall task is a huge challenge for policy-makers and the agricultural research establishment. It is also of concern in teaching agricultural students to be able to respond to future challenges. This publication is such an attempt to stimulate discussion on future options of research policy, suggesting changes of agricultural R&D for societal development in accordance with the Millennium Development Goals.

I owe a great deal of gratitude to many persons, above all to the farmers visited in the three countries over the long term of the empirical studies. Many others have generously assisted me in various ways during field visits, in stimulating discussions and by reviewing draft texts. Although space will not allow me to mention all of them, I wish to recognize a few key persons at critical stages in this exercise, such as Professor Nazeer Ahmad, Dr Richard Brathwaite and Dr Aldwyn Tangkai of the University of the West Indies; Ato Mulugeta Amha, Ato Haile Lul and staff members of the Ethiopian Science and Technology Commission, Ato Tekele Gebre and Ato Betru Haile of the former Arsi Rural Development Unit in Ethiopia; Professor Endashaw Bekele, Dr Eng. Shifferaw Taye and Ato Guluma Fekadu of the Addis Ababa University; Mr Lars Olof Larsson at the Swedish Development Cooperation Office in Addis Ababa, Professor Sten Ebbersten of the Swedish University of Agricultural Sciences, MD Leif Ekman, Uppsala Community Health Centre and Dr Thore Denward, TD Foradling Ltd. Finally, I wish to express my gratitude to my publisher for helpful advice and editing, making this book more readable than would otherwise have been the case.

June 2006   *Bo M I Bengtsson*

Contents

viii

Global Agriculture and Revisited Resource-poor Farmers

Introduction

Globally, impressive changes have taken place since World War II. Economic growth has increased global output fivefold. In 1950, the average man lived 46 years and 70 per cent of the population in developing countries was illiterate. Five decades later, life expectancy was 67 years and only 20 per cent of the younger generation was illiterate. Between 1990 and 2002, average overall incomes increased by about 21 per cent and the number of people in extreme poverty declined by approximately 130 million (UN Millennium Project, 2005a). Numerous countries in all regions of the developing world have proven that success is possible (FAO, 2004). There is significant progress in East Asia and there have been positive trends in South Asia. During the last few years, growth rates were 9 to 7 per cent, for instance, in China, India, Malaya, Hong Kong, Singapore, Chile and Argentina. Actions by strong governments have provided education, health and certain incentives for development. This has led to positive changes also at the village level, as demonstrated in India (Strandberg, 1997; Lejonhud, 2003). Still, some 850 million people are undernourished worldwide. Sixty-five per cent of the world's poor live in Asia, and Sub-Saharan Africa has faced great difficulties for quite some time. Half the people in the world now live in democracies. Nevertheless, only some of the 80 countries classified as democracies are considered to have reached a level of full democracy.

In the industrialized world, the annual per capita consumption has steadily increased during the last 25 years. According to estimates by the World Wealth Report, the number of millionaires globally increased to 8 million between 2002 and 2003, a growth rate of 13 per cent in the United States, 3 per cent in Europe and 12 per cent in China. The individual rather than family has become important. One fifth of the poorest of the global population accounts for 1 per cent of all private consumption. Increased consumption alone does not necessarily imply "development" in societies characterized by poverty and great disparities. Poverty and disparity remain at global, national and local levels and occur both across and within countries. In Africa, annual per capita consumption has declined, now being 20 per cent lower than in 1975. Overall global growth has not alleviated poverty for the majority of people in developing countries, many of which are still characterized by poverty, food scarcity, unemployment and other problems.

Some forty years ago, world hunger, resulting from population growth, was pointed out as the major problem. That prophecy of doom proved wrong. Global lack of food is not the major constraint. Rather, food may be in shortage at the national or local level. Nevertheless, the world population is still growing, albeit at a less dramatic pace than in the 1960s. At that time, environmental pollution was emerging as a danger. Side effects of industrial chemicals started to appear, such as death from industrial mercury poisoning in Japan in 1953 and mass death of fish in Mississippi in the United States in 1964. Still, grim prophecies of environmental disaster were not welcome, although global human impact on the environment was not a new idea (Marsh, 1864). Today, environmental pollution is high on the political and scientific agenda. Since agriculture and environmental issues are closely integrated, the future use of land resources will become much more important than current discussions seem to indicate.

Over the past decades, the nation-states have given attention to growth, mainly growth of industry. Consumption is becoming an ideology since basic human needs have been more than fully met in the highly industrialized countries. But industrial consumables are of less value to people who must first meet basic needs and have little or no cash. It is not only a matter of access to a sufficient quantity of food; people also expect food to be safe, palatable and reasonably priced. Methods of food production have changed dramatically in the industrialized countries after World War II. The issues have become more complex and food safety is increasingly becoming more critical than food security. Food safety and security is a complex area, calling for insights in a range of subject matters, such as biology, economics and social and cultural aspects. Not only do food and health go together today, they are also closely integrated with industrial and environmental issues. Globalization makes it even more relevant to consider these issues not only at the level of developing and industrialized countries but also at national and local levels. Improved efficiency of the industry has also led to increased unemployment in the Western world. To those with jobs in a "modern" sector, their tasks have become more demanding, often resulting in increasing sick leave. A modern lifestyle with a different set of values is one component of globalization, now being transferred to communities in the South.

Globally, the agricultural sector is large. Almost half the world population is still working or seeking work in agriculture, a reduction from 63 per cent in the mid-1980s. Then, only 10 per cent of the population in industrialized countries was working in agriculture, a reduction from some 38 per cent in 1950. In the least developed countries, the agricultural sector now accounts for some 80 per cent of the labour force and about 50 per cent of the gross domestic product (GDP). In spite of this fact, most policy-makers have not considered agriculture, forestry and fisheries as vital productive resources for growth during the last two decades. In the South, most governments have given low priority to these sectors, although agriculture and integrated environmental issues are critical parameters for any sustainable society. In the North, the farming population is small. There, political attention has focused on agricultural subsidies, agricultural trade or major crises such as mad cow disease (bovine spongiform encephalopathy or BSE).

Whether or not we have passed the limits of the earth depends partly on how we wish to interpret available statistics. Over time, there have always been critical voices, arguing that crises may follow in the pursuit of "modern" development. But global statistics do not provide the entire truth. One issue is who shapes the relevant global policies and with what tools. The outcomes of the UNCED process in the early 1990s are chiefly used in a restrictive way, applying some kind of timetable of measures. Even if well intended, these measures alone may negatively affect both future productivity and income. The political and scientific challenge is to formulate a multidimensional approach to capture growth but avoid major negative consequences. This makes the task complicated, since we usually grapple with one issue within one sector at a time. The analysis requires both a discussion of global threats and empirical evidence from farmers at the field level. This synthesis, a feature not commonly applied in agricultural research, will be attempted here.

Revisiting Farmers

The following discussion addresses major issues concerning global food security and food safety, agrarian change, agricultural development and the orientation of agricultural research to benefit the less privileged of the world. The analysis is also relevant in assessing the realistic potential of the Millennium Development Goals (MDGs) agreed to in 2000 by the heads of states and the governments of almost all the countries in the world and to be achieved by 2015. The empirical data were collected from repeated field visits to resource-poor farmers in three different countries (Bengtsson, 1966, 1968, 1983). In Trinidad, the primary focus was on the cultural practices of edible aroids grown by small and resource-poor producers in isolated villages in 1965. That was an early approach with a poverty perspective prior to the current focus on the MDGs. The study was part of my studies in tropical agriculture at the University of the West Indies. The original field visits in Ethiopia were done within a regional rural development project (Chilalo Agricultural Development Unit) some 200 km south of Addis Ababa. My initial consultations with these farmers in a feudal society started in 1967 in order to better understand existing plant husbandry and cultural practices. In 1980, more than half the same farmers were revisited in both Trinidad and Ethiopia. An equal number of new farmers were contacted. In the same year, I contacted all active and former farmers at Gramanstorp, a former parish in the province of Scania in southern Sweden. The parish had a very high degree of tenancy and the majority of farmers had small acreages, well-known features of many farming communities in the developing world. Here, the primary focus was on the adoption process of agricultural innovations and overall agrarian changes covering the period after World War II up to 1980.

In late 2003, more than half of the same aroid farmers were revisited in Trinidad. In early 2004, more than half of the same Ethiopian farmers were revisited and so were all remaining Swedish farmers once consulted in 1980. In addition, an equal number of new farmers were met in both Ethiopia and Trinidad. In total, 222 different farmers have been consulted.

The research approach permitted a long historical perspective for the same farmers, most of whom had little or no access to productive resources. Most of them were tenants with small areas of land. Major analytical tools included resource-poor farmers, a structural framework, poverty orientation, agricultural research and extension, agrarian history and research policy. Questionnaires and detailed discussions with farmers were supplemented by field observations. The same framework has been applied for a comparative analysis. It is *not* a comparison of three agricultural systems but focuses on one level of inquiry: that of policies adopted and applied for agricultural research and development and the variations in patterns of impact.

One should be aware of the difficulties with revisits in science. After many years, the anthropologist Clifford Geertz revisited places in Morocco and Indonesia. They had drastically changed–but so had he (cited by Horgan, 1996, p. 157). This suggests that there are limitations when actual findings are going to be interpreted, emphasizing a need for supplementary empirical facts. Over time, people change their minds. A villager does not move from a point A to our point B. Instead, people move towards a new society, albeit influenced by outside forces. Over time, I have changed and so probably have the visited farmers. Since the research focus has been confined to policy aspects of agricultural research and dissemination of such technology, I still believe I have captured the major features of the policy changes and trends over quite some time. The values of the past will not suffice, illustrating another challenge of future agricultural science.

What role can future agriculture and agricultural research play in reducing the number of deprived people up to 2015 and beyond? What should be the direction of future agricultural research? This relates to reflections on how agricultural universities may become more target-oriented and how science aid might be made more effective. The longitudinal approach is of specific value, in contrast to prophecies of doom. In one way, such prophecies may suit specialized scientists. If a little discovery is made, it is easier to present a new crisis, and request more funding. With reference to history, however, current events may look worse initially than in a longer perspective.

The need for proactive agricultural sciences for future development is critical. It is not confined to technical knowledge but also includes the ability to make synthesis to provide some guidance towards the future. Farmers themselves are good in synthesis in their production. But in agricultural science much of this competence has gradually been lost in an emphasis on specialization rather than combining these two characteristics. In this way, the applied contextual framework may provide some input on how to educate students on future agricultural R&D, which is now faced with huge new challenges. Above all, the future role of science in agriculture and other natural productive resources is much underestimated with reference towards the 2050s. It has even been predicted that "our descendants will see population increase level off to a point where it can be handled by advances in agronomy which–under the pressure of population growth and the need to exploit new and previously under-used environments–will replace medicine in the next century as the life-saving wonder science of the world" (Fernandez-Armesto, 1995). If this is true, what changes are required?

Agricultural Development Towards Modernization

Civilization is Rooted in Agriculture

The agricultural revolution started simultaneously in Asia, China and Meso-America. Domestication took place only when absolutely necessary, since agriculture required more labour than hunting. Dogs and pigs were domesticated before people planted crops and practised settled agriculture. Edible plants were collected and plants domesticated. Expansion into new regions was followed by diversification. Long ago, the people of the North and South American continents had a very sophisticated agriculture, experimenting with the potato for different climatic conditions, soils, humidity and solar radiation (Weatherford, 1995). They used guano and practised crop rotation and insect control. They had collections of some 3,000 varieties of potato and had first-hand knowledge of plant pharmacology. Similarly, Asian farmers have been cultivating rice long before the birth of Christ.

Technical developments in agriculture were pushed by early specialization. An elite took command of the agricultural surplus, feeding growing religious and political groups. The growth of settled rural societies and towns led to more defined territories and control of the productive resources. Distinct classes emerged but so also did a development of pottery, weaving and many of the arts. The environment was altered by human changes, sometimes destructive and leading to the collapse of civilizations. It has been argued that pre-Columbian Indians changed the American landscape more than the Europeans and their descendants did after Christopher Columbus (Denevan, 1992). Studies of ancient farming practices around Lake Patzcuaro in Mexico have shown high rates of erosion. Even the hunter-gatherers were not passive toward their environment. In general, it is thought that the collapse of societies has been mainly due to environmental disasters but other factors have also been influential, such as climate change, hostile neighbours and the inability of rulers to master emerging threats (Diamond, 2005).

To fully understand the process of agricultural development, it is crucial to recognize the role of European expansions to new continents. In 1612, the first British settlers in tropical climates entered Bermuda. When they arrived in Barbados more than a decade later, tobacco became their first cash crop (Masefield, 1972). Ten years later, a Dutchman introduced sugarcane. In 1777, cotton varieties were tested in Tobago. The first settlers arrived in Australia in the late 1770s, bringing housecats and foxes and later decimating the wildlife of Western Australia. In northeast Brazil, the Europeans successfully used large areas for extensive cattle ranches throughout the 16th and 17th centuries, even though the region was affected by drought. People moved easily and could settle on large farms. At that time, those areas were considered rich (Prisco, 2000). While the Europeans were interested in cattle and milk production, the aborigines practised rain-fed agriculture, planting cassava, maize and potatoes.

Between 1750 and 1930, Europeans settled in new places around the world and the population in those places increased more than 14 times. Charles Darwin believed that wherever the Europeans had trod, death seemed to pursue the aboriginal. Adam Smith considered the discovery of America and the passage to the East Indies by the Cape of Good Hope the two most important events in the history of mankind. Marx and Engels considered that those events opened up fresh ground for the rising bourgeoisie. At any rate, this expansion led to growth, which was considered natural and good, and led to a transformation of both nature and culture, thus influencing agricultural research and development. Even today, this expansionistic vision thrives and economic development is the organizing principle.

Religion is another feature of relevance to both civilization and agriculture. In reality, it is panhuman because it is a response to the sacred. There are probably very few people who find nothing sacred, although different people have various definitions of sacredness. All religions share some basic principles, including the one that the richness of life on the planet is greatest when the diversity of life forms is greatest (Pepper, 1996). The belief that man could manipulate and control nature was rooted in both Christianity and Jewish tradition. Nature has a spiritual dimension also in Islam. It should not be mastered by man but cultured. The old Greeks recognized an authority well beyond man. Although they had no sacred book, they held it necessary to the read the Book of Nature. To them, science was centred on values, searching for social norms within existing rules of nature and within limits that man should not surpass. The same principle (Yin and Yang) is found in the Chinese culture. To the Hindus and Buddhists, the purpose of life is not to serve the state or companies but to meet your own higher Self and work for a divine plan of harmony in the universe. According to the law of Karma, any incident is a result of many causes from both the past and the present. In consequence, unemployment is simply a bluff, a concept emanating from the framework set by industry.

In a philosophical context all jobs are important, a conception often neglected in "modern" development.

A desire to find a way to deal with nature is the common origin of both science and religion. But science is for a limited number of people, while religion is for the masses. Religion is rooted in faith and concerns itself with human beings. Science concerns itself with the objects of observation. But most people believe in science, thereby placing it on a similar footing as religion. Whereas few people are prepared to die for a scientific truth, many may die for a dogma involving sacred values, such as in a religion. Therefore, science, like religion, is also about power.

From Agricultural to Industrial Culture

Initially, agricultural development was based on solar energy and human power. With industrialization, the muscle power of people and animals was replaced by steam, electricity and fossil energy. New technology was seen as a tool for improving the standard of life. Industrialization could beat nature and basic human needs could be met. This was called progress. Time became important and the clock became critical. Progress was achieved by producing more than one's competitors in a shorter time and with better distribution. But the clock was less suitable to agriculture. Weather and seasons were more important, and still are, in rain-fed agriculture. Nevertheless, "modern" agriculture, still based on biology, is supposed to be managed as an industry. This view has implications for the worldwide transformation of agriculture, in particular for millions of resource-poor farmers.

Some have argued that past industrialization was partly financed by the wealth accrued by colonial conquests. Above all, industrialization spread through the early European and American expansions. Corporations played an influential role as early as the 15th and 16th centuries, when the British Crown chartered corporations with a new provision: limited liability. State-sponsored corporations absorbed risk-taking and promoted exploration and settlement of the New World. The liability of stockholders was limited to the investments they made in the ships, whereas earlier societies were dependent upon their own surroundings, using renewable resources. Nowadays, commercialization has put nature even more out of sight of most people. This process has also been named "development".

The Breton Woods Agreement in 1944 assumed a world infinitely blessed with natural richness. Today, that Agreement has become a global financial system and a global governance institution. Banking, industrial technology and trade have become well integrated into a global economy. This system includes agriculture and is independent of a socio-political system based on the nation-states. Instead, transnational corporations (TNCs) and privatization are key concepts. The TNCs can move around, searching for the cheapest labour. They are now multinational in contrast to colonial times, when they were operating solely within the colonial power. Nonetheless, the trend of expansion remains, either to ensure profits or even to ensure survival as a corporation.

Modernization: Progress and Decline

To many people, modern times in the Western world started with the clock, conceived as a model of the universe. Much later, Francis Bacon stressed the importance of the compass, the art of printing and gunpowder as necessary conditions for new times. Interestingly, these innovations were known in China long before, without noticeable consequences for an emerging era. Modernism has one humanistic root based in the 16th century and one scientific, starting around the 1630s. Paradise on earth was seen as central to modernism of the Western world. In discussions about the process of modernization, J A de Condorcet (1795) denied any dangers of overpopulation. Rather, wealth was to increase, leading to economic equality. Much later, it was claimed that modernism searches after "irreducible and stubborn facts and abstract principles" and has a "distinctive faith that there is an order of nature which can be traced in every detailed occurrence" (Whitehead, 1926). Views and concepts of Western Europe influenced the process of modernization, being both Eurocentric and imperialist.

Certain developments in European agriculture during the 9th and 10th centuries illustrate real technical progress through innovations such as horseshoes and horse collars on animals for traction. The introduction of the wheel-plough brought more grasslands into cultivation, resulting in increased yields. Since the farms could produce a surplus, the population grew, requiring even more food. But at the end of the 13th century, a growing population led to shortage of land, forcing many European farmers to turn to marginal lands. Subsequent consequences were erosion, environmental degradation and increased tree cutting for wood and charcoal and to clear more arable land. Water pollution was observed close to rivers, in particular from tanneries. This pattern of land degradation and pollution is now seen in many developing and industrialized countries.

The question is whether technical progress leads to a "good" and happy life. The Enlightenment aimed at combining the use of science and technology for progress with happiness and freedom for humankind. The basic idea was that policy-makers should turn religious promises into a profane fortune. This was more than progress. In agriculture, progress has been confined to increased yields per hectare and more tons per working hour. But such measurements do not tell the complete story, since technical progress often gives rise to unexpected, negative effects on the environment and even on human health. But the term *progress* has seldom been questioned, although alternatives have been suggested, Koestler, for instance, wrote "I mistrust the word progress, and much prefer the word evolution simply because progress, by definition, can never go wrong, whereas evolution constantly does and so does the evolution of ideas" (1964, p. 525). It has even been argued that the idea of progress was raised to a level of a scientific religion (Stent, 1978).

Progress is positive and based on value. To a certain extent, values are subjective and may be neither true nor false. The subject of progress is collective or social. It is about people, a nation or humankind. It is distinct from change, growth and development, all of which can be based on facts. Criteria for human progress can be defined only by those who assess their own situation, such as an individual farmer in Ethiopia, Trinidad or Sweden. Thus, the concept of progress is confusing. The value of a goal is changed to a value of the means to achieve such a goal. This turns values into objects. But values are important in life. They stir us and are responsible for the direction of social behaviour. If values change, social behaviour will change. To the Greeks something could have an instrumental value, being relevant to human beings, or an intrinsic value, independent of its practical usefulness. Such a value was common among Swedish farmers in the 1940s, who attempted to hand over their farms and land to their children in a better condition than when they themselves had started farming. Today, most Swedes may not consider nature sacred but believe nature is greater than them, implying that God may have moved out from the church to nature. Likewise, millions of farmers in tropical regions know well their immediate environments, their plants and animals, which serve as their basis for survival. With current views on modernization, economic profit is becoming the sole value, resulting in a transformation of culture itself. Such value changes may not take place without conflict.

Progress can also be identified through social matters, such as law for the people or administration, again making progress a thing and not a value. This quantification of progress in Western society took place when science became separated from the church. A modern point of view is the hope that people and society will be even more satisfied if free to follow Immanuel Kant's maxim of rationality rather than authority. Without objective measurements, however, such a view is based on faith and a disclosure of the false myths of progress is the greatest service an intellectual can perform (von Wright, 1993). The belief that everyone is going to make progress is illusory. As a general rule, three out of ten employers at any

Swedish working place are exposed to alcoholism, permanent pain or social maladjustment. Rather, there is a need for a culture to teach people how to handle not only successes but also defeats. Otherwise, those who do not achieve success will see themselves as failures. Equally, this principle may apply to adopters and non-adopters of innovations in agriculture. Progress and decline go together. Perhaps enlightenment would be a better concept than progress (Liedman, 1997).

Reacting to emerging ideas of mineral fertilizers, Karl Marx spoke long ago about both progress and decline, arguing that a short time horizon of large-scale industry in a capitalistic agriculture would lead to soil destruction due to nutrient depletion. The innovation implied, however, that nutrients could be replaced, thereby supporting a growing population. A century later, a similar debate appeared in the United States, now featuring organic or ecological agriculture (Rossiter, 1975). This highlights the need for better insight of agricultural policy-makers and improved higher education in agricultural sciences about the interdependencies between progress and decline. During the last few years, a number of minor academic courses have been introduced to attract undergraduate students both in Sweden and elsewhere. They focus on the new potentials but less on long-term side effects. They do provide universities with increased government funding and undergraduates with a piecemeal approach but fail to provide a synthesis of social relevance. Can such knowledge count as progress in agricultural sciences?

Decline is opposite to progress and carries a negative direction. It is not cyclic. Usually, new technical solutions give rise to new problems. But people have always had expectations based on a belief in technical innovations. One example includes the process of distillation introduced from the Middle East to Europe in the 15th century. In the 18th century, expectations arose from opium, used for centuries in the Middle East. Around 1900, cocaine, known for thousands of years in South America, was introduced in North America and Europe. Today, global media serve as powerful tools for a quick spread of old and new pharmaceuticals, agrochemicals, hormones, antibiotics and other substances. In most cases, their possible long-term side effects are not tested initially. Synthetic oestrogen (diethylstilbestrol or DES) is an old example. In 1957, the Grant Chemical Company advertised its use during pregnancy to produce bigger and stronger babies. It was also recommended to farmers as a feed additive for fattening their cows, chicken and other livestock. Over time, however, the experience of DES on humans showed it could cross the placenta, disrupt the development of the baby and have serious effects that did not become evident until decades later (Colborn et al., 1997). This is only one example of a technology that creates new problems.

Manageable Information

During the last few decades, conventional industry has changed. Mechanical processes have turned electro-mechanical and electronic. Material and mechanical technology are replaced by intellectual and immaterial technology. Other features are standardization, miniaturization and digitalization. New industries (glass, fibre, steel and plastic being produced from new types of raw materials) require improved communication. The powerful tool is manageable information. Almost all Swedish citizens have access to the Internet. In Ethiopia, there are currently some 30,000 Internet connections. In many developing countries, centres of telecommunications are still in urban areas. On average, low-income countries have only three telephone lines per 1,000 inhabitants compared to 400 in more advanced countries. One out of 40 Africans has access to telephones and/or mobile phones. According to the website Africa Online, one out of 200 Africans used the Internet in 2002. Above all, about 2 billion people on the planet have no access to electricity; in rural areas of India even today, half the villages are without electricity.

Information technology has been central to new findings in biotechnology. Its potentials influence both moral and values. Old, sharper lines between technology and nature are blurred. Another emerging area is nanotechnology. Its use in food and food packaging is estimated at a value of some US$ 20 billion in 2010. It has even been argued that humankind is disappearing and new control methods are necessary, although the concept of post-human is not defined (Fukuyama, 2002). The current knowledge society in the Western world is characterized by a far-reaching individualization with no boundaries. Effective communication will offer the greatest power. It offers new opportunities to organize society but may also lead to less desirable attitudes. Consumption and trademarks become critical as identity in life. Social capital is eroded because of commuting and double jobs. Family ties are becoming less important. Information societies may become so dynamic and mobile that people will lose their roots. Individuals may be less concerned with the common good. The new generation in high-income countries has freedom and good education and can make choices, often preferring self-advancement over a commitment to improve their society and the world. They are spoiled and feeling lost at the same time. Genuine socialization of neighbourhoods and communities will most likely be reduced. Fewer new things will contribute to the cohesion of society. Such a trend may have global implications as part of modernization.

There are great differences between a dot.com company and a small community, parish or village in many developing countries. With current trends in a capitalistic society, IT will largely serve the interest of capital.

Global networks are central in the hands of major economic players, involving the global corporations. The real question is how IT can be of immediate relevance to 2 billion poor people of the world. The demand for IT is active, as shown by an example from Laos, where an innovative bicycle-powered computer was tested in 2003. The bicycle charges a car battery that supplies power to the computer. When the battery is fully charged, the cyclist can stop pedalling and use the computer. One minute of cycling allows one minute of computer work. With the diminishing power of the nation-state, hopes have been expressed that corporations might gradually accept more social responsibility, a kind of "pluralism of feudalism" (Drucker, 1994). But societies are configurations of nature and continue to have an impact on natural resources and human capital. Humankind is gradually more and more removed from being part of nature. The neglect of the biological aspects fosters an illusion that we control what we create.

New forms of communication have increased service jobs in affluent societies. A greater worry is increasing global unemployment, reaching its highest levels since the Great Depression of the 1930s. Worldwide, there are more than 800 million people unemployed and underemployed, equal to the number of under-nourished people. What options can policy-makers offer them? At present, most young people live in Islamic countries. This group of some 300 million are 15-29 years old, ready to search for a better life in their own society or elsewhere. What values are they prepared to work for? In the countries of the Organization of Economic Co-operation and Development (OECD), the number of unemployed has increased by some 10 per cent since 1980. Today, unemployment in the European Union (EU) has reached almost 10 per cent, far from its target of the Lisbon strategy of 2000. Since the global corporations will continue to look for the cheapest labour for industrial jobs, this might imply more unemployment or declining salaries in highly industrialized countries. Would this imply decline only or may it also lead to new values and more equitable distribution of global wealth?

Development

Conventional language has usually equated development with modernization. The theory of development appeared in the 1700s when French philosophers discussed social change. Through Charles Darwin, the idea of development was linked to that of survival. The whole was more developed when its parts were more and more differentiated and had specialized functions. The more structurally specialized and differentiated a society, the more modern and, therefore, developed and progressive it would be. Individualism and self-advancement have been equated with the notion of general social progress. More than a century ago, a Swedish bishop, Esaias Tegnér, stressed that peace, potatoes and vaccines helped Sweden to develop as a country. Later, a Swedish poet, Nils Ferlin, argued that stone, fire and earthworms were instrumental for evolution.

After World War II, the basic idea of development promised that human beings could tap all resources in order to satisfy all their needs. However, as needs are satisfied, wishes are gradually expressed as needs. Thus, technical innovations and change must be put to use to respond to wishes. If innovations respond to wishes, they may cause demand. Since wishes are without limits and resources are restricted, there arises a steady hunger for more, or greed. During the Middle Ages, this was generally avoided in the Western world through strong religious beliefs. Over time, there were exceptions, for instance the expansion of the fen drainage system in Europe from the 17th century, which resulted from the greed of landlords and benefited only them. It has been described as a striking example of the effects of early capitalist agriculture on ecology and the poor (Merchant, 1982).

The Western model has advanced capitalism as progress itself. It aims at accumulating capital and depends on sales in the marketplace for profit. By definition, the labour force cannot afford to buy all the things it produces. The workers are not paid the full market value of their products. To make products more cheaply than competitors, industries must be "efficient", so labour must be even more "productive". This leads to minimizing costs in relation to selling price. But most social needs are seldom met in affluent societies because it is not profitable to meet them. Goods and services become commodities on the market. People are, however, separated–by an anonymous market–from the true understanding of their relations with other actors and with nature. That market sets the rules but the poor have no purchasing power. The more the capitalist market grows, the less long-sighted its outlook. Capitalism is fundamentally not sustainable because the natural environment constitutes one of its means of production. According to the Indians of the Onandaga tribe capitalism has no rules so it cannot govern. Therefore, the text of laws will not suffice to rule without acceptance of the moral principles the laws used to be based upon. What can serve as future collective values or moral principles and can they be global?

So far, materialism has been overemphasized in most of the debate on development, neglecting cultural patterns, life-styles and values. Still, these parameters constitute a reality, as illustrated by the need for beauty in Guatemala, where poor people have attractive and colourful clothes. A reminder of the old concept of development turning towards heterogeneity is found in Mexican maize varieties. Before the Spanish entered Mexico, there were some 150 maize varieties in cultivation. They were grown in mixtures well suited to various microclimates. The maize flour from one area had a characteristic colour that distinguished it from maize grown in other villages. In Western societies, farmers and their animals used to form the landscape. Today, we still wish to enjoy such a landscape but the animals are gone and the land is forested or left fallow. We wish there were no dangerous waste but we seldom agree on where to put it. Entropy will grow both as regards waste from increasing numbers of industrial products and the waste of human energy from increasing numbers of the unemployed, sick, poor and undernourished. Any change in these trends calls for a recapturing of values that make consumers feel associated with the process by which their food is actually produced.

Today, development is driven by economic integration and political fragmentation. Growth is the key. Growth pertains to expectations about more, whereas sustainability refers to balance and limitations. Many observers maintain that development would be acceptable as long as it observes laws and limits of nature. The term "sustainable development" sparked a revival of old basic ideas, but economic growth and genuine sustainability are contradictory concepts. When the definition of sustainable development is extended to include "human capital", those concepts become even more politically contentious. Nonetheless, attempts have been made to show how the efficiency of technology and organizations could be increased fourfold–by industry (Lovins et al., 1996). But the norm of Factor 10 was meant only for the reduction of carbon dioxide and not for natural resources or agriculture. Although growth is extremely important in the South, it may simply lead to even more consumerism in the North. Rather than simply more consumerism, people in affluent societies may consider looking for more substance in the meaning of life. Already, the conventional concept of development in the North is slowly undergoing a change in values.

It is doubtful whether a major change in attitudes is realistic unless there is a major global threat or crisis that forces policy-makers to take tough action. Free trade is now seen as a fundamental avenue for progress. In fact, this notion dates back to the time prior to colonization, when capitalism developed partly by the expansion of markets. Even then, rapid communication was necessary since the outposts in the colonies had to be defended against those limiting their rights to free trade. This gave rise to a defence industry. Today, IT provides quick access to new markets, illustrating the current need for the World Trade Organization (WTO) to defend the rights of free trade.

To most people, development is still the post–World War II description of socio-economic progress on a model of the industrialized nations. But the terms "developing" and "developed" are hardly useful. Development is a process that expands the freedoms that people enjoy (Sen, 1999). Internal action and spiritual freedom are significant components. Life-styles are not only about material wealth but also about differences in patterns of life and degrees of freedom. Agriculture cannot be excluded if attention is to be given to billions of people in the South. Welfare for one might be quite different from welfare for another and is thus not a collective concept. Politics alone, and capitalism, can hardly make the change. Good social structure, for example health care and schooling, contribute to welfare and development in spite of a low GDP. This requires democracy, full participation of women in society and human rights as well as human duties. Social change actually happens constantly, but for more rapid social change the power structure of those who benefit from present socio-economic arrangements must be challenged. This is embedded in most international rhetoric calling for political commitment for efforts to the poor and disadvantaged groups of the world. For instance, political leaders agreed at the World Food Summit (WFS) in 1996 to reduce by half the number of starving people by 2015. However, the number of starving people only decreased by 6 million per annum during the 1990s, a pace at which it will take 60 years to reach the political target, according to the Director General of the FAO.

Some people argue that development might not be planned through sensible governance. The unexpected may give rise to new possibilities. Since there will be surprises, there is a need for scientific backup to counter-balance possible negative effects and surprises. If a change means real progress, all should agree to what is better and for whom. An old but quite interesting and unconventional measurement of development used to be common among the Swedish farmers visited in this study. It was related to the daily farm wage and the price of one litre of alcohol. In 1929, they were exactly the same, that is, Swedish Crown (SEK) 2.60.[1] In 1960, the price of alcohol was SEK 28 and the daily wage SEK 30. In 2004, the figures were SEK 273 and SEK 450, respectively. These simple value changes implied a sense of progress. We need to specify (1) the change, (2) whether it leads to measurable, genuine improvements, and for whom and (3) overall consequences, and for whom? Do the final results constitute life improvement for a majority? These questions are highly relevant in agricultural development and research for future food security and food safety.

In general, the power of the nation-state increases with progress in science and technology. But that may not measure real progress. When growth has passed certain limits, its value may not be the same because of

---

[1] SEK 1.00 = US$ 0.14.

environmental or social side effects. Thus, there is no direct linkage between progress in science and technology and increased social happiness. When the computer is shut down, the basic philosophical questions about life still remain, as they did for past generations both in the North and in the South. Art may give some guidance on how to get a good life. If we ignore certain things, we can spend more time on reflection. This is relevant also for teaching and research at agricultural universities. How can the agricultural universities and their graduates produce new knowledge and better help humanity and the poor? The challenge is, recognizing the current power structure, to focus on those with greatest needs. One extreme is city-dwellers in mega cities or shanty towns or those living in rural areas of low-income countries. Here, basic human needs ought to be a priority. For the future, this task is critical, since science and technology spread with little humanity during the earlier European expansion (Liedman, 1997).

Towards the Future

The growing strength of the TNCs may not cease over the next decades. They may even challenge the existence of international organizations. Global governance through a World Parliament through the United Nations is unrealistic, even if there might be power shifts within the UN Security Council and other UN reforms. Within one or possibly two decades, China and India will probably become the major political and economic challengers to the United States. In 1945, there were 74 independent nation-states and now there are more than 190. For the foreseeable future, the nation-states will probably continue to lose power. The EU and the new African Union (AU) might be future actors for certain decision-making, although they are also distant from the realities of their constituencies. If central political power cannot handle problems in the sub-regions, future decision-making will become more regional and closer to the inhabitants. Another alternative may be the formation of regional groupings, probably based on populism rather than patriotic nationalism.

A few years ago, a future with eight or nine different civilizations was predicted, religion being a uniting element (Huntington, 1996). Western interventions were seen as one reason for instability and potential global conflict, the Chinese and Islamic civilizations being prime challengers to the Western type of civilization. This prediction is in contrast to earlier optimism, stating that the only remaining trend was for liberal democracies (Fukuyama, 1992). That belief was rocked by the events on 11 September 2001 at the World Trade Centre in New York. A more pessimistic view of the future was recently formulated with global threats to humankind emanating from new potentials of biotechnological research (Fukuyama, 2002).

An interesting comparison has been made between present times and the Roman Empire (von Wright, 1994). Prior to its collapse, migrating people from the East attacked Europe and so did religious groups and sects from within. After the Roman Empire, there was an interregnum of several hundred years, followed by the Middle Ages, when religion became strong. Nation-states started to grow. All this led to expansion and new technology, influencing the thinking and values of people. Today, most of Europe has similar kinds of pressure from an internal unemployed proletariat, providing some basis for further thoughts in the longer time perspective. Like the Roman Empire, the Chinese Empire was based on expansion of territories. In contrast, the Spanish and Portuguese empires were based on the expansion of trade. This applies also to the recent US hegemony, characterized by its military importance, financial markets and a special American culture (Bacevich, 2002). Except for the recent US dominance, all the other changes took place over centuries.

In the past, the approach was to reorganize society so as to produce more effectively. Life is not possible in a post-agricultural world. The issue is how the agrarian sector best can serve to reorganize society in light of both past experiences and new opportunities. This calls for vision and ideas on the shaping of societies without destruction of the environment and human beings. Is genuine agricultural sustainability possible? What is it? What are the options in order to integrate poor and hungry populations with access to their productive resources? Do we need to maintain all known species of diversity, or how many? What about climatic variations and how much climate change is acceptable, and in what time perspective? Against what norms and values should development effects be measured, and by whom? To both urban and rural dwellers these issues require a discussion of a future vision, global threats and the future role of agricultural research for development.

Process of Adoption of Agricultural Innovations, Changes of Cultural Practices and Technical Change Among Revisited Resource-poor Farmers over almost Four Decades

Aroid Farmers in Trinidad

Trinidad and Tobago

In 1498, Christopher Columbus landed in Trinidad. Tobago was first inhabited in 1632 when a Dutch company sent some settlers. In 1802, Trinidad became a British colony and so did Tobago in 1814. Together, the two tropical islands formed a joint Crown colony in 1889. They were granted self-governance in 1956, got independence in 1962 and became a republic in 1976.

The current population of about 1.3 million has increased marginally since 1980. The population density is about 230 persons per square kilometre. Total available land area is about 500,000 ha, some 60 per cent being forested. In the 1980s, the oil industry boomed and only 3 per cent of GDP emanated from agriculture, forestry and fisheries. About 16 per cent of the population was active in agriculture, a level that has now dropped to a small percentage. The government started to give some attention to food production in the mid-1960s, although it has imported food for almost a century. In the 1960s, root crops constituted some 40 per cent of all food crops in Trinidad and Tobago. They included edible aroids belonging to the *Araceae* family, namely eddo (*Colocasia esculenta*, var. *antiqourum*), dasheen (*C. esculenta*, var. *esculenta*) and tannia (*Xanthosoma sagittifolium*) (Purseglove, 1974). Root crops are well suited to environments of the humid and sub-humid tropics. They constitute one important source of carbohydrate and are rich in Vitamins A and B. In the Pacific, *Colocasia* species are known as taro, a symbol for cultural, physical, economic and political revival among native Hawaiians. According to the FAO, *Colocasia* sp. ranks 14th among vegetables consumed worldwide with an annual production of almost 6 million tons. Taro is grown in some 30 countries and has recently been promoted in South-East Asia to reduce dependence on rice.

In Trinidad and Tobago, the main use of aroids is the starchy corm, whereas the young dasheen leaves are used as spinach and in soups. The production of tannia is much lower, mostly in home gardens. Average aroid production in Trinidad and Tobago is on the decline from some 10,000 tons in the mid-1960s. World production of edible aroids is approximately 10 million tons.

Relying on its oil and gas exports, the government continues to give low priority to agriculture. In 1969, the per capita national income was equivalent to US$ 630. Ten years later, it had increased to US$ 3,390 and reached US$ 8,580 in 2004.[1] At present, some 11 per cent of the population are reported to live in extreme poverty.

Selected Areas and Farmers
In 1965, I visited resource-poor farmers at three locations in Trinidad, considered to be representative for dasheen, eddo and tannia, respectively. This focus allowed an early poverty perspective, aroids also being given little research attention. Biche is situated in the lowland area of the Nariva Swamp on the eastern coast, about 70 km from Port of Spain. The area is characterized by natural forest. Major crops are cocoa, coffee, rice, coconut, dasheen and vegetables. Most of the farmers had moved into that area prior to the 1960s. The Caroni area is close to the capital. On this flat land, sugarcane and rice used to be common crops in the 1960s. Now, various vegetables are grown there, eddo still being the major aroid. Sans Souci and Toco are more isolated areas, located in the northeast some 90 km from the capital. Most land is under natural vegetation and forests with cocoa, banana and coffee as major crops. Tannia is cultivated in the protected valleys of a rather hilly hinterland.

In 1965, 10 farmers were contacted at each location. In 1980, 17 of these farmers (57%) were revisited. The majority of the others were deceased, two had moved and two could not be traced. In addition, 19 new farmers were approached. In late 2003, 17 farmers (55%) were revisited. The majority of the others were deceased, one was hospitalized, one had moved to Port of Spain and two farmers were not traceable. In addition, 12 new farmers were consulted. Several farmers (17%) were again revisited. In total, 61 different aroid farmers were consulted. In 1965 and 1980, the majority of aroid farmers were 40 years or older. In 2003, the average age of revisited farmers was 71 years, and some of them were over 80 years old. The average age of newly contacted farmers was 53 years.

Farm Size and Aroid Acreage
In 1965, it was difficult to estimate the average farm size but new attempts were made in 1980. Based on discussions with the farmers, the average

---

[1] 1.00 TT $ = 0.16 US$

farm size was about 4.0 ha at Biche, 3.5 ha at Caroni and 2.7 ha at Toco. In 2003, the farmed area remained almost the same for revisited farmers at Biche, although newly met farmers had almost doubled their farm size. At Caroni, the average farm size had declined to 2 ha. This was explained by its proximity to the capital and land scarcity. The average farm size at Toco was difficult to assess.

In 1965, the average aroid crop comprised about 1.6 ha of dasheen, 0.5 ha of eddo and 0.5 ha of tannia. Up to 1980, there was a slight increase of the acreage of eddo and a slight decline of tannia. But there was a marked decline of dasheen (0.9 ha). Over the next 20 years, this decline continued for both eddo and tannia. In 2003, the average acreage of eddo was 0.3 ha. Moreover, two former eddo growers were now cultivating dasheen for leaves at Caroni. At Toco, one tannia farmer had turned to large-scale production (1.2 ha), whereas all the others had reduced their acreage and now confined it to home gardens. In 2003, the average acreage of dasheen had increased among newly visited farmers (1.4 ha) compared to 1980.

In 1965, no detailed information was gathered on land distribution by tenure. Squatting was very common at Toco and Biche. In 1980, one fourth of the farmers at Biche remained squatters, one third leased their land and one third were owner-cultivators. At Caroni, almost two thirds of the farmers leased their farm. Two farmers had purchased their farm prior to 1945, whereas eight farmers (22%) had done so after 1960. In 2003, lease of land was normal practice at Caroni. A major problem for revisited farmers was that their applications to the government for formal lease arrangements in the late 1960s had not yet been acted upon. Again, the number of owner-cultivators was highest at Biche. Except for one major tannia grower, squatting was probably still common at Toco.

Some Features of Aroid Cultivation and Technical Changes

Soils and Soil Preparation
Wet soils have generally been considered best for dasheen. In the 1960s, dasheen farmers preferred running water to stagnant water, the latter resulting in anaerobic conditions and rotting of the corms. But running water had the disadvantage that it transported snails, damaging the crop. This problem remained in 1980. Flooding was, however, not considered a problem in 2003 probably because newly visited farmers had purchased their own water pumps in the 1990s. Running water was considered to be less significant to dasheen.

Ever since 1965, farmers at Caroni mentioned a need for ground limestone as a soil ameliorant for their cultivation. In 1980, more than one third of them emphasized the importance of ground limestone. This might be explained by the fact that the Bejucal clay soil of the eastern fringe of the

Caroni swamp is acid and requires lime. Farmers at Toco or Biche did not find limestone a problem. In 2003, limestone was no longer a concern for most farmers at Caroni, who used chemical fertilizers.

Over time, the cutlass and the hoe have been important tools for aroid farmers. In 2003, the cutlass remained the commonest tool for soil preparation and land clearing for tannia. In 1965, no special tillage was given to dasheen. In 1980, two dasheen farmers hired a tractor for soil preparation. At Caroni, tractor hire for ploughing was common practice in both 1965 and 1980. Twenty-three years later, half the number of farmers cultivating dasheen and eddo had purchased a tractor of their own. The cutlass and the hoe were still used by some of these farmers for weeding and mounding.

Hillside planting of tannia has long been practised at Toco, farmers arguing that inter-planting did not increase erosion. In the 1960s, most tannia farmers cut the grass to cover the soil as mulch and did, not burn it. Most eddo and dasheen farmers used herbicides to kill the vegetation before cutlassing. Then, the trash was burnt. At Caroni and Biche, soil erosion was not identified as a problem in 1965. Fifteen years later, one fifth of the farmers at all the three locations found erosion troublesome, a view maintained also in 2003. However, a few farmers at Caroni and Toco claimed that soil erosion had declined, making reference to a lower rainfall over the Northern Range Mountains during the last three decades.

Planting Material and Planting
Aroids are propagated by vegetative means. Seeds are rarely set by the plants, although there have been reports on seed production from Hawaii and Dominica. Since the mid-1960s, there has been no exchange of planting material. All visited farmers used their own genetic material. A majority of the dasheen farmers planted corms, whereas a few used a sliced portion. Tannia farmers usually used sliced corms or large corms. In 1980, the majority of farmers used large corms of eddo and dasheen as planting material. In 2003, small corms, if healthy, were acceptable to all eddo farmers.

The identification of local names of aroid cultivars was most prominent among eddo farmers. In the past, most of them claimed they planted "Chinese eddo". In 2003, all newly visited eddo farmers identified their cultivars as "White Stem" or "Ordinary". Tannia farmers rarely mentioned the name of their cultivars. Dasheen farmers did not identify names of their planting material in 1965. One farmer referred to "White and Black" in 1980. Twenty-three years later, one farmer emphasized "White Stem" dasheen as fast growing, while another cultivator argued that "Red Stem" was superior for corm production. In spite of these specific names, it is unclear whether or not the varieties had distinct genetic differences.

A constant feature of aroid cultivation in Trinidad is that all visited farmers paid great attention to the phases of the moon when planting their aroids. This practice was common in 1965 and remained up to 2003. Farmers observed the moon directly or followed advice from the MacDonald's Farmers Almanac. The small booklet, printed in the United States, gives planting dates for various crops on the basis of astrology. Both in 1980 and in 2003, this almanac was particularly popular among eddo growers. In 2003, the majority of farmers reported planting aroids in dark nights.

In 1965, dasheen was planted about 90 x 90 cm apart, a spacing that farmers had reduced to 75 x 75 cm in 1980. For tannia, spacing was much more variable and wider than for dasheen in the mid-1960s. The majority of eddo farmers preferred a spacing of about 90 x 45 cm. No further changes in spacing were noticeable for any aroid crop in 2003.

Both in 1965 and in 1980, farmers at Caroni were hoe mounding their eddo, usually one month after planting. This was not common in tannia or in dasheen. In 2003, all newly met farmers at Biche were mounding their dasheen crop and mounding by eddo farmers was less common. The tannia crop was not mounded at all.

Irrigation

In 1965, no farmer used irrigation for aroids. This had changed in 1980, when half the farmers at Caroni did so, using the large drainage and irrigation system established for sugarcane and rice. Since then, maintenance had deteriorated so it could not be used at all in 2003. A few farmers had tried to use their own diesel pumps but had difficulties even in getting water because of clogged canals. In 1980, two dasheen farmers had bought their own pump. Twenty-three years later, irrigation was practised by one fifth of the dasheen farmers, all of them newly visited.

Crop Rotation

In 1965, interplanting aroids was common practice. Tannia was grown in the forest together with banana and cocoa. Most tannia farmers also mixed in a few dasheen plants. At Biche, dasheen was interplanted with cassava and cocoa. At Caroni, half the farmers planted a few stands of dasheen and sweet potato with their eddo. Fifteen years later, eddo was no longer interplanted with sugarcane. A few eddo growers had pure stands while others grew eddo together with pigeon pea, maize or cassava. At Biche, half the farmers had almost pure dasheen stands under cocoa trees.

There was no crop rotation system for the dasheen and tannia. They were generally grown on the same piece of land for 2-3 years and 3-6 years, respectively. Then, a period of fallow followed for several years. The majority of eddo growers claimed they practised a rotation sequence of eddo-vegetables-fallow, although some grew eddo without fallowing.

In 2003, interplanting aroids was still a practice among revisited farmers (40%). All newly met aroid farmers planted pure stands. A mono-crop approach to dasheen and eddo was now normal. At Caroni, there was no sugarcane and rice production but farmers grew vegetables and aroids without fallow. Most farmers at Biche planted dasheen without fallowing, making reference to their use of mineral fertilizers.

Organic Manure and Mineral Fertilizers
Farmers have long tried to intensify aroid cultivation. In Guinea, people placed rotten leaves and green grasses around the aroids. Farmers in Hawaii burned growing weeds around the crop to release the nutrients, a practice of tannia farmers around Toco in the mid-1960s. In 1980, all farmers applied organic manure to eddo, an increase from one fifth in 1965. This was mainly a result of increased costs of mineral fertilizers in the 1970s. Cow manure was applied at planting or at mounding. No organic manure was applied to dasheen in 1965 or in 1980. This had changed by 2003, when one third of the dasheen farmers used chicken manure. At Caroni, both bagasse and chicken manure constituted the organic manure applied by half the eddo farmers. In 2003, organic manure was used by one third of all the aroid growers and was somewhat more common among newly met farmers. This was a slight increase since 1965.

In 1965, the use of mineral fertilizers was common among one third of the farmers but mainly on eddo at Caroni. The most widely used fertilizer was sulphate of ammonia. In 1980, the use of fertilizers had increased. Both urea and a compound NPK fertilizer were now common on vegetables at Caroni. All eddo farmers applied fertilizers; only a few dasheen farmers and one tannia farmer did. Increasing costs of the mineral fertilizers after 1974 led one tenth of the farmers to stop using them on aroids. In 2003, half the farmers visited applied fertilizers on aroids. As in the past, all growers did so at Caroni. They broadcast both urea and Blaukorn, a compound NPK+MgO fertilizer (12:12:17:2). Since 1965, the rate of adoption of fertilizer use had declined among revisited dasheen and tannia growers. In 2003, a minority of revisited farmers (41%) applied fertilizers to aroids; 65 per cent applied fertilizers to other crops. The use of fertilizers on aroids was less common among revisited farmers (41%) than among the newly met farmers (62%).

Weeds and Weed Control
The majority of aroid farmers were able to identify the most serious weeds in their cultivation, using local names. This ability was most prominent at Toco and Caroni. The most noxious weeds were *Ischaemum rugosum, Rottboellia exaltata, Commelina elegans, Paspalum fasciculatum* and *Momordica charantia*. In 1980, one tenth of the farmers argued that the number of weed plants had decreased, mainly at Caroni, where Paraquat had been used for several years. Two farmers reported a new weed, although that plant could not be identified. One third of the farmers

claimed that weeds had increased in number, mainly as a result of fertilizer use and unsatisfactory soil preparation. In 2003, the majority of farmers considered weeds a persistent problem in their aroid cultivation. No new weeds had appeared.

In 1965, all aroid crops were hand-weeded. In 1980, farmers used herbicides only at Caroni. In 2003, half the aroid growers applied Paraquat or Gramoxone, again mainly on eddo. Herbicide use was more common among newly met farmers (62%). The marginal increase of herbicide use since 1980 is probably explained by increasing costs, although some farmers also argued that agriculture today was easier because of the existence of agrochemicals. In 2003, the cost of daily labour for hand weeding was a very serious constraint.

Pests and Diseases

Aroids have traditionally been regarded as having no serious pests or diseases, not even major storage pests. In 1965, dasheen farmers reported snails as their major pest, a nuisance persisting in 1980. Snails were not a major problem in 2003, since dasheen farmers had found a way of controlling the running water. Instead, a black beetle, probably *Ligurus ebenus*, had become a major pest. This insect was described as a problem on eddo already in 1965, its frequency increasing on both dasheen and eddo up to 1980. Mites, aphids and grasshoppers were identified as pests of some importance on eddo. At Toco, farmers did not report any serious pest on tannia.

Only a few farmers mentioned leaf diseases as a problem. Most likely, this was leaf blight on eddo, the most serious disease on aroids in the Pacific. In 1980, it was mentioned by one fourth of the eddo cultivators, who claimed that the incidence and damage by leaf blight had increased. In 2003, there was no confirmation that leaf blight was a major disease.

Few farmers applied control measures to prevent attacks by pests and diseases. In 1980, farmers at Caroni reported the use of insecticides on eddo. They were introduced in the late 1960s. A minority of the aroid growers (28%) applied insecticides in 1980; none did in 1965. In 2003, the use of insecticides on aroids had further declined (13%), the chemicals being confined to eddo.

Yield Levels of Aroids

In 2003, two thirds of the visited farmers stated their aroid yields had declined since 1980 in spite of their use of mineral fertilizers. One eddo farmer maintained his yield had increased over time. Average yields were much lower among revisited farmers, while newly met dasheen farmers reported less serious declines. Also, the corm size of dasheen had decreased, a fact highlighted by farmers even in 1980, referring to declining soil fertility.

In 1965, dasheen farmers estimated their yields in bags per acre, equivalent to some 8,000 to 10,000 kg/ha, the second harvest being slightly higher. In general, this was in line with agricultural research findings, reporting dasheen yields of some 12,000 kg/ha at the first harvest and about 8,000 kg/ha at the second harvest. Still, results of field experiments had shown that a good crop of dasheen might yield 15-20 tons/ha at the first harvest. In 1980, farmers at Biche reported an average dasheen yield at 6,600 kg/ha for the first harvest. Information on a second harvest was scarce and unreliable. This low figure must be viewed in a broader context. At that time, many farmers were preoccupied with another crop with a much higher profitability. In 2003, corm yields of dasheen ranged from 6,000 to 10,000 kg/ha.

In 1980, eddo farmers reported an average yield at about 11,500 kg/ha in the wet season and 14,000 kg/ha under irrigation. This was a substantial increase from about 5,000 to 6,000 kg/ha in 1965, when the highest yield reported was about 9,000 kg/ha. In 2003, the average yield of eddo was estimated at 10,000 to 12,000 kg/ha.

In 1980, farmers at Toco stated the average yield of tannia at about 2,600 kg/ha for the first harvest, a figure slightly lower than in 1965. For consecutive harvests, there was a yield decline of about 200 kg/ha for each harvest. In 2003, the major producer obtained tannia yields of 2,000 kg/ha. Other farmers reported lower yield levels.

Agricultural Extension, Research and Experimentation

In 1965, a majority of aroid farmers complained that no agricultural extension agents visited them. In 1980, one third of all the aroid farmers were regularly visited on their farms. There were great differences between locations. Visits were rare at Biche and Toco (7% and 13%, respectively). At Caroni, the corresponding figure was much higher (83%). One third of all farmers contacted their closest extension office at least once a year for the use of subsidies for agricultural inputs. In 1965, the majority of farmers (80%), except most farmers at Biche, expressed a desire for advice on aroid farming. Few farmers participated in field days and demonstrations but one third listened to commercial agricultural radio programmes. Although the programmes did not inform them about aroids, farmers found them useful.

In 2003, extension officers visited only a few farmers (13%) at Caroni. One tenth of the farmers at Biche and Toco could recall such visits in the past. Only one fifth of the farmers at Biche and Caroni expressed interest in receiving technical advice. There were no agricultural demonstrations, field days or radio programmes of relevance to aroid cultivators.

On the whole, aroid farmers were sceptical as to how agricultural research might help them. Several of them questioned why the research establishment had not produced improved cultivars of eddo since such work had been done on vegetables and rice. Since the mid-1960s, one farmer had adopted a new cassava variety developed by the University of the West Indies (UWI). Research on aroids in Trinidad has received very little attention. It was not until the 1960s that the first mineral fertilizer trials were conducted on aroids. Only once had a mineral fertilizer experiment on eddo been conducted in a farmer's field at Caroni. No field trials had been carried out at Toco and Biche, ecologically very distinct from each other and from land at Caroni. Years ago, an agricultural scientist had approached a farmer at Caroni for discussions and exchange of information but this was not sustained. In 2003, there were no field trials on aroids at any of the three locations.

Already in 1980, eddo farmers had emphasized the need for agricultural students at the UWI to visit their farms to experience field problems. This happened once. Farmers at Caroni stressed specifically the need for more field research on their soils, research on blight on eddo and the design of labour-saving, mechanical innovations. Small-scale machinery and equipment was demanded to suit their farming situations. In 2003, eddo farmers reiterated specific demands for a regular system of receiving updated knowledge on possible outbreaks of aphid attacks and hints about new crops. Another expectation was to be invited to meet with researchers for a dialogue and/or for agricultural training on certain topics. Aroid farmers considered their own experience important so any plans for field trials should be closely discussed with them and conducted in their neighbourhood to avoid unrealistic outcomes to them.

During the period under investigation, a few farmers had, within their limitations, been innovative on their own. In 1965, one farmer demonstrated his own very small-scale fertilizer experiment on dasheen, interplanted among his cocoa trees. Another cultivator experimented with planting material of eddo. Since the late 1990s, he had begun to grow his own eddo seed on a separate piece of land, to be well tended in order to get large and good quality corms. He made his own selections by picking corms from the best individual plants. At Biche, one farmer grew dasheen in two separate fields, one for leaves and another for corms. In the late 1990s, one farmer at Caroni had started perennial production of dasheen leaves, not previously done in the location where eddo used to be the predominant aroid.

Some Development Indicators

By 1980, all but one aroid farmer had access to their own radios. Television sets were purchased in the 1970s by 79 per cent of the owner-cultivators,

67 per cent of tenants and 50 per cent of squatters. Half the number of eddo growers had bought a private car, and only one tannia and one dasheen farmer had. No squatter had purchased a private car. The first aroid farmer to buy his own tractor did so at Caroni in 1967. At Biche, a dasheen farmer purchased the first tractor in 1975. They were both owner-cultivators. In 2003, one fifth of all visited farmers had a tractor of their own. Six farmers (four at Biche) had a pick-up car of their own for transporting their produce to various markets. All aroid farmers had acquired a TV set.

In 1980, one third of the visited farmers found that the overall situation had improved since 1965, in particular at Caroni and Biche. They noted improved schooling, better access to health clinics and some improvement of major roads. The level of profit had increased but so had the cost of living. One dasheen farmer was proud to have travelled to Mecca in 1979. It was his major accomplishment in life at a cost of TT$ 8,000. During the 1980s, the dasheen crop lost most of its former importance at Biche, where farmers made more profits by producing marijuana. During the 1990s, the cost of living continued to increase. Food imports forced several revisited farmers at Caroni into part-time farming and taxi driving. At Toco, life improvements were negligible for the tannia farmers.

Problem Identification by Farmers and Their Vision for 2015

In 1980, major problems for farmers were shortage of labour (44%), low prices of agricultural products (30%), import of aroids (11%) and low priority of agriculture to the government (9%). Twenty-three years later, more than one third of the farmers reaffirmed their view that the government was not interested in agriculture and now gave it even less priority. Because of low prices farmers at Biche and Toco predicted there would be no cocoa production within the next five years. Food imports remained a major obstacle to farmers and so did the lack of feeder roads (31%), in particular at Toco. All eddo farmers repeated previous complaints about the government's neglect in solving the drainage and irrigation problems at Caroni, a problem highlighted even in the mid-1960s.

In 2003, labour shortage was of less concern (28%) than in the past. Many farmers had decided to reduce their efforts and investments in agriculture. Labour costs exceeding TT$ 50 a day were found unaffordable. Although mineral fertilizers were readily available, increasing costs for their transportation made them counterproductive to farming in isolated areas. The extra costs for transporting a bag of fertilizer to a Toco farmer amounted to TT$ 20, excluding the cost for the fertilizer (TT$ 60-75 for a bag of urea or about TT$ 130 for Blaucorn). In considering these constraints, aroid farmers indicated that they would rather revert to garden crops for their own consumption than grow for the market.

In 2003, three quarters of the aroid farmers had lost their belief in agriculture. They predicted doom for agriculture in 10-15 years: "it will be gone". All farmers cautioned their children against farming. However, there was no other option for currently active farmers. They had great difficulty in understanding the government policy of abandoning sugarcane land, leaving it idle and leaving 9,000 more people without jobs. They also predicted that more agricultural land would become idle if current trends continued, ultimately causing social problems. Nevertheless, both food and jobs are required in the immediate future so agriculture could be one avenue. Food imports may gradually become more expensive as supplies of oil and gas decline. According to the farmers, the country faces grave difficulties beyond the next decade, in particular since the majority of the population is quite young and not interested in farming.

The aroid crop was vital as food for all revisited farmers. However, a higher price for the crop would not stimulate them to increase their aroid production. This attitude was in great contrast to both 1965 and 1980, when labour was their main constraint, preventing one third of the farmers from producing more aroids. In 2003, the option was to "grow larger or die", but unavailability of suitable land was a major problem. For long, it has been difficult to buy or even rent agricultural land at Caroni. Another option was food exports to the United States, accomplished by two farmers at Caroni, producing sorrel and dasheen. This highlights the issue of plant quarantine.

Summary of Highlights

- The use of agrochemicals was the most important agricultural innovation among visited aroid farmers, although they were not developed specifically for aroids. The use of mineral fertilizers had increased from 30 per cent in 1965 to 44 per cent in 1980 and 50 per cent in 2003.
- In 1965, no farmer used herbicides on aroids. In 1980, one third did so, and 50 per cent did so in 2003.
- No farmer applied insecticides to aroids in 1965. Only eddo farmers at Caroni (75%) did so in 1980, a percentage that declined (to 13%) in 2003.
- On average, it took about 10 years as a minimum for the majority of farmers to adopt an agricultural innovation of relevance to aroids. Time was shorter for the use of mineral fertilizers. There was no marked difference between owner-cultivators and tenants.
- As an average, 59 per cent of the aroid farmers had never had a visit by an agricultural extension agent on their farm in 1980. Their visits were uncommon at Biche and Toco but very frequent at Caroni. In

2003, only a few farmers at Caroni (13%) reported visits by an agricultural extension agent on their farms.

- In 1965, no farmer considered soil erosion a problem. Over time, farmers found that soil erosion had increased. It was reported as a problem by 16 per cent of the farmers in 1980 and 20 per cent in 2003.
- In 1965, no farmer at Caroni used the irrigation system for aroids but half of them did so in 1980. Because of poor maintenance, no farmer could use that system 23 years later.
- A major obstacle to farmers was the neglect of agriculture by their government, a situation underlined in 1965 and persisting in 2003. By then, the majority of the aroid farmers had lost faith in agriculture, some of them returning to home gardening rather than producing for the market. The neglect of agriculture by the government was the crucial problem rather than labour costs, which were considered the most serious constraint to aroid production in both 1965 and 1980.

Farmers in the Ethiopian Highlands

Ethiopia

Ethiopia has a long history. It has not, except for a very short period, been under colonial rule. In area, Ethiopia is two and a half times the size of Sweden, equivalent to 1.1 million sq m. The country has now about 70 million inhabitants, 85 per cent living in rural areas, and a population growth rate of 3.1 per cent. This is a significant increase over about 23 million people in the mid-1960s. For a long time, its economy has been based on agriculture and the land is rich in biodiversity. In the late 1950s, Ethiopia was self-sufficient in staple foods and even exported food grains. Now, some 16.6 million ha are under cultivation but almost 73 million ha are considered to be potentially suitable for production. The country has the largest livestock population in Africa. Livestock provides 30 per cent of the agricultural GDP. After coffee it is the second most important source of export earnings.

Officially, the average smallholder family is now composed of five to six members and cultivates 1.4 ha of land. The highlands constitute 36 per cent of the total land area but support 88 per cent of the population. In 1967, the annual per capita income was US$ 60, climbing to US$ 130 in 1979 but fallen to about US$ 110 in 2004.[2] About 40 per cent of the Ethiopian people are living in absolute poverty.

Selected Areas and Farmers
In 1967, several field studies were conducted in the former Chilalo awraja, some 200 km south of Addis Ababa. This region was selected for an

[2] Birr 1.00 = US$ 0.48

integrated rural development project (Chilalo Agricultural Development Unit or CADU) and financed by the Ethiopian and Swedish governments. Farmers were randomly contacted at four locations, the purpose being an investigation of existing cultural practices and crop sampling. Kulumsa is just north of Asella at an altitude of 2,200 m above sea level. Wajji is further south at an altitude of 2,400 m and with a higher degree of grassland. The two other locations were Aleltu Silosa further south and Dighellu to the east along the road to Ticho. The altitude at Dighellu is about 2,700 m.

In 1967, 43 farmers were consulted for interviews and field observations on ploughing, seeding, weeding and crop sampling of yield levels. In 1980, 25 of these farmers were revisited (58%) and an equal number of new farmers were contacted. Field investigations were repeated and crop samples were taken in farmers´ fields. The average family size was 5.7 persons. In early 2004, 26 farmers were again revisited (52%), 19 were deceased, one was hospitalized and two were not traceable. In addition, 24 new farmers were consulted. One third of the farmers from the 1967 study were revisited. In total, 92 different farmers were contacted.

In 1980, only 16 per cent of visited farmers were of Amhara origin, the remainder being Oromos. Most farmers were born in the area, except at Kulumsa. In 1980, almost two thirds of them had moved into that area. In early 2004, such movements had more or less ceased but the number of Oromos seems to have increased. More than 70 per cent of the farmers visited in 1980 were 40 years old or more. In 2004, many revisited farmers were very old, some of them in their late seventies. Some were even above 80 years of age, in particular at Dighellu, where the higher altitude seemed to influence the survival rate.

Average Farm Size
Originally, farmed land was calculated in "timads", defined as the area of land a pair of oxen would plough in one day. By practical field measurements in 1967, one timad was estimated at 2,200 sq m, a figure that was a little lower in 1980 owing to weaker oxen. Already then, farmers claimed that four large timads constituted one hectare. In 2004, most farmers used hectares, except for crops such as field peas, flax and chickpea.

In 1980, the average farm size of all visited farmers was 2.9 ha, including 1 ha of grassland. The area of grassland was smallest at Kulumsa (0.6 ha) and largest at Wajji (1.9 ha). Revisited farmers had 3.7 ha compared to 2.2 ha of newly met farmers. The difference was most likely a result of the land reform process during the Ethiopia Military Government (DURG). A more thorough examination revealed that many of the revisited farmers actually had been selected as so-called model farmers by the CADU project. Quite a few were originally landlords. Most likely, they had been able to exert power within the peasant organizations during the implementation of the land reform. Nonetheless, there had been considerable shift of power since

1967. Then, an average landlord at Wajji owned 27 ha, compared to a sharecropper with 2 ha.

In 2004, the average farmer had 2.5 ha as cropped land and about 0.9 ha as grassland. As in 1980, the cropped area was larger among revisited farmers (2.6 ha) than for newly visited farmers (2.0 ha). Also for grassland, the trend in 2004 showed that revisited farmers had larger average acreages (1.3 ha) than the newly visited farmers (0.6 ha). At Wajji, the average revisited farmer had 2.5 ha compared to 0.2 ha of newly met farmers at Kulumsa. This is another indication that the old power structure had not been fully dismantled.

Features of Plant Husbandry and Technical Changes

Soil Preparation and Farm Implements
In 1967, most farmers used the ard, the bent wooden beam with a small iron or steel point for soil preparation. Only a few farmers around Kulumsa (11%) hired a tractor. At that time, the overwhelming majority of farmers desired a better plough. In 1980, times were very harsh and all farmers used the ard. All revisited farmers had a pair of oxen of their own. For newly visited farmers this was true only for three fifths. One fifth of them had only one ox and another fifth had no ox at all. According to the Ministry of Agriculture almost one third of all farmers in the country had no oxen in 1980.

In 2004, the average farmer in the area had 1.6 pairs of oxen. Revisited farmers were better equipped with oxen. Some farmers with the largest acreages had up to three pairs of oxen. One farmer rented a pair of oxen at ploughing time, whereas two farmers had no oxen of their own. The market price for one pair of oxen was about Eth. Birr 2,500, equivalent to US$ 1,200. At Kulumsa, a minority of farmers (15%) hired a tractor for ploughing.

As a general rule, land for arable crops was usually ploughed three times, followed by seed covering. Land was ploughed once for linseed and field pea but twice for broad bean. The ploughing pattern was different for grasslands. In the past, they required five to six ploughings. In the 1960s, farmers at Wajji and Aleltu Silosa used soil burning or "guie" to get sod land better prepared and to utilize the nutrient release from burning the soil. In 1980, no farmer practised that system, although one revisited farmer used it occasionally on potato. It reduced the number of red ants but was very labour intensive. Instead of guie, some revisited farmers at Wajji had increased the number of ploughings for wheat and barley, mainly to combat weeds. In 2004, there was no significant change in the number of ploughings. No farmer used soil burning but one revisited farmer had used it in 1999.

In the early 1970s, agricultural researchers at CADU had designed, produced and launched various farm implements. In 1980, very few visited farmers (8%) used the spike-tooth harrow and the mould-board plough. One farmer had used an ox cart, but it had broken down. In general, farmers had found the plough and harrow too heavy both for them to carry to the fields and for their oxen to pull. In addition, the implements were expensive and the local blacksmiths had great difficulties in repairing them. Some farmers found the harrow effective on *Avena* weeds. Although difficult to manage, the plough was considered effective on *Rumex* weeds.

In 2004, no farmer used these farm implements. A few years ago, one farmer had stopped using the CADU plough, since it could not be repaired. However, one third of the visited farmers declared they had initially adopted the use of both the CADU harrow and plough but stopped during the Military Government period.

Crops and Crop Rotation

In 2004, wheat and barley were the major crops (80%), wheat being the preferred one. The wheat acreage had increased at all three locations south of Asella, where barley used to be the major crop. Only small areas (one or two timad) of field pea and broad bean were grown for home consumption. In 1980, cereals were cultivated on 60 per cent of the total area, an increase from 45 per cent in 1967. Up to 1980, the acreage of pulses had increased slightly, mainly at Kulumsa and Wajji, a trend that was discontinued in 2004. By 1980, linseed had disappeared from all locations except Aleltu Silosa, where it had marginally increased. In 2004, linseed was uncommon.

In 1980, a few visited farmers grew oat, rye, fodder beet and carrot as new crops. In 2004, other crops such as onion, potato and carrot were more common at higher altitudes. Hybrid maize as a garden crop was now grown south of Asella, the new varieties planted as late as in May.

In 1967, farmers had kept one third of their land as grassland. The majority of farmers (80%) kept land fallow, usually for two years (30%) or longer (24%). One-fifth of the farmers kept land fallow for one year. In 1980, the fallow period was reduced to two years as a maximum and observed by more than one third of the farmers (40%). In 2004, the grassland areas around Kulumsa had almost disappeared, resulting in fewer cattle and more exposure of the hillsides to erosion. South of Asella, the number of cattle was reduced because of shrinking areas of grassland. Except at Aleltu Silosa, farmers were well aware that erosion had increased, in particular at Kulumsa and among revisited farmers. The fallow period was now reduced to one year, except at Wajji, where farmers might keep land fallow for up to two years. Fallow was more common among revisited farmers (46%) than for newly met farmers (17%).

In 1967, crop rotation was greatly influenced by the existence of fallow land. Although there was no systematic crop rotation system, a typical approach might be as follows: fallow-barley-broad bean-field pea-linseed. In 1980, the crop rotation had become more flexible, cereals being the dominant crops. Twenty-four years later, wheat and barley were planted on a continuous basis with crops of broad bean and field pea as exceptions.

Seed

In 1967, the overwhelming majority of farmers planted their own seed for all crops, except wheat. Other sources for seed were the local market or neighbours. Half the farmers around Kulumsa (16% of all farmers) purchased improved seed of wheat ("Kenya I") from the Kulumsa Seed Multiplication Farm. In 1980, the same pattern persisted; the overall majority of farmers used their own seed. During the 1970s, many farmers had bought their first improved wheat and barley variety from the Kulumsa Seed Farm. After the first year they used seed from their own harvest. Another noticeable change was better seed cleaning of cereals both for revisited and newly met farmers. In samples from farmers´ seed, the number of weed seeds was reduced from 5 to 1 per cent. Field trials in the 1970s showed that seed cleaning had significant effects on yields.

Also in 2004, the majority of farmers (82%) continued to plant their own seed. A few farmers purchased new wheat varieties every two to three years. A minority (16%), mainly revisited farmers at Wajji and Kulumsa, claimed they were making annual purchase of improved seeds. No farmer bought cereal seeds at the local market as they commonly did in the late 1960s.

In 1967, farmers planted seed of all crops, except wheat, in local mixtures of different cultivars, an example of *in situ* conservation of land races. In 1980, farmers were growing improved cultivars of wheat (80%), barley (32%) and maize (12%). Maize was only grown at Kulumsa, often as a mixture of both hybrid and local maize. "Beka" was the only improved barley variety. It was grown together with local barley cultivars in adjacent fields as a security measure. In 2004, "Bira" was the most commonly grown improved barley. Revisited farmers (77%) stated they had planted it for more than a decade. Only a few farmers (15%) were cultivating older improved barley such as "Beka" and "Muga". Two farmers still planted the local "Arusso", commonly grown in the 1960s.

Up to 1980, several new wheat cultivars had been both adopted and rejected by farmers, including "Yaktana", "Supremo" and "Israel". "Enkoy" was claimed to be the highest-yielding wheat. "Salmayo" ranked second in popularity since it did not shatter and was a good yielder as well. "Laketch" was quite common in spite of its susceptibility to rust. Most commonly grown local cultivars were "Bokake" and "Tikur sindie". In contrast to barley, new wheat cultivars were seldom cultivated together with local ones. In 1980, one third of the farmers had returned to local wheat cultivars. Farmers argued they had no access to improved wheat seed and maintained that local cultivars were superior, providing yield stability over a longer time. In 2004, "Kubsa" was the most commonly grown improved wheat (44%). It had been cultivated for more than a decade by the earliest adopters but some farmers (8%) had just started to plant it. "Galema" was also commonly grown by almost half the farmers. It was of more recent origin, having been adopted one to four years ago. Local cultivars were still cultivated, such as "Enkoy" (16%), "Laketch" (6%) and "Salmayo".

In 1967, specific measurements were made of the seed rate in farmers' fields. The number of "kunas" per timad of different crops was weighed by scale in the field when farmers were to broadcast their seed. Thus, kunas per timad were transformed into kilograms per hectare. In 1980 and 2004, the majority of farmers expressed their seed rate both in the conventional manner and in kg/ha. In 1980, the seed rate of wheat was higher at lower altitude. It ranged from 200 kg/ha at Kulumsa to 185 kg/ha at Wajji and 140 kg/ha at Dighellu. This pattern remained in 2004. In 2004, the average seed rate of wheat remained as in the past, for example, approximately 200 kg/ha. The average seed rate of barley was 225 kg/ha at all locations, except at Wajji, where it was much lower at 175 kg/ha. Since 1967, farmers had increased the seed rate of barley by 50 kg/ha, reaching 200 kg/ha in 1980 as an average in spite of recommendations of 85-100 kg/ha by the National Institute of Agricultural Research. For other crops also, farmers used seed rates that were higher than official recommendations. Probably, they did so because of their conviction that the number of weed plants had dramatically increased and that they had to plant more crop plants to increase their competitive ability.

Organic Manure and Mineral Fertilizers
In 2004, almost all farmers (96%) applied organic manure to their garden crops, a practice taught by the agricultural extension staff. This was an increase from 1980 (80%) but less so compared with 1967 (90%). Since the number of cattle had declined, the quantity of cow manure now available must be rather small. Organic manure was applied on maize at Kulumsa.

In 1967, no farmer applied mineral fertilizers. In 1980, all farmers had tried them once and the majority (92%) actually applied di-ammonium-phosphate or DAP (18:46) to cereals. Most likely, the approximate rate of application did not exceed 50 kg/ha; it was a little higher among revisited farmers. Eight per cent had stopped using DAP, since they found it too expensive. In 2004, all farmers used mineral fertilizers, except on chickpea, field pea and flax. DAP was rarely applied on broad bean. Two farmers applied urea (on potato at Dighellu and on grasslands at Wajji). With reference to estimates by the farmers, the average rate of fertilizer application can be estimated at 100 kg of DAP per ha. If correct, this is high. The FAO has reported the average fertilizer use in Ethiopia as 16 kg/

ha on arable land (FAOSTAT, 2002). Tanzania used 6 kg/ha and Mali 11 kg/ha in contrast to 279 kg/ha in China and 365 kg/ha in Vietnam.

Weeds and Weed Control

The majority of farmers found weeds a major problem in their cultivation. The actual weed counts in the farmers´ fields supported their assessment. In 1980, an average un-weeded field had some 1,100 weed plants in competition with about 305 wheat plants and 215 barley plants per square metre. Seed covering reduced the number of weeds to about 250-300 plants in cereal crops. In field pea and linseed fields the number of weed plants was much higher than in cereals.

Farmers are well aware of the most noxious weeds and identified about 60 per cent of the most difficult ones by local names. Monocotyledonous weeds constituted about 40% of the weed flora in both 1967 and 1980. Most common were *Phalaris paradoxa, Setaria pallidefusca* and/or *S. acromelana, Snowdenia polystachia,* and *Avena* spp. Most troublesome among the dicotyledonous weeds were *Polygonum nepalense, Galium spurium, Galinsoga parviflora, Rumex* spp., *Amaranthus angustifolia, Guizotia* sp. and *Lolium temulentum.* Between 1967 and 1980, *Polygonum nepalense* and *Galinsoga parviflora* increased in numbers. But perennial weeds might also have increased because of changing cropping patterns and use of mineral fertilizers.

In 2004, no weed counts could be made in the field. Instead, farmers ranked their most troublesome weeds in the following order of importance: *Snowdenia polystachia, Galinsoga parviflora, Guizotia* spp., *Avena* spp., *Rumex* spp., *Galium spurium, Phalaris paradoxa* and *Polygonum nepalense.* There were few indications of reduction in weed infestations.

Over time, a few new weeds had appeared. At Wajji, the big-seeded *Avena* spp. and *Avena fatua* were most likely introduced during the Italian occupation in the late 1930s. In 1967, farmers at Aleltu Silosa claimed that seeds of *Guizotia* spp. were new introductions at the Sagure market. In the early 1970s, *Bidens pilosa* was identified as a new weed around Asella, whereas *Tagetes minuta* was reported to have first appeared in 1977. Recent weed introductions were "Mara" (unidentified) in linseed fields after 1980 and *Argemone mexicana,* a spiny weed reported to have spread in the area as a side effect from food aid provided during the period of the Military Government.

In the 1960s, crops were hand-weeded, once in wheat and barley, twice in broad bean, and not at all in linseed and field peas. At that time, the observed effects of hand-weeding a barley or wheat crop were reductions of weed plants by 40 to 60 per cent. In 1980, the majority of farmers (62%) affirmed that weeds had increased in numbers. Hand-weeding was still common practice. A small minority of farmers (12%) used herbicides

(MCPA and 2,4D). In 2004, almost all visited farmers (96%) used herbicides (2,4D) on their cereal crops. Most revisited farmers had started to apply them in the early and mid-1970s. Most of the newly met farmers started during the Military Government period but two farmers had adopted the use of herbicides after that period.

Pests and Diseases
In 1980, one third of the farmers declared they had no serious pest problem. Armyworms on barley could be difficult during certain years but the severity of attacks had declined since the late 1960s. Cutworms were considered an annual problem on barley by one fifth of the farmers but the majority believed that cutworms were more destructive in the 1960s. In the mid-1970s, aldrine was suggested as a control measure and was used by some farmers. In 2004, a minority of farmers (42%) reported problems with pests. They included cutworms on barley, mainly at Wajji and Dighellu, red ants on potato at Aleltu Silosa and Dighellu, and pod worms on field pea and broad bean at all locations, except Aleltu Silosa. Ever since 1967, a black beetle was reported to damage broad bean but it was found to be less important in 2004. Maize stem borers remained a problem in the Kulumsa area.

In general, farmers mentioned few problems with storage pests, except at Kulumsa, where grain weevils damaged stored maize. In the 1970s, farmers had tried Phostoxin but no one used it in 1980. Only one fifth of the farmers used insecticides in 2004; none had used it in 1980 and 1967. Another fifth of the farmers had tried them up to 1980 but stopped using them. In 1967 and 1980, one third of the farmers reported rodents as a major problem. In 2004, they were not mentioned as a serious problem.

In 1980, half the farmers reported rust diseases on both wheat and barley. Over the years, change of cultivars had been the obvious method of control, mainly for wheat. In 2004, the incidence of rust was reported to be less frequent but it did occur at Wajji, in particular. Possibly, farmers had also acted by shifting to more rust-resistant cultivars. In the past, some farmers had confused rust with the effects of cold nights. In 1980, one third of the farmers at Wajji and Dighellu declared frost to be a major problem on barley and field pea. In 2004, frost during cold nights remained a concern for farmers at Dighellu. It was unclear whether the currently planted barley cultivars had better frost tolerance than the local ones.

Crossbred Cattle
In 2004, one fifth of the farmers kept crossbred cattle originally produced by the CADU project. Revisited farmers at Wajji chiefly kept these hybrids. Some cattle were recently bought since the old ones had died from diseases. For several years, it had been difficult to get access to veterinary medicine

in Asella. At the other locations, the grassland areas hardly allowed as much feed for animals as in the past.

Crop Yields

In 1967 and 1980, several hundreds of crop samples were taken from one square metre in farmers´ fields at all four locations. Individual samples were trashed and dried and yield figures were calculated in dry matter content (DM) in kg/ha. In 2004, farmers were asked about their approximate yields and whether they had increased or declined over the years. To allow for a comparison of yield trends over time for the same farmers, using actual field data for 1967 and 1980, they have been adjusted to moisture content of 16 per cent (Table 1). This is assumed to correspond fairly well to field conditions at harvest time and be comparable to assessments made by the farmers in 2004.

Up to 1980, the average DM yields of wheat increased by 25% and those of barley by 34%. In 1980, the highest average wheat yields were recorded at Aleltu Silosa and the lowest around Wajji. Average yields of barley were highest at Wajji. The average yield increase of field peas was marginal (7%).

Based upon the estimates by the farmers in 2004, the yields of wheat had doubled since 1967 and those of barley had increased by 60 per cent. At present, wheat yields of 3 500 to 4,000 kg/ha were not uncommon in good years. At Wajji, barley yields may exceed 3,000 kg/ha. In contrast, farmers at Aleltu Silosa stated that their wet soils meant lower yields of both wheat and barley, at least a reduction of 500 kg/ha. The yields of field pea have declined by one third, probably as an effect of no weeding, no improved varieties and no fertilizers. Broad bean has been more able to compete, partly since it is regularly weeded in contrast to field pea. Thus, broad bean yields have been maintained.

Table 1. Average crop yields among farmers around Asella between 1967 and 2004 (kg/ha).

Crop	1967 (DM)	1980 (DM)	1967 (Adjusted)	1980 (Adjusted)	2004 (Estimates by farmers)
Wheat	980	1180	1140	1370	2340
Barley	1130	1660	1310	1930	2100
Broad bean	1500	1930	1740	2240	2300
Field pea	930	970	1080	1130	700

Three fourths of the farmers considered the use of mineral fertilizers and improved seed the major reasons for the yield increases. Nevertheless, one fourth of them argued that their soil fertility had declined since they could not afford the required amount of fertilizers, which were costly. If the price of fertilizers continues to rise, soil fertility will further decline, a great concern to all revisited farmers. In 1980, the revisited farmers had the highest yields, partly because many of them had benefited more from the development project. This was not the case in 2004, when the average yields were highest among newly visited farmers.

Agricultural Extension, Research and Experimentation

In 2004, agricultural extension agents were very selective, paying farm visits to about one third of the farmers. Farmers had to contact them to purchase improved seed and fertilizers. One third of the farmers did not do so. In the late 1960s, about half the farmers had been visited by an agricultural extension agent, a pattern that changed completely in 1980. Then, the peasant organizations were responsible for technical agricultural advice as a new government policy of 1975. Thus, the agricultural extension agents did not visit any farmer. Likewise, there were no field days for farmers, a sharp contrast to the early 1970s. Nowadays, field days and demonstrations are non-existent. Farmers with access to a radio declared that the programmes on agriculture, twice a week, were quite useful and even better (64%) than listening to the agricultural extension agents. In 1980, only one tenth of the farmers had listened to such radio programmes. As part of the development assistance activities in the area many demonstration plots were laid out in various extension areas up to the mid-1970s. No such plots were observed in 2004.

In 2004, less than one fourth of the farmers found the contact with an extension agent useful. Twice as many considered it less helpful. In general, revisited farmers considered extension agents less technically competent, although a few agents were found to be excellent. Four farmers claimed that they had been cheated when requested to pay in advance for fertilizers over the last two years. Besides, they were asked to pay an extra, unidentified fee. This problem was accentuated when extension agents were suddenly transferred from their area, which was not uncommon.

To some farmers, experimenting in agriculture on their own was not uncommon. In the late 1970s, one farmer at Aleltu Silosa had diluted DAP fertilizer in a large bowl of water overnight. The crop seeds were soaked and got thinly coated with fertilizer before they were broadcast. By that approach, the farmer saved some fertilizer, got quicker germination and reported the same yield level. In 1980, three neighbours had adopted his idea. In 2004, many farmers were well aware of this practice. Some additional farmers had tested it, although none was actually practising it. Using the CADU harrow as a model, another farmer had designed his own, hand-pulled harrow with 12 spikes for seed covering at Kulumsa. He used it for some years in the late 1970s, although his neighbours considered it, and him, somewhat odd. Those farmers who on their own had decided to turn to forestry and wood production had not only made economic gains but also demonstrated a more profitable land use system than previously thought of at Wajji. Recently, one farmer had planted ginger as a spice crop for the local market but realized that the price was too low.

In 2004, farmers expressed demands for new approaches. One tenth of them argued for a system of two crops a year. Moreover, they wanted both grain and milk to be locally processed in the future. Such an orientation emerged as a research priority at the CGIAR Science Council in early 2005. Other farmers (8%) desired training at the new Farmers´ Training Centre in Asella on how they might produce more improved seed, which is often in short supply. Two farmers regretted they had no access to agricultural researchers at the Kulumsa station to influence the research to deal with acute problems. This was a rather exceptional view because the majority of farmers generally did not expect much from agricultural research. One of them stated explicitly they "had never received any substantial contribution from university work".

Some Development Indicators

All revisited farmers had experienced significant improvements from increased production and higher income but they faced ups and downs. Improvements included larger and better housing conditions and education of their children. A very old farmer with "70 years of experience of Ethiopian governments" found the present one to be the best. Three revisited farmers had turned to forestry as a major source of income, making significant economic gains. One of them was able to make his trip to Mecca in 1995 as a result of his long-term investment in an alternative method of production. The trip was his major achievement in life and resulted in a donation by the Saudi Arabian Government for the erection of a mosque on his compound at Wajji in the early 2000s. But a majority of farmers argued they had simply regained by 2004 what they lost since the 1974 revolution, a view expressed by most of the former landlords.

To former tenants, the abolishment of tenancy was a major positive change, being explicitly recognized in 1980. Nevertheless, the land reform had also resulted in smaller holdings for many families, not only the former landlords. One farmer with no children had to survive on one timad allotted to him in 1974. There were positive changes also among the newly met farmers, such as higher production (50%) and better schooling and education for their children (21%). Still, a number of them felt that "change is not for farmers."

Between 1965 and 1980, half the visited farmers had bought their own radios, although several were forced to sell them to get cash during the Military Government period. In 2004, four fifths of the farmers had access to radios of their own. None had access to electricity, a TV set or a private car.

Problem Identification by Farmers and Their Vision for 2015

In 2004, farmers identified three major problems in their agricultural production: the high cost of agricultural inputs (62%), the unavailability of mineral fertilizers when required (62%) and the increasing land tax (38%). One farmer found "that costs of inputs are not for development". A farmer at Wajji stated that "expenses are high when using a new technology—my fallow system gave about the same result so it was better". By 1980, high costs of fertilizers were found to be a major constraint (40%). At that time, the price of DAP had already tripled to Eth. Birr 90 from Eth. Birr 30 in 1974. Since then, the price of DAP had further escalated, reaching Eth. Birr 320 per 100 kg in 2004. In contrast, the price of 100 kg of wheat had increased from Eth. Birr 25 to just Eth. Birr 100-130 in 2004.

The land tax is paid as a lease of government-owned land. It had increased from Eth. Birr 8-20 per ha during the Military Government period to some Eth. Birr 160 per ha in 2004. Other problems included shortage of arable land (28%) and grassland (24%), too many weeds (16%) and short supply of improved seed and herbicides (8-10%). Many farmers argued that the agricultural extension agents could be more effective through dialogue and not just serve as administrators. About one tenth of the farmers mentioned deteriorating quality of drinking water over the last decade, lack of an organized system of hiring combiners at harvest and tractors for ploughing, and waterlogged soils at Aleltu Silosa. Finally, farmers claimed that market follow-ups were non-existent in contrast to work done by CADU some 30 years ago.

Twenty-three years ago, the major problems were delays in helping farm families in which the husband was at the warfront (24%), lack of oxen (24%), poverty and lack of food (12%), shortage of land (10%), and shortage of improved seed, feed and veterinary medicine (10%). The most technical problem was too many weeds (46%). Interestingly, the interrupted agricultural extension activities, replaced by peasant organizations, were not considered a major problem, except by a few (6%).

In 2004, all farmers were well aware of emerging development problems within a decade. Two thirds of them mentioned population pressure, eventually leading to crowded land even for housing. Increasing costs of production and land scarcity are major political issues for the immediate future. With a growing population, farmers were convinced there might be hunger or poverty (20%), even if they were to sell more cattle and plough up more grassland (22%). These views were more prominent among newly met farmers, some of whom stressed they were not so interested in agriculture. Rather, they felt new jobs must be created outside agriculture, in particular for the young generation.

Several newly met farmers (12%) advocated a future with fewer children and one wife only, a way of life that children now were taught at school. In fact, they argued that schools had an important role to teach new skills not only to children but also to active farmers. In general, revisited farmers considered current school teaching too focused on basic sciences rather than on agricultural sciences and the practical professions needed by the country in next two decades.

At two locations, one third of the farmers expressed strong reservations about the government resettlement policy. They did not wish to move to the lowlands. They argued that the government must first conduct research there on both medical and agricultural problems before people were asked to move. Otherwise, they would not be able to survive. The farmers considered any such movement "a death penalty". In contrast, a Presidential announcement in May 2004 had declared a voluntary resettlement programme to ensure food security was becoming largely effective, being based on both the interest of farmers and feasibility studies. Two million Ethiopian Birr was allocated for 60 research programmes on peasant land in eastern Ethiopia. New seeds of maize, sorghum, potato, wheat and barley, being drought-prone and resistant to pests and diseases, were distributed.

Whereas all farmers realized the need for increased food production, they would contribute actively to increased production only if there was no further increase in the land tax and the price of fertilizers. Even so, one fifth of the farmers would not increase their food production. The strongest reservations were found among revisited farmers (50%). These complications are serious obstacles for increased national food production in the future, a huge task for millions of farmers. Ethiopia achieved quite large national food grain harvests in 1996, 1999 and 2001. During 2004, farmers harvested about 14 million tons of food grain. This was 24 per cent higher than in 2003 and more than the national needs of some 12.5 million tons. This issue is not confined to Ethiopia but a challenge to African nations, where agriculture must annually grow by 6 or 7 per cent.

Summary of Highlights

- Agricultural research started in the Chilalo area in the late 1960s with a focus on surveys, meteorological observations and field trials on mineral fertilizers, crop cultivars, crossbred cattle and work on farm implements.
- Although a number of agricultural innovations were made available in the late 1960s and early 1970s, the use of fertilizers and improved seed remained the major ones. Some farmers have adopted crossbred cattle and tried vegetables.

- The use of mineral fertilizers increased from nil in 1967 to 92 per cent in 1980 and to 100 per cent in 2004.
- In 1967, no farmer used herbicides but they were used by 12 per cent of the farmers in 1980. During that period, they had been tried once, by at least one third of the farmers. In 2004, herbicides (2,4 D) were used by 96 per cent of the farmers.
- Insecticides were not in use in 1967 or in 1980. During that period, they were tried by less than 10 per cent of the farmers. In 2004, insecticides were used by 20 per cent of the farmers.
- One third of the farmers had adopted new farm implements between 1970 and the 1980s but stopped using them. In 2004, no farmer used the CADU plough or harrow.
- Since the mid-1960s, the acreage of wheat has been greatly expanded at the expense of grassland areas. This has led to a reduced number of cattle and less fallowing.
- Erosion increased both at Wajji and at Dighellu but particularly at Kulumsa, where all land was under cultivation in 2004.
- Yield levels of wheat and barley have increased. In 2004, wheat yields had more than doubled since 1967 and those of barley had increased by 60 per cent. There was a yield reduction by one third for field pea.
- On average, it took seven to eight years for the majority of farmers to adopt improved wheat seed and the use of mineral fertilizers in the late 1960s and early 1970s. Adoption time was shorter for fertilizers. Within four years, 68 per cent of revisited farmers adopted fertilizers, compared to 40 per cent of the newly visited farmers.
- Up to 1974, farm visits by an agricultural extension agent had been more common for revisited farmers (72%), who used to be model farmers in the development project, than for newly met farmers (8%). In 1980, the agricultural advisory role was solely handled by the peasant organizations. In consequence, agricultural extension agents seldom paid visits to farmers (4%). In 2004, the agricultural extension agents made selective visits to one third of the farmers.
- In 2004, the three major problems for farmers were the high cost of fertilizers, their unavailability when required and an increasing land tax. High costs of mineral fertilizers were identified 23 years ago. Additional concerns included shortage of both arable land and grassland areas and short supply of improved seed and herbicides.

Swedish Farmers under a High Degree of Tenancy

Sweden

Sweden is a small country of about 450,000 sq km. It has a population of 9 million, an increase from 7 million in 1950. About 2 per cent of the population is active in agriculture, compared to 55 per cent in 1900. Today, some 2.6 million ha are cultivated or under fallow, a decrease from 3.5 million in the 1950s. Forests cover half the area or some 23 million ha. Since the 1920s, the forest resources have almost doubled from 1.8 billion cu m in growing stand. In the mid-1960s, every single Swede had about 3 ha of forested land at his or her disposal. Sweden is an industrial market economy. The per capita income has more than doubled since 1979, reaching about US$ 38,500 in 2005 after a decline in the early 1990s.

Agriculture and forestry employ about 400,000 people and comprise about 9 per cent of the gross national product (GNP). About 100,000 persons are employed by the Swedish forest industry and the food industry employs some 60,000 persons. The Swedish Federation of Farmers owns almost half the Swedish food enterprises and foreign companies control about one third. Together with products from forestry, cars and pharmaceutical drugs are significant exports mainly to Europe and the United States. About 15 per cent of total Swedish export goes to Asia but some also goes to Africa and Latin America. Imports from these continents account for about 11 per cent. The food export has increased by 150 per cent since Sweden joined the EU, vodka being a major item. In 2003, the export value of forestry products was almost four times that of food exports Starting in 2001, paper pulp had to be imported from Indonesia and other countries.

Selected Area and Farmers
The province of Scania in southern Sweden is almost the size of the island of Trinidad. It accounts for about one third of the national primary crop and animal production, equivalent to almost 45 per cent of the value of the total food industry. During the last 40 years about 13,000 ha of the arable soil of top quality has disappeared due to urban development, roads and more than 50 golf courses. Non-farm activities are now regaining importance to many farmers. In 2001, there were some 10,000 farms in this province, although agro-industry is expected to require only some 1,000 farms by 2010.

Gramanstorp is a former parish situated in the northwest of the province of Scania. Agriculture has been practised for a very long time and farming expanded in the 10th century. The province has been under both Danish and Swedish rule, ultimately becoming Swedish territory in 1658. In the neighbouring parish, one early industrial activity, a paper-mill, was established in 1573, followed by a brewery (1877), a dairy (1893), a leather factory (1893) and a brick-works (1898). Except for the paper-mill, these industrial activities have now closed down. In 1906, the parish was connected to a railroad system, which was replaced by buses in 1953.

In 1980, all former and active farmers of the parish were approached for interviews and field investigations. The majority of the farmers were born in the parish or had lived there most of their lives. About one third of the tenants had moved to Gramanstorp between 1925 and 1980. Up to the 1960s, many tenants had additional income from a variety of sources, such as carpentry, slaughtering, mail distribution, milk collection, blacksmithing, daily work at the estate, ice-cutting for delivery to restaurants in winter time, chemical weed control and wood cutting. In 1980, farming was the single source of income for all but two farmers. The average household was composed of 2.5 persons. This was in contrast to 4.6 persons in 1925 and 4.3 in 2004.

Within the parish, there are seven distinct sub-locations, three on the plain land and four with more forested surroundings. In 1925, the parish had 112 farms with a tenancy rate of 90 per cent. Their size ranged from small (< 10 ha) and relatively small to medium size and large (> 50 ha) farms. In 1980, there were 19 farms, the tenancy rate still being high (58%). In 2004, three farmers remained active in agriculture. Two of them had formed a limited company, whereas one was a tenant. In all, 69 different farmers were consulted about technical changes in their agriculture.

Farm Size and Terms of Lease
In 1980, the average farm size was 93 ha, including one very large entailed estate. Eight farms had more than 50 ha. Exclusive of the large estate, the average farm size was 41 ha. This was a doubling since 1925. In 2004, the average size of the three remaining farms was 593 ha of arable land. In addition, the largest farm had 1,300 ha of forested land. Today, the average acreage of a Swedish farm is about 40 ha.

Around 1900, the terms of lease began to be based on cash instead of day-labour obligations. With one exception, all the tenant farms belonged to one major landowner. Up to the 1940s, there were few changes of the terms of lease per hectare between different farm sizes. This changed in the 1960s, stimulated by the policy of the Swedish Government. The terms of lease increased significantly and the tenants of small farms had to pay a higher lease per hectare than those of larger farms. Between 1925 and 1975, the terms of lease increased by almost 1,200 per cent. In contrast, prices of wheat and barley increased by about 200 per cent and that of nitrate of soda by 150 per cent.

Technical Changes and Adoption of Agricultural Innovations

Crops and Crop Rotation
In the mid-1930s, tenants started to cultivate sugar beets, previously grown only by a few of the large owner-cultivators. A few tenants on the larger farms tested white mustard. Otherwise, the most common crops included rye, barley, wheat, oats and leys with clover. All farms had dairy cows and horses. In the early 1940s, oilseed-rape was introduced as a new crop. In

1948, a small tenant was the first adopter of marrow-stem kale. At that time, some owners and tenants were cultivating timothy and clover for seed on a contractual basis. Contract farming of carrots, beetroot and other vegetables was introduced in the late 1950s. A decade later, four tenants on the smaller farms had begun producing strawberries and black currants for a commercial market. In 1972, a tenant of a larger farm introduced broad bean as a new crop. The most recent new crop introduction was the cultivation of canning and freezing peas on contract with the Findus Company.

By 2004, cereals (wheat and barley) were the predominant crops on the plain land. Farmers bought certified seed of all cereals every year. The sub-locations close to the forested land had turned into young forest or permanent pastures. In 1980, the majority of farmers bought certified seed of rye and winter wheat every year. For barley and oats, they used their own seed for the subsequent three or four years.

In 1980, there was no systematic crop rotation because of the expansion of cereals at the expense of leys in the 1960s. When the government policy made milk production less profitable, farmers turned to continuous cultivation of cereals. Prior to that, most farmers grew two-year leys as part of a seven-year rotation with two cereal crops. It was also a necessary combination in order to produce grass for the dairy cows. In principle, the same crop rotation system was in use from the 1930s up to the late 1950s: ley-ley-barley/oats-fallow-winter wheat-sugar beet-barley. In 2004, two farmers were fully dependent on milk production, cultivating crops for ensiling and cereals as feed for their dairy cows. This cultivation did not follow a strict crop rotation system.

Organic Manure and Mineral Fertilizers
Up to the early 1960s, cow manure was applied on potato, sugar and fodder beet. In 1980, farmers with pig and milk production used manure on potato, cereals and leys. In 2004, the two dairy farmers applied organic manure primarily on winter wheat.

Nitrate of soda and other fertilizers came into common use after World War II. In the 1940s, fertilizer spreaders were introduced on the larger farms at Gramanstorp. They were gradually adopted according to farm size, the small producers purchasing second-hand ones. One fifth of the farmers, mainly the owner-cultivators, invested in soil mapping and analysis of nutrient content during 1940-1980. A small minority did it twice or three times. In the early 2000s, the remaining farmers had recently invested in a new soil mapping and soil analysis.

Up to the mid-1970s, nitrogen fertilizers were applied at higher rates than the national average. In a national context, the use of nitrogen fertilizers had increased from 30 kg/ha in 1960 to 80 kg/ha in the mid-1970s. In the early 1980s, rates of fertilizer applications had begun to drop below official recommendations. At that time it was believed that a reduction of fertilizer use by 25 per cent might reduce yields by some 10 per cent. A reduction by 50 per cent would imply a yield decline of 20-25 per cent (Jonsson, 1982). Around 2000, mineral fertilizers were applied to three quarters and manure to one quarter of the Swedish acreage. National average nitrogen application was about 100 kg/ha in the early 2000s.

Weeds and Herbicides
Up to the 1940s, the most important weeds at Gramanstorp included *Circium arvense, Sinapis arvensis* and *Chenopidum album*. For the next two decades, *Stellaria media* and *Galium aparine* remained the major nuisance. In 1980, farmers complained mainly about *Chenopidum album, Polygonum* sp., *Matricaria perforata* and *Elytriga repens*. The weed flora of spring-sown crops was mainly composed of dicotyledonous weed plants. Prior to herbicide sprayings the average field had about 200 weed plants per square metre, competing with some 200 plants of wheat and 340 of barley. Field observations showed that herbicides reduced the number of weed plants by more than 70 per cent. Over time, certain weeds have declined in numbers and new ones have appeared. During the 1990s the monocotyledonous weeds have increased, particularly *Elytriga repens* and *Apera spica-venti*. As a new weed, *Avena fatua* was first observed at Gramanstorp in the late 1960s due to the rapid expansion of cereal production and the joint use of combine harvesters. *Lamium purpureum* was on the increase up to 1980. Even in 2004, one farmer argued that *Avena fatua* was a new weed to him.

In 1949, both herbicides and insecticides were introduced at Gramanstorp. On commission by the Svalöf Farm School, one tenant was given this responsibility. This approach led to quite a rapid adoption of herbicides. In 1980, the majority of farmers (92%) used herbicides, a rate of adoption in line with national statistics. All large cultivators did so within a decade, as did some small farmers (50%) and medium-size farmers (85%). In 2004, herbicides were in common use on crops. For environmental reasons they could not be applied in forest production, which was much regretted by one farmer who was responsible for 1,300 ha of forests.

Agrochemicals against Pests and Diseases
In contrast to herbicides, the adoption of insecticides was slow. In 1980, a minority of farmers (46%) used insecticides. The majority of the large farmers did so but small farmers (7%) and relatively small farmers (22%) were reluctant. They argued the chemicals might damage honeybees and other useful insects—a signal of "the green movement". In 2004, insecticides were commonly applied, if needed, mainly to control aphids.

Since 1985, the national use of herbicides, fungicides and herbicides has been reduced by 70 per cent. In 1995, the Swedish Government introduced an environmental programme and the Parliament decided that the use of poisonous chemicals in agriculture should be reduced by one tenth by 2001. In terms of active substances, however, their use increased, one reason being the priority given to winter wheat through the EU subsidy system.

Mechanical Innovations

The majority of farmers at Gramanstorp adopted a range of labour-saving agricultural machinery, a process that again started after the end of World War II. As an exception, large farmers purchased the iron-wheel tractor in the mid-1930s. The milking machine was introduced in the mid-1940s. Saving manual labour, it spread to a majority of farmers over a decade.

In 1956, all owner-cultivators had a tractor with rubber wheels; only half the tenants did. In 1980, every farmer had at least one tractor. In the early 1950s and onwards, half the farmers on large and medium-size farms adopted the towed combine. The time taken for its adoption was roughly 8 years with little difference between tenants and owner-cultivators. It was usually bought second hand at auction. In 1958, the first combine harvester was introduced, gradually being adopted by all farmers between the late 1950s and early 1970s. Again, the majority of farmers bought this machinery second hand. In 1980, most farmers (79%) had their own combine harvester. All farmers had them in 2004.

The first grain drier using cold air appeared in 1960. Six years later, the first drier using hot air was installed, followed by a second one after 10 years. In 1980, all large farmers and all medium-size farmers except one had grain driers of their own. More recent examples of adopted innovations at Gramanstorp include the no-tillage system, which has been practised since 2001. Other examples are the use of mini-tractors and large grain driers. The rotary system for dairy cows was introduced on the largest farm in 2003. It was considered superior to a milking robot since it carried less economic risk, required less maintenance and repair and involved a shorter milking time. The rotary system was designed long ago but has mainly been used in the United States and New Zealand. It appeared in Sweden first in the 1970s, whereas the milking robot was launched around 2000.

For most of the 20th century, agricultural machinery was mainly produced by or with active involvement of the Swedish manufacturing industry. Nowadays, there is no national manufacturer of tractors and combine harvesters and only one Swedish manufacturer of milking and dairy equipment.

Crop Yields
Between 1925 and 1980, the yields of winter wheat at Gramanstorp tripled and those of barley more than doubled. Thereafter, crop yields have increased at a lower pace. During the last decade, the farmers considered yields of winter wheat to have increased by 750 kg/ha, an increase they did not find profitable. This overall trend among farmers is even reflected at the national level. From an average yield of winter wheat in Sweden of some 2,100 kg/ha in 1945, it more than doubled to 4,800 kg/ha in 1980 and is now about 6,100 kg/ha. In the late 1880s, average wheat yields were about 1,600 kg/ha (Mattson, 1978). The early increase was modest or some 500 kg/ha between the 1880s and 1945.

Table 2. Average yields of wheat and barley between 1925 and the early 2000s (kg/ha) according to farmers at Gramanstorp.

Crop	Year or period for estimations			
	1925	1950	Late 1970s	Early 2000s
Winter wheat	1700	3200	5200	6250
Barley	1600	2400	4000	5000

Milk and Pig Production
In 1980, 8 of 19 farms had dairy cows. The average farmer reported an annual milk production of about 6,200 kg per cow, a significant increase from 2,500 kg in the 1930s. One farm kept 280 cows, whereas the average dairy farm had 23 cows, an increase from 11 in the early 1950s. In 2004, the two dairy farms had an average number of 215 cows, the larger having more than 350. The average cow produced 9,750 litres of milk, a fourfold increase in 75 years.

In the early 1930s, Sweden had 1.8 million milking cows, a figure that has now declined to some 400,000. Likewise, the number of milk producers has declined from 200,000 in 1960 to less than 9,000 in 2005. The average dairy cow produces some 9,000 kg per annum. It took 45 years to increase the average annual milk production from 2,500 to 3,500 kg. Such an increase is now believed to take less than 10 years, a trend also anticipated by the visited farmers.

In 1980, 11 farmers at Gramanstorp raised pigs with an average herd size of 310 pigs. Three farms produced annually more than 600 pigs. In 2004, no farmer produced pigs.

Agricultural Extension, Research and Experimentation

Up to the 1980s, the majority of farmers (90%) claimed the agricultural extension staff had not visited them on their farms. The others, mainly those on large and medium-size farms, had themselves contacted agricultural extension officers for advice. One third of the farmers found radio programmes on agriculture to be of general interest, although seldom of practical use. Field demonstrations by private companies and the county agricultural society attracted many farmers. In 2004, the farmers were in close dialogue with the county agricultural society on crops and the society of artificial insemination. Although recognized as a source of technical information, the Internet was seldom used. It was too time-consuming compared to the rapid service farmers could get via their mobile phones.

In the past, sales representatives of private firms had played a major role by annually visiting three fourths of the farmers to provide agricultural machinery, seed, agrochemicals, paint and other products to both tenants and owner-cultivators. Over time and with the declining number of farms, the representatives stopped visiting each farmer. When the private local grain trader closed down in 1973, farmers had only one option, to turn to their nearest agricultural cooperative society. A similar situation prevailed in 2004 but that society was located far from the farmers.

In the 1950s, the county agricultural society had started to conduct field trials on some farms in the area. The experiments focused on application rates of fertilizers and crop cultivars. For advice on crop cultivars, farmers found the seed catalogues of the plant breeding institutions the best source of information together with advice by the local grain trader. In 2004, farmers gained new knowledge from various sources, including dialogues with other farmers living outside of the former parish.

Between 1940 and 1980, farmers themselves were quite innovative regarding both new technical solutions and for the modification of purchased agricultural machinery in close consultation with local blacksmiths. One tenant farmer invented the Bjersgard hurdle in the 1950s and even acquired a patent. Using shorter hayrack stakes and a single strand of barbed wire placed in a notch on the top, a tractor with a front loader could replace manual labour. No small farmer adopted that innovation. This system was later replaced when the forage harvester was introduced. In 1960, another tenant built the first grain drier at Gramanstorp, using the cold air technique, based on a design by the National Testing Institute for Agricultural Machinery. In 2004, there was no example of an innovation produced by the farmers themselves.

Some Development Indicators

A telephone system was first installed in 1910 on a few farms in the nucleus of the Gramanstorp village. Electricity for light was available to these farms in 1918 but larger farms outside the village got electricity in 1925-1931. The majority of farms got electricity installed after 1941 and the last ones had to wait till 1953. Radios appeared in the mid-1930s but were not commonly adopted until the 1940s. The installation of telephones on the farms started in 1940 and was completed in 1954. The first television set appeared in 1957. In 1980, all farmers had a TV and a private car of their own. In general, the adoption process for these development indicators took longer than most agricultural innovations. For agricultural use, personal computers appeared on the farms around 1990 and so did mobile phones.

Problem Identification by Farmers and Their Vision for 2015

In 2004, farmers found costs of crop production a major constraint, as well as the increasing set of rules and regulations through the EU. A technical problem is shortage of personnel in animal production, which farmers presumed will persist for quite some time. At the same time, some 8,000 Swedish persons born abroad, and often unemployed, have education for work in agriculture, forestry and land use systems. Environmental concerns in forestry were identified as obstacles to a balanced production. Bio-energy was seen as a potential avenue for future land use. Farmers doubted the effectiveness of the Swedish Federation of Farmers, since it has an ambiguous role as the spokesman for all Swedish farmers. This is not possible, since 15 per cent of them are responsible for 85 per cent of the total production. Still, the major producers have no real power over the food they produce, according to farmers at Gramanstorp.

As to the future, the farmers envisaged continued specialization in both crop and animal production. Competition among farmers will accelerate, they feel, leading to further emphasis on economics. The number of farms will continue to decline, although there might gradually be a limit to farm size, partly due to the environmental regulations, for instance regarding a maximum number of cows per farm. The recent expansion of the EU in 2004 means an additional 4 million farmers to the existing population of 7 million farmers. This will definitely change Swedish agriculture, although the farmers will continue to get financial support equivalent to SEK 6.6 billion. In 2004, the Swedish contribution to the EU was about SEK 24 billion. Future development of agriculture requires a sound national capacity to produce a steady supply of fuel, inputs and raw materials and feed, the ability to repair agricultural machinery and irrigation pumps, and easy access to spare parts and other essentials. These are all national requirements in developing countries as well.

Summary of Highlights at Gramanstorp

- In 1903, agricultural research started in the county of Kristianstad, where the former parish of Gramanstorp is situated. The initial focus was on fertilizers. For many years, most field trials (75%) were devoted to studying the use of mineral fertilizers. The first field trials on fertilizers in the parish were laid out in 1927. A few trials were conducted on the larger farms with the best soils. It was not until the mid-1950s that more diversified field research activities were started to include trials on crop varieties, herbicides, drainage and crop rotation. Few trials were laid out at Gramanstorp and only some experiments on crop varieties were carried out during the last two decades. In 2004, there were no field trials in the former parish.

- A focus on field trials of mineral fertilizers was a national trend up to the early 1960s. Three quarters of the Swedish field experiments dealt with plant nutrients. The trend gradually declined and reached 20 per cent in 1981. The national service for agricultural field experimentation ceased in the late 1970s and was transferred to the College of Agriculture.

- For a long time, tenants living in the northeastern part of the parish had cleared forested land for cultivation, a process that some of them continued up to the late 1940s. When tenants started to give up farming in the late 1960s, some of these previously forested areas were gradually reforested, turned into grassland or even left fallow in the late 1990s. On the plain land, the acreage under cereals increased, particularly for wheat.

- Between 1925 and 2004, the yields of winter wheat more than tripled and those of barley almost did the same. In the early 2000s, cereal yields at Gramanstorp were close to national averages.

- For the majority of farmers, average time for adoption of a variety of agricultural innovations was 8-10 years. Rate of adoption was sometimes based on farm size, sometimes on ownership. New crop varieties were generally adopted more quickly. In the late 1970s, it generally took 2-3 years after a wheat or barley variety had been accepted in the national list of crop varieties. In the 2000s, all farmers bought certified cereal seed every year.

- Up to 1980, the agricultural extension agents visited only 10 per cent of the farmers, mainly those with large and medium-size farms. To all categories of farmers (in total 75%), the sales representatives of private companies paid regular, annual visits up to the mid-1960s. In 2004, the remaining farmers had regular contacts and visits upon request by technical advisers.

- The former parish is characterized by a dramatic decline in number of farms. From 112 farms in 1925, the number was reduced to 19 in 1980. The rate of tenancy has long been high (90%) and remained so in the mid-1950s. In 2004, there were three farms left in the former parish, two of which operated as private companies. Nationally, the rate of tenancy was 45 per cent.

- No specific attention was directed to the research problems of resource-poor farmers, for example, the tenants with small holdings but constituted the large majority up to the late 1980s. During a period of 70-75 years, this category of farmers vanished.
- Starting in the 1930s, the majority of farmers gradually joined different farmers' cooperative associations. In 1980, this was more common among larger than smaller farmers. In 2004, all farmers were members of agricultural producer cooperatives.

Brief Comparative Analysis of Policies for Agricultural Research and Extension

Adoption of Agricultural Innovations

As a general conclusion from the field studies in the three countries, it took generally about 10 years for a majority of farmers to adopt a variety of agricultural innovations. The most easily adopted agricultural innovations were the use of mineral fertilizers and improved crop cultivars. Time taken for adoption of crop cultivars was somewhat shorter. Average time for adoption of the use of herbicides was also about 10 years but more than 15 years on aroids in Trinidad. There were no herbicides specifically designed for aroids. It should be noted that dicotyledonous weeds were most frequent in Sweden (92%) and Ethiopia (60%) and less so in Trinidad (20%). This has implications for the type of herbicides to be designed under different ecological environments. The period for adoption of the use of insecticides was much longer than for herbicides. They were never fully adopted in Ethiopia or in Trinidad (Table 3).

Table 3. Percentage of resource-poor farmers adopting agrochemicals between 1965 and 2003/4.

Agrochemical	Period	Trinidad	Ethiopia	Sweden
Fertilizers	1965/1967	30	0	100
	1980	44	92	100
	2003/2004	50	100	100
Herbicides	1965/1967	0	0	72
	1980	33	12	100
	2003/2004	50	96	100
Insecticides	1965/1967	0	0	10
	1980	28	0	22
	2003/2004	13	20	100

Mechanical innovations were more common in Sweden; the time taken for adoption was again about 10 years (the Bjersgard hurdle, tractor on rubber tyres, fertilizer spreader, forage harvester). It took somewhat less than 10 years for the milking machine and the AIV ensilage method (named for A I Virtanen, its inventor) to be adopted. Small farmers adopted the milking machine more rapidly than the large farmers, who were quicker in buying the combine harvester. Adoption time for towed combines was more than 10 years. Up to the late 1960s, local blacksmiths were instrumental in modifying purchased technology for Swedish farmers. They were not involved as a resource in Trinidad or in Ethiopia, lacking adequate tools to be helpful. In Trinidad, tractors were the major mechanical innovation, the adoption taking much longer. In Ethiopia, the majority of farmers rejected the introduced CADU farm implements. The plough was too heavy, a fact already encountered in the development of a plough for Swedish farmers and their oxen in 1800. But labour remains a constraint, explaining why the Sasakawa Global (SG) 2000 now promotes animal traction, for instance in Uganda. A training programme for more than 850 farmers on a multipurpose tool bar has been carried out to use more than 250 new animal training kits (Foster et al., 2002).

The conclusion on adoption time is consistent with other examples, for instance the spread of hybrid maize in the United States. Maize was first grown in the 1930s, and farmers planted about half of the US maize hybrids by the mid-1940s. If time for research design and field experimentation is added, one can anticipate a total time of about one and a half decades for R&D to reach a majority of farmers in the field. This has significant implications for how to reach millions of resource-poor farmers up to 2015 in line with the targets set by the MDGs. Many observers, such as the UN Millennium Project, argue that technology is available. Field records, however, have shown that the process takes time and available technology is often not appropriate, for various reasons, to the major target group of farmers. In a not so distant future, fertilizers may not remain a cheap agricultural input as they used to be. Second, farmers need agricultural innovations other than fertilizers and improved seed. Third, much agricultural research is directed towards certain crops not always of high priority to resource-poor farmers. Research by the private business sector in agriculture focuses mainly on crops and animals for a commercial sector. Another factor is that adoption of innovations is affected not by technology alone but also by various aspects of government policy, which may or may not be conducive to development.

The overall context in which the adoption of innovations takes place needs to be considered. It took 8-10 years for the majority of adopters to accept fertilizers in Ethiopia. In Trinidad, the time span was about the same although shorter for fertilizers on eddo at Caroni. By the late 1940s, sulphate of ammonia was introduced by the plantation industry in the sugar belt in Trinidad and distributed to small sugarcane producers at Caroni. A decade later, it was commonly applied to cocoa, banana and coconut but it was applied much later to food crops. The first eddo farmer applied fertilizers only in the early 1960s. At Gramanstorp in Sweden, nitrate of soda from Chile was first used around 1925 on a few farms. Other fertilizers came into more common use in the mid-1940s. When fertilizer prices increased, as in the 1970s, farmers stopped using them or applied much less than recommended rates, a common feature in all three cases. Likewise, a general feature of field trials on fertilizers was the search for optimum levels, although farmers seldom applied such rates.

National fertilizer consumption in Ethiopia has increased substantially from about 31,000 tons in 1981 to 230,000 tons in 2000. Still, it is estimated that only 20 per cent of Ethiopian farmers apply fertilizers, a tenfold increase from 1995. In Sub-Saharan Africa, however, fertilizer consumption stayed the same over the last 15 years, whereas in Asia it more than doubled. For a long time, the World Bank supported fertilizer subsidies but no private sector supplier could be involved in the follow-up work during the Ethiopian Military Government. In addition to a parastatal, AISCO, private importers of fertilizers handled one third of the imports to Ethiopia. One reason for quick adoption of fertilizers on eddo in Trinidad was proximity to the fertilizer factory.

Between 1967 and 1980, the use of improved seed by Ethiopian farmers increased from 16 to 78 per cent for wheat and from nil to one third for barley. In 2004, however, the farmers complained about a shortage of improved seed of their choice, in particular of wheat. In fact, the Kulumsa research station had difficulties in producing sufficient quantities of certified seed for the Ethiopia Seed Enterprise. This demand is also illustrated nationally since the use of improved seed tripled from the mid-1990s to the late 1990s. Today, there is also a private seed company, the Ethiopian Pioneer Hi-bred Seeds. Seed production is of great concern in the whole of Sub-Saharan Africa. During the last 15-20 years, most governments closed down their public seed companies, hoping the private business sector would take over. Still, improved seeds are not available to small producers with regularity and in the quality and quantity they want. In 2004, the SG 2000 even proposed a private sector Smallholder Foundation Seed Service.

No improved genetic material was available to aroid farmers in Trinidad. Still, the International Institute of Tropical Agriculture (IITA) has identified and refined a treatment to induce flowering in edible aroids. Thus, the bottleneck of selective hybridization and genetic improvement of these crops may be removed. In the mid-1990s, the Technical Advisory Committee of the Consultative Group on International Agricultural Research (CGIAR) reviewed research on roots and tubers, noting that the consumption of roots and tubers, particularly yam, was increasing in Africa. But the committee found no need to change mandates of the CGIAR centres or add aroids to the CGIAR research agenda. This position is to be questioned when CGIAR recently has agreed to a focus on the MDGs. In its recent priority setting for 2005-2015, the new Science Council has not explicitly included crops such as edible aroids.

In the early 1980s, Swedish farmers at Gramanstorp adopted new crop varieties of wheat 2 or 3 years after they had been accepted in the national list of certification. Improved seed was made available long ago, starting with selections in the early 1900s, resulting in early yield increase of some 25 per cent for wheat. A range of crop varieties appeared thereafter through plant breeding. Sometimes, large areas were covered, such as for "Starke" winter wheat (90%), first released in 1959, or "Kosack" in the 1990s. Today, crop varieties change even faster. Since seed production is now confined to one Swedish plant breeding company, expanding towards the international market, its future crop varieties may have to cover wider ecological regions and markets. This will provide more space for local seed companies, such as Scandinavian Seed and TD Foradling Ltd.

Great ecological variations at the sub-locations in Trinidad (3), Ethiopia (4) and Sweden (7) call for consideration of a more distinct focus to suit the actual climatic conditions of resource-poor farmers. Variations in altitudes in Ethiopia within short geographical distances and the ecological differences between plain land and more forested areas in Sweden are good illustrations. This relates to agricultural research in general and specifically to plant breeding for quality seed. Globalization of seed production will not help ecological precision of new varieties. This would be of great concern in many countries and may gradually result in a need for new seed producers targeting local markets.

Access to land is one factor influencing the rate of adoption of agricultural innovations, often most pronounced in the early phase. There was little difference in the rate of adoption of the use of fertilizers among aroid farmers who leased their farm or who were owner-cultivators. Among the owner-cultivators, there were more non-adopters of insecticides than of herbicides. In Ethiopia, one third of the tenants adopted the use of fertilizers compared to two thirds of the landlords during the first 4 years after their appearance. That difference disappeared completely over a span of 7 years. Prior to the revolution in 1974, only one tenant had used herbicides, compared to 43 per cent of the owner-cultivators. No tenant had tried insecticides or tested improved farm implements, actually designed for the resource-poor farmers and of little use to the feudal landlords with large farms. At Gramanstorp in Sweden, a similar trend was noted. Within 5 years, one third of the tenants and half the owner-cultivators adopted the use of herbicides. After 10 years, all farmers had adopted that practice. For most mechanical innovations, the initial rate of adoption was more frequently based on farm size than on ownership of land.

These few examples illustrate that the rate of adoption of new technology must be measured in terms of consequences for different categories of farmers, in light of the local power structure and the time horizon. Land tenure is a major issue in many countries. Customary property regimes support mobility, and the presence or absence of a title to land in Africa affects the use of land-improving investments or productivity. Thus, development efforts must be seen in the context of ownership and in a longer time perspective than usually applied. Then, it may be politically more acceptable for a government with both economic and social concerns to accept some initial, though adverse, effects to some farmers rather than enforce all farmers to the same level of poverty, as demonstrated by the Ethiopian Military Government in the 1970s and 1980s.

Agricultural Extension

New varieties of crops and new animals change the behaviour of farmers who adopt new ideas. With the introduction of technical changes, agricultural researchers also create social changes. In consequence, the fundamental assumptions of the so-called diffusion model in agricultural extension have proved dubious. That model was mainly confined to farm-level changes and a notion that there was relative equality among farmers with respect to technical change. Since the agricultural extension staff in Ethiopia was too small and it was given insufficient resources, that system of agricultural extension developed a specialized clientele of model farmers, disliked by the Ethiopian Military Government. On the other hand, the peasant organizations were short of technical expertise. The agricultural extension programmes in Sweden and Trinidad and Tobago had the same approach.

In all countries, few farmers were actually visited on their farms. For a number of reasons, the agricultural extension agents did not travel to the more isolated areas in the 14 sub-locations. Thus, it would be reasonable to make better use of agricultural radio programmes, even commercials as in Trinidad. They provided useful information to farmers. They can be further developed and would be less expensive than a large agricultural extension staff in the field. This may be a new task for the private sector. Major suppliers of agricultural innovations could establish a radio-based news agency similar to the one launched by the World Association of Community Broadcasters, with its focus on development issues.

Furthermore, farmers can be used as the principal agents of change, such as the Swedish tenant who introduced herbicides at Gramanstorp, most likely stimulating rapid and high rate of adoption of their use for all categories of farmers. In 2004, one Ethiopian farmer suggested that he and other farmers could produce improved seed if trained how to do so. This illustrates a potential among resource-poor farmers, illustrated for instance by the "peer-to-peer" training programmes in Cameroon. Farmer training is crucial, must be appropriately designed on priority areas and must include women. A transfer of responsibilities from a less efficient agricultural extension service to farmers ought to include work for them in conducting simple on-farm trials and recording the most acute problems of their fellow-farmers.

Recently, one myth affecting extension training and delivery in Africa was highlighted, namely that only science-based technologies are worth communicating to farmers (Opido-Odongo, 2000). Many farmers are innovators and further develop the technology provided to them by tapping additional sources of information. This is why simple demonstration plots may be effective, as was done almost 40 years ago by the CADU project, probably facilitating the rapid adoption of fertilizers by Ethiopian farmers around Asella. Nationally, the FAO Freedom from Hunger Campaign had launched such an Ethiopian demonstration programme in 1967. Nowadays, SG 2000 is repeating this approach, having established almost 400 Extension Management Training Plots as "standards of excellence plots" (Sasakawa Africa Association, 2003). Similar approaches can be used to demonstrate new cultivars, the use of other inputs or cultural practices. A possible disadvantage might be an initial delay in the adoption process. The key factor will be that the new technology should be timely and locally available to those farmers wishing to try them.

Further financial investments in the conventional agricultural extension service are justified only if staff are better trained, better paid and more mobile. Often, there are too many supervisory staff in offices for report writing and too few staff members in the field in constructive dialogue with farmers. Effective and relevant training must consider the various roles that farmers expect from them. One interesting initiative is a mid-career extension programme developed at the Alemaya University in Ethiopia. It is a practically oriented BSc programme for only 2-2½ years instead of the conventional 4-5 years. The course is targeted to those with at least 5 years of field experience. Students share their insights with the university staff, who have more theoretical knowledge. At its start in 1997, it experienced strong resistance and various challenges but soon considered quite successful (Gebrekidan, 2000). In 2005, this programme at Alemaya University had involved a total of 247 Ethiopian students, out of whom 191 had graduated (Sasakawa Africa Association, 2006).

Agricultural Research and Experimentation

For quite some time, there has been a false assumption that the research establishment and development agencies are well aware of the major problems of farmers. This has been accentuated by the lack of dialogue between agricultural researchers and farmers. In the three communities, the identification of relevant research problems by the agricultural extension service seems to have been almost non-existent. In 2003/2004, there was no feedback from farmers in Trinidad or Ethiopia to the research establishment at the UWI or the Kulumsa Research Station, respectively. Thus, agricultural researchers select the topics of research. Those topics may or may not be of relevance to the resource-poor farmers, fertilizers and improved seed being good examples. Farmers may have more to offer. For example, in Colombia, scientists allowed local farmers to choose among 20 new bean varieties. Because of their better knowledge of their terrain and personal interest in increasing yields, their selections out-performed those chosen by the scientists by 60-90 per cent. In Africa, farmers have recently proved to be very knowledgeable about the best agroforestry species. Early involvement of farmers in the research process makes it possible to develop improved practices and speed up the process of adoption of new technology. Through this approach, together with field demonstrations, new technology may spread on its own merits and not require an agricultural extension agent to contact individual farmers. Such an approach requires more specific research methodologies to cope with ecological variations, greater precision in research and new experimental designs with much more research in farmers' fields.

Early agricultural research in Trinidad did not focus on small producers or food crops. In 1891, a Botanical and Agricultural Department was set up as an extension of the Botanical Garden, which had been founded as early as 1819. A Department of Agriculture was established in 1909 but it was not until the early 1960s that a Food Crop Breeding Unit was created. In 1969, the Tate and Lyle Sugar Research Station of Caroni Ltd. refocused on problems of the local sugar industry, since its international activities were transferred to the United Kingdom. A Regional Research Centre started in 1965, gradually becoming a part of the Faculty of Agriculture of the UWI. When the UK Government reduced its contribution to that Centre in 1975, it was transferred into the Caribbean Agricultural Research and Development Institute (CARDI), which was supposed to work in close collaboration with the Faculty of Agriculture of UWI. It was to transform the agricultural sector to be internationally competitive, applying a market-driven approach to research. This was hardly a move towards a focus on resource-poor farmers. Since 1999, the financial contributions by the Trinidad and Tobago Government have been reduced by one fourth. In the early 2000s, research on root crops concentrated on sweet potato, cassava and yams, but not aroids. It is very doubtful whether such research orientation is being designed within the framework of the new University of Trinidad and Tobago launched in 2005. Although it plans to refrain from a disciplinary research approach, its role is more industrially oriented, it being designed to "discover and develop entrepreneurs".

In Ethiopia, official agricultural research is of recent origin. The Institute of Agricultural Research was founded in 1966. At the same time, research and experimental work was started by the CADU project around Asella. Although the institute was supposed to have about 140 agricultural staff members, it never reached that target and had great difficulties in the 1970s due to high staff turnover and shortage of funds, vehicles and even petrol. In the 1970s, the Plant Genetics and Resources Centre was formed and so was the Ethiopian Sorghum Improvement Project. Today, the Kulumsa Research Station is part of the Ethiopian Agricultural Research Organization. It has five substations covering the lowlands (1,650 m) up to the highlands (3,000 m). It has a staff of 42 researchers, compared to only two at CADU in 1966. Its major activity is the production of breeders´ seed for the Ethiopian Seed Corporation. In the short term, vegetable seeds and temperate fruits are also planned. So far, forestry has been given little attention. During 2003-2004, some 200 agricultural research projects were executed within Ethiopia, although the budget for 2002/2003 experienced a cut of 26 per cent (Alemaya University, 2003, p. 11). A significant portion of the agricultural research is financed through development assistance.

Swedish agricultural research dates back hundreds of years. An early handbook with monthly agricultural tasks was published in Swedish in 1643, also making reference to the effects of the moon on crops. After the foundation of the Royal Swedish Academy of Sciences in 1739, agricultural literature became more frequent. During the 1700s and the early 1800s, much of the research literature was contributed by owners of estates, large farms, clergymen and schoolteachers. To a large extent it was based on observations and experiences in the field or in barns. Scientists took over in the late 1800s and the early 1900s when specialized research institutes were also established. The 19th century saw an increased production of agricultural literature. The introduction of compulsory primary schooling in 1842 led gradually to a higher literacy rate. At the same time, it was argued that it was not a scarcity of information but rather a surplus that prevented farmers from finding the most useful information (Larsson-Kilian, 1916). Inputs from individuals continued to be substantial. In the late 1880s, a private initiative by Per Jönsson Rösiö led to a farm school and pushed for a new enthusiasm for agriculture. One of his pupils, Nils Larson started a farm school in the province of Skane around 1900. That training gave special attention to involving small farmers and those living in low-potential areas, an orientation not advocated by the official Swedish agricultural research. Such a focus is now important in international agricultural research for development.

Private initiatives were crucial for the development of agriculture at Gramanstorp and in southern Sweden: the first farm schools, the first import of fertilizers, the production of farm machinery, the establishment of a fertilizer factory and a private plant breeding station, all came prior to government actions. In Trinidad, the private sector played a great role in agricultural R&D, producing fertilizers and other agrochemicals, making their access relatively easy, even for resource-poor farmers. This was not the case in Ethiopia, which had a small private sector in agriculture.

In the 1940s, major efforts were made to elevate Swedish research to international level, although agricultural research was confined to domestic issues for quite a few decades. In 1945, the Swedish Council of Forestry and Agricultural Research (SJFR) was established. It was to provide public research funds to scientists in agriculture, forestry and veterinary medicine. In the early 1980s, it started to fund research on alternative methods of production. The notion of organic or ecological agriculture was politically seen as an initial and possible step towards a sustainable agriculture. Since then, the debate on ecological production has been lively. Low-input agriculture, as practised in many developing countries, was seen by many observers as unrealistic and a waste of resources. Nevertheless, the Swedish agricultural research establishment has, so far, failed to show other options for change, except business as usual. Under new leadership in the mid-1990s, the SJFR was reluctant to spearhead a change, confining itself to recommend to the Swedish Government that ecological production should be seen as "a niche" (SJFR, 1996). After Formas took over the tasks of SJFR in 2001 it has turned more international, even expressing interest in the Challenge Programmes of the CGIAR. For quite some time, there has been a declining trend in funding by the Swedish Government of agricultural research and field experimentation; it is anticipated that this work will be taken over by the private sector and the farmers themselves.

One special aspect relates to research linkages and exchange with the CGIAR institutes. Since the inception of the International Livestock Center for Africa (ILCA), some of whose activities are now carried out by the International Livestock Research Institute based in Nairobi, Ethiopia has served as its host country. A new plough for oxen was designed and tested in the field, similar to the one produced by CADU. Comprehensive reports on its impact are scarce. Today, the International Livestock Research Institute (ILRI) has certain activities in Ethiopia and so has the International Food Policy Research Institute (IFPRI), starting in 2004. It is unclear to what extent Ethiopian livestock research and production has benefited from its collaboration with ILCA and later ILRI over more than three decades. By 1980, Ethiopian partners had established research contacts with the International Maize and Wheat Improvement Center (CIMMYT), IITA and the International Crops Research Institute for the Semi-Arid Tropics

(ICRISAT). Through cooperation with the International Center for Agricultural Research in the Dry Areas (ICARDA), new lentil varieties were developed and are now grown on some 4,000 ha in the highlands. Over the years, 10 plant collection missions have been carried out in the country with support from the International Plant Genetic Resources Institute (IPGRI), collecting cultivated and wild species totalling more than 2,700 samples (E A Frison, 2004, personal communication). In 2005, IFPRI and Alemaya University set up a Center for Agricultural Research Management and Policy Learning for Eastern Africa. In Trinidad and Tobago, research contacts have been established between the former Faculty of Agriculture of UWI and CIMMYT, IITA, ICRISAT and the International Center for Tropical Agriculture (CIAT). The specific outcomes of these collaborative arrangements are less known.

The Nordic Gene Bank, created in 1979, has maintained close contact with the IPGRI. It has also served as consultant to the Swedish International Development Authority (SIDA) in its financial support to plant genetic resources. In the past, contacts between the CADU project in Ethiopia and Swedish research institutions promoted an exchange of plant genetic resources of primarily wheat, barley and chickpea. These institutions include both public research departments and the plant breeding institutions (the former Weibullsholm and Svalöf Seed Association, merged into Svalöf Weibull AB, renamed SW Seed in 2004). Access to this wealth of Ethiopian germplasm is considered to have been very useful for Swedish plant breeding work. Exchange of plant genetic resources has also included other CGIAR institutions.

Agricultural Policy

During almost three decades of rhetoric, donor agencies and national governments in developing countries have given lip service to the need for poverty eradication. On the whole, little attention has been given to agricultural research policy and the consequences of agricultural research for different categories of farmers. In developing countries, very few social institutions can counter-balance negative consequences of the introduction of new technology. Social relevance and responsibility must be part of agricultural research policies if poor farmers are to be assisted in accordance with the MDGs. This requires genuine dialogue between agricultural scientists and policy-makers to identify major problems, some of which are researchable and others non-researchable, simply requiring political action.

In Trinidad, the lack of an overall transparent and coherent government policy on domestic food production remained a major concern for aroid farmers over four decades. In spite of official statements, the government has not given attention to agriculture and resource-poor farmers and those living in isolated areas. Over the years, farmers have questioned food imports since food crops including aroids could easily be grown locally. Such local production would have provided employment and contributed to food security. The current policy with a focus on natural gas and petroleum implies that few farmers will remain when the sources of fossil energy start to drop.

With Swedish development assistance to Ethiopia in the 1960s, the Government of Ethiopia was ready to give more attention to agriculture. The feudal structure collapsed when the Military Government decided on land reform. The reform was implemented by peasant associations in less than a decade. With a doubled population, this move has led to a multiplication of the number of small-scale farms and Ethiopia faces difficulties in food security immediately and in the next decade. This has given rise to strategies for food security and development and policies have been outlined in the Sustainable Development and Poverty Reduction Programme. In June 2003, a New Coalition for Food Security was set up and a technical group created, keeping in mind an additional 1.8 million people being added each year. The objective is to turn around the food insecurity in a three-year framework, a huge challenge.

At Gramanstorp in Sweden, the problem of tenancy was solved over time. The government policy anticipated that large-scale agriculture should deliver cheap and safe food for consumers, as did policies in the United States and Denmark. Until the 1960s, the size of Swedish agricultural production was determined by domestic consumption needs. Such a policy meant no expansion of agriculture. In the early 1980s, environmental concerns were becoming more central, anticipating new methods of production and increased attention to food quality and animal health and care. When presenting its bill on sustainable fisheries and agriculture to Parliament in 1997, the Swedish Government stressed ecological aspects and a need to change the agricultural policy of the EU towards environmentalism and biodiversity conservation and to decrease nutrient losses and the use of agrochemicals. Also, the government proposed a further increase of ecological production in Sweden. In retrospect, one can seriously question whether the original intentions and the rhetoric of the 1997 bill have brought about major changes. Two years later, a new bill related to agriculture was accepted by Parliament. Within a generation, 15 national targets should be met regarding climate, air, use of chemicals, quality of ground water, forests and landscapes. Specific goals were identified for the period up to 2010 but it was also recognized that there would be difficulties in reaching the targets set in 1999 (Miljoradet, 2002). In reality, that bill simply suggests that the process of environmental degradation will continue, but at a slower pace.

Up to the 1990s, Swedish industry provided employment and the number of farms declined. Between 1950 and 1970, almost 1.3 million people left Swedish agriculture for industry. Agricultural policy has been directed to get the cheapest food, although this has disregarded aspects such as biodiversity, food safety, social and environmental consequences, including growing unemployment. Cheap food is based on fossil energy imported at a relatively low price. Such a policy orientation is irrelevant for agricultural research and development in a majority of developing countries with a large agricultural sector. It also requires rethinking of "modern" agriculture. Even though Swedish farmers produce 2.4 times more food per hectare today than they did in the 1920s, they accomplished this by using 13 times as much energy (Rydberg and Jansen, 2002). A tractor tilling a field burns fuel containing about 67 per cent more energy than the hay a horse would require for doing the same job. A steady increase in the price of fossil energy must radically influence future agricultural research approaches. Another fundamental question is how the overall approach in Sweden has global implications for the international agenda, both within the EU and in the developing countries that are the subject of the MDGs. The issue remains for a vision on sustainable agriculture and land use in Sweden.

After Sweden joined the EU, its policy on agriculture has been further changed. In the late 1990s, the EU had 128 million ha of arable land with about 7 million farms. Agriculture accounts for about 6-8 per cent of the GNP of the current Union. The budget for agriculture is Euro 43 billion, almost half the total EU budget and amounting to about 0.98 per cent of the joint gross national income of all member countries. The EU agricultural policy of 2005 means that Swedish farmers will get continued support. Nonetheless, the structural problems of European agriculture have not been solved. Agricultural subsidies remain a problem in the United States also. When these problems are resolved, they will have a drastic effect on future agriculture. As in Sweden, biodiversity is scarce on farms with specialized production. This challenges the current thinking on large-scale agriculture as the only avenue. The average US farm size grew from 59 ha in 1900 to 176 ha in 2000 (USDA, 2001).

There is a similar pattern in Sweden and guidelines and directives on future Swedish agriculture and land use are no different, except for a political focus on ecological agriculture. Recent studies and investigations have been confined to isolated aspects rather than a comprehensive examination of options for future policy of land use. This is probably explained by the fact that two different ministries are responsible for agriculture and forestry, respectively. Using 1995 as a basis, four different scenarios for 2021 were elaborated by a study group of the Swedish Environmental Protection Agency. They were confined to a modelling of crop production but none turned out to be very useful. In another report commissioned by the government it was concluded that 10,000 farms would be required to feed Sweden in the future, anticipating conventional agriculture and based on maximum yields and efficiency in economic terms (Myrdal, 2001). In addition to the supply of food and fibre, three alternative objectives were identified for agriculture: (1) biodiversity and landscaping, (2) new products such as tourism and (3) ecological production and efforts to avoid further pollution of soil and water, all of these were to be given government financial support. Although these supplementary objectives might be justified per se, the basic problems remain, since a "conventional approach" hardly can be classified as "sustainable", meet aspirations stated in the political rhetoric or meet new challenges. Rather, the suggested objectives are questionable in light of the higher price of fossil energy, shortage of land for bio-energy and the closing down of electricity-producing nuclear reactors.

Both at Gramanstorp in Sweden and around Asella in the Ethiopian highlands, cereal production increased over four decades. In Ethiopia, this led to less fallowing, reducing the grazing areas and thereby the number of cattle. Up to the late 1990s, cereals were not grown at the expense of the acreage of broad bean and field pea. Shortage of grazing land was noted already in the early 1970s, also a national trend. According to the Central Statistical Authority of Ethiopia, areas cultivated under major crops, particularly cereals, by the smallholder sector expanded from 5.5 million ha in the 1980s to 9 million ha in the mid-1990s. The acreage under wheat more than doubled and that of maize more than tripled. Areas under barley, sorghum, pulses and oilseeds remained more or less the same. To some revisited farmers at Wajji, forestry had turned out to be a more economical land use than crop production. This highlights a need for considerations of future land use, including agroforestry rather than analysing agriculture or forestry in isolation.

At Gramanstorp in Sweden, the cultivated area declined by one third, some of which was left fallow. During the 20th century, arable land in Sweden decreased by about 1 million ha. Forest plants were planted on one fifth of this land. The area under natural meadows declined by half. Nature has been less diversified and there were 1,953 endangered species in Sweden in 2000. The number of family farms decreased by one third from 1986 to 2001, with a current trend of three farms disappearing daily.

As regards aroids in Trinidad, there is both a declining trend in production and a change in the overall pattern of land use. Recently, sugarcane land was abandoned. Only one sugar factory remains; a similar situation has developed in Barbados. This reduction of sugar factories increased transport costs of raw cane so smallholders had become insignificant suppliers in 2003. In 1957 there were 15 factories, and today there are two. A similar trend is taking place in Europe, where the number of sugar industries, based on sugar beet, has declined by one third to just 150. This trend will intensify with a new EU policy on reduced subsidies to this crop.

Another aspect of policy relates to adequate channels of communication between farmers and policy-makers on marketing, the producer cooperatives in Trinidad and Sweden and the peasant organizations in Ethiopia. By 1965, aroid farmers were already experiencing difficulties with the Central Market Agency and formed a Food Crops Association. Because of internal conflicts, that association did not last very long. In 2005, the Trinidad and Tobago Government considered a new legislation to ensure good governance of the cooperative credit unions. In Ethiopia, the cooperative movement initiated by the CADU project has ceased to operate. Although the peasant organizations still exist, most farmers were quite sceptical about their efficiency as a channel for purchasing inputs and as a means to actively approach their government on policy matters.

In Sweden, there is a growing concern among farmers about declining access to policy-makers. Since the 1930s, the agricultural producer cooperatives have played a major role in Swedish agriculture. The Swedish Federation of Farmers (LRF) is the central organization, with some 159,000 individual members and cooperative companies, such as dairies, slaughterhouses, forestry companies, Lantmannen and national organizations. AgroEtanol produces ethanol and feed from some 100,000 ha. The companies used to operate exclusively within Sweden. For many years, the agricultural producer cooperatives expanded and became highly centralized on the basis of economies of scale. One example relates to dairies. In 1960 there were 357 Swedish dairies and gradually the numbers declined under Arla Foods.

Starting in 1991, Lantmannen took an international approach, tying up with the Nordkorn Company, a partner in eastern Germany. But the combination of grain trade and real estate led to economic deficit due to ignorance about the fierce competition from some 60 other companies in the same region. This situation was in contrast to Sweden, where Lantmannen had almost a monopoly. More recently, LRF has become involved in forestry and international business activities and the Swedish forest companies have turned global. Moreover, LRF is now facing a number of strategic, financial, and organizational issues. It must decide whether to represent large- or small-scale farmers. The Swedish food processing industry may have to continue to merge with other industries. If so, the agricultural cooperative organizations will increasingly be forced to operate as global corporations for profit. Alternatively, the food companies may decide to serve local producers within a looser network with delegated power and responsibilities at units based around the country in regions where major food items are produced. Alternatively, a new organization may appear to serve the needs of small farmers and/or ecological producers, a reminder of the pioneering days of the Swedish agricultural cooperative movement. Then, the cooperative ideas and ideals were central to compete with and constitute an alternative to the private sector. Today, globalization and international market forces challenge these ideals and old values of the cooperative movement in a global context. Another factor is the new EU agricultural policy of 2005, implying that all farms will get support regardless of how much they produce, based on production figures of 2000-2002.

The LRF is increasingly turning to international activities. In 2000, the Swedish Arla Foods merged with the Danish MD Food. The new Arla Foods has some 65 per cent of the Swedish and 93 per cent of the Danish market. In Sweden, the remainder is in the hands of seven other Swedish dairies. Arla Foods markets some 500 products in 23 countries. In 2003, Arla Foods merged with the British Express Dairies. It has also been operating in Haiti, dumping milk powder with export subsidies from the EU. Its plans to merge with the Dutch Campina were stopped in early 2005 because of doubts among Swedish dairy producers. If the merger had succeeded, with about 21,000 dairy farmers as members, it would have become the second largest company in the world after Nestlé AG. Such a merger had meant a reduction by half of Swedish dairy farmers, now numbering 6,000. Arla Foods entered the Chinese market in late 2005 in a joint venture with the China Mengniu Dairy Company. It is to produce milk powder for the Chinese market, although the long-term target is export of milk powder from Sweden and Denmark. In the same year, LRF sold its shares in the Pieno Zvaigzdes dairy in Lithuania.

In practice, the small Swedish milk producers are unfairly treated since they are required to pay an extra fee for each milk collection. Another disadvantage is the closing down of 17 Swedish dairies due to increasing costs for transportation. However, other studies indicate that local food production in Sweden may annually reduce the requirements of fossil energy equivalent to the production of eight nuclear power reactors. The centralized approach by Arla Foods also implies that farmers have only one company for delivering their milk, as in their days of serfdom. During the past five years, every third Swedish dairy farmer delivering milk to the former Arla has been forced out of production.

Since 1996, meat imports to Sweden have doubled. Swedish Meats, the largest company, accounts for almost 60 per cent of the slaughter and one third of processing and meat packing in the country. It has fewer than 30 slaughterhouses, a reduction from 289 in 1960. There are seven private slaughterhouses, a reduction since 2000. Swedish Meats has had plans for a merger with Danish Crown, which annually slaughters 24 million pigs compared to about 3 million in Sweden. Some 80 per cent of the Danish pig production is for export. The merger has been postponed with reference to the relatively higher incidence of salmonella in Denmark. Nevertheless, the overall policy of Swedish Meats means a further reduction of the number of slaughterhouses to less than five in the near future, with greatly increased costs of fossil energy for transportation. Together with Swedish Meats, LRF owns 40 per cent of the shares in Polish Sokolow SA and 20 per cent of the shares in Finnish HK Ruokatalo. In late 2005, Swedish Meats announced that it was moving some of its production from southern Sweden to Poland to reduce costs, in spite of higher transportation costs. Its search for a merger will continue.

Today, a chicken is ready for slaughter after 35 days in contrast to 70 days in the mid-1940s. In turn, this has led to weaker legs and increased risks of damage to the animals. Chickens are mainly produced in large-scale production facilities. In 2003, serious problems were identified such as unsatisfactory hygiene and damaged chickens through handling in Swedish slaughterhouses. Whereas there are very few slaughterhouses in Sweden, there are about 5,000 in Austria, 9,000 in Germany and 3,000 in Italy. The small private Swedish slaughterhouses pay high costs and fees for regular inspections, approaching some 25 per cent of the price per kilogram of meat. Still, the Organization of Small Farmers argues that meat in Austria is cheaper than in Sweden. Another major problem for small slaughterhouses are the specific demands set by the wholesale industry and the food chains.

Global Trends—Threats or Challenges to Future Food Security and Food Safety

The Same Old Alarm Bells?

Around 1960, the prophets of doomsday argued that existing resources of fossil energy and coal were to diminish over "a few more decades" (Borgstrom, 1966, p. 282). This proved wrong but a looming water scarcity was rightly pointed out for the United States and other countries. Today, water scarcity is high on the international agenda. Also, the Club of Rome proved wrong when stressing the limits of growth in 1972. Problems of erosion had become more apparent on all continents. Early evidence was presented, linking cancer to environmental changes due to chemicals. Gradually, the political debate on environmental issues picked up through the Report by the Brundtland Commission and later through Agenda 21. Although Agenda 21 was universally accepted at the UN Conference on Environment and Development (UNCED), it has remained a document with little follow-up action. At that time, the Union of Concerned Scientists warned humanity that "a great change in our stewardship of the Earth and the life on it is required, if vast human misery is to be avoided and our global home on this planet is not to be irretrievably mutilated." In 2001, a similar message by scientists was communicated through the Amsterdam Declaration. A year later, the World Summit on Sustainable Development in Johannesburg again highlighted all the problems but provided little prescription for practical actions.

In general, action has been marginal. The world is still facing some of the same issues identified in the mid- or early 1960s, and even earlier. A process of gradual change of natural resources has been going on for a long time. Over the last three centuries, the total global areas of forests and woodlands diminished by 19 per cent and those of grasslands and pastures by 8 per cent (Richards, 1990). But the decline in global forest cover since pre-agricultural times is estimated at 50 per cent. Grasslands have been converted into crop and pasture land, having an impact on soil fertility. Revisited farmers in both Trinidad and Ethiopia emphasized declining soil fertility and increased erosion. There is more evidence of cancer due to new life-styles. A steady increase of chemicals in daily life has also posed both direct and indirect threats to humans and the environment. For instance, PFOS (perfluoroctan sulphonate) is predicted to become a major problem, being highly persistent in the environment and found in fishes and whales at high levels. Unlike DDT, it accumulates not in fats but in proteins. More studies have linked chlorination of drinking water to bladder and rectal cancer and in some cases to cancers of the kidney, stomach, brain and pancreas (Steingraber, 1997). This refers back to an early questioning of the need to chlorinate drinking water in the United States. Today, 7 out of 10 Americans drink chlorinated water compared to one-third consuming non-chlorinated water in 1940. By 1974, scientific findings had already pointed out negative effects on the ozone layer due to the chlorofluorocarbons (CFCs). High levels of ozone in the lower atmosphere will damage plants, legumes in particular, and cause skin cancers to humans. Two out of three Australians reaching the age of 75 might have skin cancer, a likely effect of the thinning ozone layer. In a recent report to the United Nations, projections indicate that the hole in the ozone layer at the Arctic might be healed in 50 years if all partners adhere to the requirements of the Montreal Protocol of the Vienna Convention, which came into force in 1987. More recent alarms include food contaminations observed within the EU and the long-term environmental effects of pharmaceuticals, so far tested only for their acute effects.

Some major global trends are briefly highlighted below. In various ways, they relate to future food security and food safety. They constitute major challenges to both the political and scientific establishments, requiring decisive actions rather than new mega conferences. Some of them are acute threats for the immediate future.

Poverty

Thirty years ago, about one third of the world's people were seriously poor by their own or anyone's standards. About 800 million people lived in absolute poverty with an income of one US dollar a day. Today, women constitute some 70 per cent of some 1.1 billion absolute poor of the world. When the boundary is set at two dollars a day, that number will be around 1.6 billion. But there has been an overall decline in global poverty during the last decade, mainly due to effects in some developing countries in Asia. Indonesia reduced the incidence of poverty from 60 to 20 per cent in one generation, although recent events have broken this positive trend. In India, poverty remained at 55 per cent from 1960 to 1974 but is now estimated at 25 to 29 per cent. Likewise, China has significantly reduced the incidence of absolute poverty from 270 million people in 1978, although there was again a small increase in numbers in 2004. In Latin America, the percentage of poor has dropped but the number of poor people in 2003 had increased by 10 million in numbers over the last decade. Only in Sub-Saharan Africa has poverty continued to increase.

In spite of advice by the IMF that countries increase their exports and in spite of privatization, domestic budget cuts and deregulation of the financial markets, developments have not taken place in Latin America. Moreover, changes in a global economy can be quick. By 2000, the number of poor people in Argentina had increased by 6 million, almost one third of the population. In 2002, with drastically falling salaries, the percentage of poor had almost doubled and was equivalent to some 20 million people.

The World Bank has argued that the number of poor people may decrease further, assuming successful WTO negotiations. This will require better terms of trade on textiles and agriculture for developing countries, reductions of subsidies to agriculture, writing off of old debts and a doubling of development assistance. If so, some 300 million more people may earn more than one US dollar a day by 2015. In terms of money, a solution to the poverty of the world would cost less than one third of the subsidies given to agriculture by the OECD countries.

There are not only gaps between rich and poor nations but also widening gaps within democracies. In India, some 100 million are world consumers, while 800 million are left out. In Europe, about 15 per cent of the population lived below the poverty line in the mid-1990s. In the United States, some 30 million people live in food-insecure households and about one fifth of the children were living in poverty (EPI, 1999). Moreover, the gap in US incomes has been widening. According to the US Economic Policy Institute, the salaries of chief executive officers (CEOs) increased by 79 per cent between 1989 and 2000. A CEO made 24 times his average employee's wage in 1965, 71 times in 1989 and 3,000 times in 2000. Compared to the mid-1970s, income gaps have also increased statistically in Sweden. The salaries of Swedish CEOs increased by 217 per cent between 1988 and 2003. In the 1990s, 20 per cent of the richest Swedish people owned 90 per cent of the total wealth, a figure roughly the same as in the United States (Folster, 2003). In fact, there are increasing income gaps also in developing countries, for instance in Vietnam and China.

In conclusion: In spite of a range of political agreements and international declarations on poverty eradication, poverty remains a fundamental threat. Political commitment has been lacking at the national level. Gaps have widened not only between developing and developed countries but also within societies. Globalization, in terms of a growing interconnectedness of the world, holds a promise of reducing poverty and hunger. But poverty has more dimensions than economics. It will not be eradicated simply by more funds for development assistance. To make a change, current power structures must be challenged, nationally and internationally. Since about

75 per cent of 1.3 billion poor people are dependent upon agriculture, directly or indirectly, efforts to reduce poverty can no longer neglect investments to this sector in developing countries with a large agricultural sector. The political issue is not only about agricultural production but goes much beyond that, to purchasing power and incomes.

Rate of Population Growth

In the 1960s, population growth was considered a major threat to humankind, leading to the establishment of the UN Fund for Population Activities (UNFPA). The current growth rate of the world's population is 1.6 per cent a year. Globally, this means an annual addition of another Mexico. This rate of growth was about 2 per cent in the late 1960s. Then, Kenya had 5 million people, now it has some 30 million. Nigeria had 22 million people and now it has 126 million. In 2004, the world population was 6.4 billion. Estimates by UNFPA indicate a world population of 8.9 billion in 2050, similar to estimates by the World Bank. Other projections indicate that, within two generations, four out of five women will give birth to two children at most. Women in more than 60 countries have less than 2.1 children, the critical level. This would imply a maximum of 7.5 billion people in 2030, and then a decline. Using another method of probabilistic forecasting it has been argued with 60 per cent probability that the world population may not exceed 10 billion before 2100. Most estimates indicate, however, that the world may have approximately 2 billion more people around 2020. Then, nearly half the world population will live in South Asia and Sub-Saharan Africa. In Sub-Saharan Africa, the population will more than double, reaching more than 1 billion.

Discussions on population issues can be oversimplified. Given certain conditions, an increase of population is an asset in economic growth. India has continued to develop despite a doubling of population. Its agricultural production has increased at about the same long-term growth rate over 20 years. In Egypt, most of the people live on 4 per cent of the land. On a traditional diet, Egypt could feed its people but food is also grown for livestock to suit a small elite. Meanwhile, the distribution of land has become more inequitable under adjustment privatization schemes, favouring large landholders and increasing the number of landless peasants. Instead, people move into urban areas. This set of issues is of global concern.

Prior to UNCED, governments and international decision-makers were called upon by academies of the world to take action and adopt an integrated policy on population and sustainable development. The goal was improvement of quality of life, requiring zero population growth within the next generation. There have been few actions, though. According to UNFPA, government commitments to population issues at the UN Conference in Cairo are short of about 50 per cent of the budget, mainly because of failure by developed countries. However, the Chinese government has made a serious attempt on birth control. In 1979, the fertility rate was 2.8 per cent; in the early 1990s it was approaching 1.9 per cent. Some observers have found such measures of birth limitation draconian. The World Population Plan of Action in 1974 stated that all couples and individuals "have the basic right to decide freely and responsibly the number and spacing of their children and to have the information, education and means to do so." As a contrast, Thomas Aquinas agreed long ago with Aristotle that the number of children generated should not exceed the resources of the community. If needed, the law should ensure this since overexploitation of the natural resources would lead to poverty and chaos. In fact, without its one-child policy, an additional 240 million more people would have been born in China over the last two decades (Feeney and Feng, 1993). With a few exceptions this policy persists in urban areas. In 2004, it was changed in the city of Shanghai and on the island of Hainan, accepting two children per family. With 18 per cent of the population older than 60 years there are serious concerns for the future social security system. In the rural areas, a second child may be accepted after seven years and more than one child is also accepted in minority groups. By 2050, China expects 100 million people older than 80 years, a tenfold increase from 2000. This may also influence the one-child policy at the national level.

Some time ago, conventional ethics and morals were challenged by the proposal of a strategy on Health in a Sustainable Ecosystem (King, 1990). By saving children through large health programmes one may just prolong their misery in poverty, hunger and under-nourishment for a few more years. That view was argued against by WHO. Improvements in child health would be balanced by the deaths from HIV/AIDS, a trend that proved to be correct in many societies. Finally, one should not forget that the population-ecology crisis cannot be solved by rationality but "only with truths that stir the flaccid collective will by touching the sacral core of human willing" (Maguire, 1998, p. 24).

In conclusion: For the foreseeable future, population growth remains an acute threat in several parts of the developing world. There, efforts to merely increase food security will be futile if population growth is neglected. Ideally, freedom of choice in reproductive matters must be the norm but personal rights may need to be curtailed by requirements for the common good, as politically demonstrated by Chinese leaders. The misery caused by HIV/AIDS and other emerging epidemics may over time balance population with available natural resources. This balancing act is an issue for national policy-makers and also applies to a need and duty to stimulate birth of children in countries with aging populations.

The HIV/AIDS Epidemic and Emerging Diseases

In 1981, the first scientific report on a new disease appeared and five years later the virus was named HIV/AIDS. In 1987, WHO presented its first official figure of 64,000 cases. Already at that time, the French Foundation France-Libertés presented a report with alarming figures on the HIV epidemic in Africa. In larger cities, 7-30 per cent were infected. By 1992, the epidemic had its focus in developing countries.

Today, HIV/AIDS annually kills some 3 million people and infects 5 million. In late 2005, almost 39 million people were infected, three quarters of them living in Sub-Saharan Africa. In total, some 20 million people have died worldwide. High increases of infection are now observed in India and Central and Eastern Europe. With some recent exceptions, the situation is still worsening in Sub-Saharan Africa. Every fourth South African woman in the age group of 20-29 years is infected. Almost 40 per cent of the population of Zimbabwe and Botswana have HIV/AIDS. In Ethiopia, about 3 million people are infected and in China 1 million. In the Caribbean region about 2.4 per cent of the population has been infected. Without serious actions to combat the disease, the death tolls in the coming decade may exceed the deaths by the Black Plague in Western Europe.

The HIV/AIDS epidemic has serious implications not only for health but also for food security. African agriculture and food production has been hard hit by this epidemic in addition to older diseases such as malaria. Between 1985 and 2001, about 7 million agricultural workers died from AIDS in the 27 most affected African countries according to the FAO. In certain regions of Ghana, Zimbabwe and Uganda, some 10 per cent of agricultural extension workers have died (Spore, 1999). One effect of AIDS/HIV relates to the loss of local knowledge, for instance of traditional agricultural seeds. The loss of a male head of a household reduces cash crop production, while the death of a female affects grain production. By 2020, some 16 million more deaths are predicted for Africa, requiring livelihood support interventions. It is a task for several ministries, not only those of health or agriculture.

Like other diseases, HIV/AIDS will produce resistant viruses; one out of four new diagnoses of HIV/AIDS are showing resistance. Between one third and two thirds of AIDS patients in the United States have developed resistance to existing drugs. A resistant virus will stay on at low levels in the lymph nodes of the body, ready to return whenever the patient stops taking the drug. Recycling of drugs will not be effective. Recently, cheaper medicines have been made available in the poorest developing countries, cutting annual cost per person from US$ 10,000 to US$ 140. Still, this is unreachable for poor people. For quite some time, there have been great hopes for a vaccine. It takes a long time from the availability of a vaccine to a point at which its full impact can be judged. It took more than 70 years to achieve full global control of smallpox. Also, scientific experience shows that if less than 92 per cent of a population is vaccinated, the particular disease will continue to be prevalent (Dagens Medicin, 1999, p. 69). It is critical to bear in mind this time factor when discussing impact by vaccines against epidemics such as HIV/AIDS and malaria.

New diseases may give rise to additional concerns. Today, the health hazards of one part of the world easily appear somewhere else. Increased trade with food is one route. Urbanization augments the problem. Centralized systems of food and water in large urban areas can amplify the impact of otherwise modest microbial outbreaks. In "modern" production, animals are kept in larger numbers and in close quarters. Their diets have changed from hay to grain and often imported feed. When ruminants are fed fibre-deficient rations, microbial ecology is altered and animals may become more susceptible to metabolic disorders and in some cases to infectious diseases. This situation is gradually turning out to be a global issue. Imported contagious livestock diseases can be a major problem in highly industrialized countries, for instance the United States, where the most severe ones have been eradicated. A vaccine for them may be non-existent or seldom used.

In 1991, a virus (PCV-2) attacking the immune defence of pigs was found in Canada and later spread to Europe and Asia. Latently existing in most pigs, it causes Post Weaning Multisystemic Wasting Syndrome. Outbreaks started in the United States in the 1990s. The disease is frequent in Denmark. The first case in Sweden was reported in 2003, and the number of cases reached almost 40 in 2005.

In 1997, a three-year-old boy in Hong Kong died from an infection of the virus H5-N1. It was not known to infect humans but spread through bird faeces. It is a virus against which most human beings have no defence. Some 10 per cent of all birds were infected at the Hong Kong chicken markets, leading to slaughter of some 3 million birds to stop further infections. In 2003, about 31 million chickens had to be slaughtered in the Netherlands and the virus also killed one veterinarian. Some researchers hinted it was a recurrence of the influenza epidemic in Europe in 1918. In early 2004, the virus again appeared in southern Vietnam, Thailand, South Korea, China and Turkey. In Thailand alone, 25 million chickens were slaughtered. Altogether the virus up to early 2006 has killed some 120 persons worldwide. It has been observed in 48 countries at least, including India, all the EU countries and some African countries. There are suspicions

that the virus may mutate to a human form and be transmitted from animal to man. Then, it may become pandemic.

In 1999, another virus never found before in the western hemisphere appeared in the State of Florida USA (Weiss, 2002). It was spread by mosquitoes. Like HIV/AIDS, it was classified as a US national security threat. Stemming from the West Nile district of Uganda, it killed some 2,000 people. This Ebola virus is killing humans rapidly. Another new epidemic, caused by a bacterium (*Mycobacterium* sp.) has recently been reported from 26 countries, mainly in West Africa. By reducing the immune defence system through a toxic product it affects muscles and bones. In late 2003, the outbreak of an acute respiratory disease (SARS) in China was initially reported by the *Southern Metropolis News* and later followed by an official report to the WHO. It spread quickly within China and other Asian countries, the United States and Canada. SARS caused death and created panic worldwide. Through classic methods of quarantine and isolation of the infected patients, it has been kept under control so far.

In conclusion: HIV/AIDS will remain a serious threat in many countries for years. Since costs for medical treatment are quite high, this alternative is hardly realistic for most poor people in a short-term perspective, unless provided free. The epidemic is reducing population growth in parts of the world, affecting current and future food production. As a result of globalization, increased food trade and more travelling, new diseases may appear. They pose a lurking threat to future agriculture and food safety, particularly in countries with large-scale farming operations. A first step in controlling them is the enforcement of strict quarantine regulations in all countries.

Biological Diversity

An Untapped Diversity of Food Plants
Biological diversity is the total variability within all living organisms and the ecological complexes they inhabit. They include the ecosystem reflected in the number of species, the different combination of species and genetic diversity or the different combinations of genes within each species. Although we do not know exactly the number of species in the world, various estimates indicate a range from 10 million to as many as 100 million. According to the University of Texas (USA), some 1.7 million species have now been classified, but it plans to map all species over the next 25 years. There are possibly about 5 million insect species, of which some 70,000 species are known. They may be pests but serve also as a food source for birds, amphibians, fish and reptiles. Out of more than 250,000 known plant species, about 10,000 to 50,000 are edible, but only 7,000 are nowadays used for food. About 120 species are cultivated but just some 20

species provide 90 per cent of all the calories consumed by humans. Nine species provide more than three quarters of our food, with rice, wheat and maize accounting for the major part. For the long-term future, there are probably great possibilities to find new crop species for food production rather than an exclusive focus on the nine major ones.

Losses of Biodiversity Related to Agriculture, Fishery and Forestry

Over the last decades there has been intensive debate about the loss of species. There are considerable differences in the rate and extent of losses. They vary for species, ecosystems and areas. In general, the average abundance of species has declined in the last three decades. In 2003, researchers argued in *Science* that some 20 per cent of plant species of the world were threatened by extinction. Globally, 58 fish species have become extinct in 30 years. During the 20th century, about 1,000 animal races disappeared, according to the FAO. About 1,500 animal races are threatened by extinction. Figures such as these have led to a debate in which preservation has dominated rather than a search for potentials, for instance demonstrated by Inbio in Costa Rica, which reported some 700 previously unidentified species in 2003. The Global Census of Marine Life discovered three new fish species a week during 2001 and 2003.

In agriculture, new crop varieties have been adopted over large irrigated areas and in endowed rain-fed regions. This has led to loss of landraces. Since the early 1960s, more than 1,700 varieties from more than 100 rice breeding programmes have been released, about 75 per year since 1980. Moreover, the origin of 1,709 modern rice varieties in Asia can be traced back to almost 11,600 traditional varieties (IRRI, 1998). These developments should be viewed in the context of the total number of existing rice varieties, estimated at more than 140,000, and the rice germplasm available in the rice gene bank at the International Rice Research Institute (IRRI). As a result of a new breeding technique ("wide crossing"), recently improved varieties from CIMMYT are reported to have about the same level of genetic diversity as wheat varieties that were grown before modern plant breeding. The technique, introduced some 15 years ago, incorporates genetic material from wild relatives of wheat that have not intermixed with cultivated wheat varieties for many hundreds of years.

Some observers have argued that around 30,000 species of the tropical rainforests annually become extinct. The World Conservation Union and the Worldwide Fund for Nature have jointly reported about one tenth of 80,000 to 100,000 known tree species to be endangered. Others consider this an exaggeration (Lomberg, 2001). Tropical forest species are more subject to loss than temperate ones. Deforestation has been more significant in certain regions of the world, in particular during the 20th century. In the Philippines, forests covered one third of the country in the 1950s. Today, less than half of that forest remains, most of it degraded or young second-growth forest. Initially, logging companies exploited the forest. Later, small farmers cleared it, mainly under population pressure. Likewise, one tenth of Madagascar used to be covered by forests, which were gradually lost in the early 1990s to make way for rice. Such trends are not a new phenomenon: in China, when the water buffalo was domesticated some 3,000 years ago, large forested areas were cut down to get arable soil, and erosion and ecological decline followed. As a practical approach to global forest biodiversity, a set of about 100 of the richest sites, strategically located and covering 3 to 5 per cent of the tropical forests of the world, could conserve the great majority of all tropical biodiversity (Sayer et al., 2000). Other sites could be managed by people to meet human needs and requirements for timber. Furthermore, reports on forest destruction are sometimes exaggerated, as illustrated by one example from El Salvador. In earlier reports, it was claimed that only 5 per cent of the forest was left because of overpopulation and poverty. During the last two decades, El Salvador has become greener, the forest having grown almost 40 per cent between 1992 and 2001 (Hecht et al., 2006). The relatively dense forest now covers some 60 per cent of the country.

An emphasis on the preservation of existing species neglects the natural changes within ecological systems. Some species disappear and new ones will expand. Biological diversity is about balancing the positives and negatives of a system. Furthermore, species extinction has been a trend in a geological time perspective. During the last 500 million years, the planet has been exposed to about ten major catastrophes, leading to extinction of species and biotopes. Life on earth was dramatically changed some 250 million years ago and most living organisms above the bacteria are thought to have become extinct. The cause was not a single one but a combination of stress factors wiping out some 90 per cent of all species in the oceans and 70 per cent of vertebrates on land (Erwin, 1996).

Until recently it was assumed that the recovery of species depended on the size of the catastrophe. It has been argued that it takes 10 million years for a new organism to develop and reach its maximum after extinction (Kirchner and Wil, 2000). In contrast to the past, current threats to the globe are now mainly due to humans and their quest for "development". The issue is what exactly can human beings actually control in a long-term perspective, and how long is that? About 100 BC, the oyster lagoons in south China were a basis for pearl fishing. After 200 years they had been completely exploited. That industry collapsed and so did the economy of the region. Pearl fishing was banned for a period, allowing the oysters to come back together with increasing wealth about 150 AD (James and Thorpe, 1994). Recently, there are concerns about environmental pollution of the coastal regions of the Baltic Sea and from the Gulf of Mexico to Chile.

A special case is the US Chesapeake Bay Program. Although restoration efforts started in 1983, a full recovery is not in sight. The oysters, which were abundant in 1948, are almost lost and the crabs are at historic lows (Horton, 2005). Nevertheless, crab and oyster dishes are still on the menus; they are supplied by Asia and the states of Louisiana and Texas.

There is a basic contradiction between the second law of thermodynamics and the evolutionary law. The former says that the universe must tend towards increasing disorder and simplicity (entropy), while the latter suggests the reverse. Over time, evolution is producing forms of increasing order and complexity. However, the law of increasing entropy has given scientific support to the belief in a decline or setback. All energy is transformed into thermal energy and the universe may collapse at a final temperature. Thus, evolution is just a temporary movement within limited space and time. If so, the thermodynamic law is as relevant for the belief in decline as the evolutionary theory is for the belief in progress (von Wright, 1993).

Access to Genetic Resources for Food and Agriculture

Three crops are known to have existed in both the Old and New World: the common gourd, sweet potato and the coconut (Purseglove, 1974). Otherwise, crop movements and plant collections are ancient. Some 3,500 years ago, Egyptian rulers started to bring back plants after military expeditions. The lesser millets (*Setaria* and *Panicum* sp.) from central and eastern Asia reached India before 3,000 BC. Sorghum, finger millet, cowpea and pigeon pea originated in Africa but arrived early in Asia. Triploid cultivars of *Musa* originated in Malaysia, reaching East Africa around 1,000 BC. Christopher Columbus brought sugarcane, citrus and other crops to the Caribbean. He carried back maize, sweet peppers and other crops to Spain. *Eucalyptus* was introduced to Europe from the tropics. Cassava and maize in West Africa originally came from the Americas. Rice, wheat, barley, oat, soybean and pea have spread from Asia to other continents. Potato, cassava, groundnut, bean and maize have come from the Americas. Plant genetic resources have flowed in many directions and the spread has taken quite some time. Today, no country is self-sufficient in plant genetic resources.

A number of tropical crops have originated outside Vavilov's centres of origin, such as the oil palms and sugarcane in New Guinea and *Oryza glaberrima* in West Africa. After domestication, crops have spread to a diversity of environments. Optimal conditions are often in another hemisphere. In expanding populations, the number of naturally occurring mutations will be increased, in particular when crops are introduced to new and variable environments. This adds to the gene pool. Today, major economic crops are grown with maximum production in areas far from their regions of origin. One example is rubber, the seeds of which were smuggled out from Brazil and planted in Malaysia. Other examples include soybean, which originated in northeast Asia but is today mostly produced in the United States; cocoa, which originated in South America but is grown in Ghana; vanilla, which originated in Mexico but is grown in Madagascar; coffee, originating in Ethiopia but grown in Brazil. This principle also applies to many monocotyledons. It reinforces the need for plants to be exposed to maximum competition and an exchange into new areas, rather than simply being preserved and confined to their local habitats. This requires both time and easy exchange of plant material if new plants are to be used in future food production.

Maintenance of Plant Genetic Resources

In the past, botanical gardens played a significant role in the maintenance of genetic resources and in plant collections, particularly in the tropics. Gradually they lost their importance until a decade ago, when they became more broadly based botanical resource centres. They can serve as a resource in education and contain living collections, mostly of wild species, in contrast to dried seed kept in agricultural gene banks. There are some 1,700 botanical gardens worldwide (Hawksworth, 1995). The IPGRI has estimated their number to be significantly less but estimates that they hold some 3.5 million samples in their collections (IPGRI, 1999).

With new crop varieties, landraces have been replaced. This may not always imply that the gene pool has been completely lost but just that it has been partly integrated through plant breeding. Landraces are constantly evolving. In farmers´ fields, biological diversity is a dynamic process through active selection, natural mutations and influx of new genes. This dynamic is not captured in gene banks, although they were considered the best way of conserving plant genetic resources in the late 1970s. There are more than 1,400 gene banks in the world. A major part of these genetic resources is deteriorating because of poor maintenance. In China alone, some 178,000 accessions of the 10 most important crops have been collected and distributed throughout the country (Weidong et al., 2000). Most of them were preserved in gene banks but about 16,000 were held in field gene banks. Globally, the number of minor species in total holdings is small, amounting to about 7 per cent (Padulosi, 1999). Only 10 per cent are wild species, which are often neglected in plant collections. However, they can provide useful data, as illustrated by wild species of *Cucurbitaceae* in Kenya (Njoroge and Newton, 2002).

Regional gene banks have also been established. Most of them were set up to save costs and were often initiated by donors to help developing countries. In the mid-1980s, SIDA planned for the support and creation of a regional gene bank in southern Africa. The well-intended idea was seen in a perspective of 20 years. Initially, the former Swedish Agency for Research Cooperation with Developing Countries (SAREC) expressed grave concerns referring to lack of national government commitment, scarce national research capacity for effective regional cooperation and unclear division of labour and responsibilities between the new regional gene bank, the Nordic Gene Bank and national activities. A national gene bank was more likely to attract political support, showing national identity. Furthermore, free access of plant material required national legislation. The patenting of plant genetic material was another emerging issue. As consultant to SIDA, the then IBPGR simply responded, "There is no reason to believe that any government will succeed in establishing a national unit." It is now realized that many regional gene banks have not been working because of political constraints (Frison et al., 2002). In addition, this problem has been accentuated by an earlier UN concept of sovereign rights, now endorsed by the Convention on Biological Diversity (CBD). Restrictive legislation on access to germplasm has been put in place, for example by the Andean Pact countries. A similar lesson has been learnt in a Project on Development of National Biosafety Frameworks by the United Nations Environment Programme and the Global Environment Facility. An external entity cannot impose regional cooperation. The countries themselves must do so at their own pace in response to their own needs (UNEP, 2006).

At the international level, the gene banks of the CGIAR currently hold some 600,000 accessions of staple food crops, forage and agroforestry species. But the number of accessions has grown marginally since the early 1990s. As a result of the CBD, the number of plant collection missions has declined. In the past, they often resulted in over-duplication of samples. Now, rationalization of these collections in most gene banks is urgent to complete the characterization and evaluation of germplasm in current collections.

The use of new research tools of biotechnology implies there may be less need for gene banks of the current design. Rather, there might be a need for gene and/or DNA libraries to trace genes and gene complexes that have been introduced into specific crops. Today, researchers are focusing on DNA fingerprinting of cultivars for accurate comparison of accessions between countries. Brazil has recently opened a gene bank to preserve the DNA of its endangered plant species. The private Philippine Sugar Research Institute Foundation Inc. (Philsurin) has produced DNA fingerprints of its sugarcane germplasm collection. A third example is the Library of Plant Genes Germinates at the University of North Carolina (USA). Its database

(Phytome) on 39 plant species allows researchers to ask complex questions from some 730,000 protein sequences in more than 25,000 protein families. Through the base, one may simultaneously compare the genetic maps of multiple species and predict the gene content in regions of plant genomes that have not yet been deciphered. Developments such as these give rise to pertinent questions regarding where and under what control the genetic resources should be handled, calling for a clear division of labour between the public and private sectors.

From Free Access and Exchange to Political Bureaucracy

Over centuries, there has been free access and exchange of genetic resources for research cooperation, the basic genetic material being considered to belong to humankind. This changed suddenly with the CDB placing ownership of genetic resources under national sovereignty. The political move, with good intentions, was to assist developing countries in which most genetic resources of the world originated. The assumption was that they should be able to trade these resources, get increased development assistance for Agenda 21 and prevent the private sector from taking the profits of the advancements made possible through biotechnology research. So far, increased funding for the developing countries has failed and so have noticeable effects of the principle of benefit sharing. The development of an international law on access and benefit sharing of genetic resources will be both complicated and problematic. The challenge for an access and benefit sharing regime is to resolve the contradictions between the CBD and the Trade Related Intellectual Property Rights Agreement (TRIPS). Currently, the agreement is undermining the implementation of the access and benefit-sharing provision of the CBD by allocating the control of genetic inventions to patent holders. Negotiations are calculated to last for several years according to reports from the meeting of the CBD Ad Hoc Open-ended Working Group in early 2005. Another issue is whether or not the CBD is retroactive and what direction the discussions in the International Treaty for Plant Genetic Resource for Food and Agriculture (ITPGR) are advocating. In retrospect, there might have been a miscalculation by the CBD architects, neglecting historical and technical features of the spread of plant species. In mid-2006, 102 countries had ratified the treaty. It remains to be seen whether the treaty simply means that an additional administrative and political layer has been added. Besides, it needs to cooperate with the Commission of Genetic Resources for Food and Agriculture, the Global Crop Diversity Trust and the third party beneficiary with reference to a core budget of US$ 2.8 million approved by the first session of the Governing Body of the ITPGR in mid-2006. The budget provision by the Regular Programme of FAO is relatively small (US$ 1.1 million) and US$ 1.7 million is still to be funded.

One outcome of UNCED and the CBD was a trusteeship agreement between the CGIAR and the FAO in 1994. It was important since the CGIAR holds about one tenth of the accessions or entries of the world's crop gene banks. Essentially, this allowed the CGIAR to continue to have a say in its genetic resources through the independent boards of trustees of individual CGIAR centres, but under a political umbrella. To the FAO, this move was a step towards the inclusion of the CGIAR gene banks under an international agreement. It is planned that the new treaty will govern the "in trust" germplasm under new provisions. The Multilateral System for Access and Benefit-Sharing applies to 64 major crops and forages. The treaty is to protect farmers' rights including the protection of traditional knowledge and a right to participate equitably in benefit sharing. However, this will not be the case for farmers in non-ratifying countries, some of them hosting CGIAR Centres. Several other issues remain (see further under CGIAR, pages 237-241). Recent data on the genetics of one million microbes of the Atlantic water outside Bermuda highlight significant prospects, but microorganisms are outside the remit of the treaty. On intellectual property rights (IPRs), the treaty specifies that they cannot be claimed for "genetic resources for food and agriculture, their genetic parts or components in the form received". This may turn out to be a poor defence against privatization. Some governments are not signatories to the CBD and some may not wish to join the new ITPGR.

The private sector has supported the CBD in principle but questioned its implementation. The issue of IPRs has become more explicit. Although that debate was initiated by some international NGOs it was closely controlled by the private sector and continues on both access and IPRs. Restrictions in sharing germplasm are going to be a major problem, in particular for plant quarantine regulations. It is important to share genetic material without administrative and political complications.

In conclusion: General trends in biodiversity indicate some serious threats. However, some new plants and underused genetic material may turn out to be practical alternatives for the long-term future. The past policy of a biased preservation of existing situations implies stagnation. The new political conflicts on genetic resources further complicates an already difficult situation, requiring further international negotiations to accomplish constructive arrangements on benefit sharing, in particular at the international level. Conventional gene banks should be maintained if operating efficiently. Gene mapping and biotechnology research will require new kinds of gene registers or "gene libraries". New approaches have to be developed to provide good quality germplasm *in situ* to farmers and local groups.

If public funding of agricultural research continues to decline, the role of agricultural transnational corporations will grow in importance with regard to access, maintenance and use of all genetic resources. This is amplified by more frequent use of IPRs and the potentials of a complete gene mapping of major crops and animals. Active involvement of the TNCs, which should be considered relevant participants in future decision-making processes, may bring about more benefits for society as a whole than preventing them from attending the international negotiation table. In spite of different agendas, the private sector and the public sector may complement each other; this potential needs to be more constructively considered by both governments and the NGOs.

Towards an End of Cheap Fossil Energy

Oil, gas and coal provide about 85 per cent of all global energy. Today, world production of oil is about 24 times the production in 1925. With annual global demand increasing by 2 per cent, consumption will double in 35 years. Twenty years ago, there were few private cars in China, which are estimated to have increased from 5 million in 2000 to 24 million in 2005 (World Watch Institute, 2003). Up to 1993, China was self-sufficient in oil but its annual import is currently increasing by 9 per cent. Between now and 2020, China will need extra energy equivalent to the entire energy demand of Western Europe today.

Although only one quarter of the world population lives in the North, it consumes more than 70 per cent of the global fossil fuels. The EU is dependent on fossil fuels for up to 80 per cent of its energy consumption. Europe imports about half of its oil. Projections indicate that this share will reach some 75 per cent in 2011 and 90 per cent in 2025. As regards agriculture, fossil energy is estimated to account for some 3 per cent of total consumption, although there are variations. The percentage is even smaller in developing countries. But agriculture has not been part of the mindset of most energy planners.

Since the mid-1960s, the number of new oil resources has globally declined. One fifth of the current oil production comes from oil fields found 60 years ago. In 1991, the United Nations concluded that the remaining oil reserves of the world would last 75 more years. At that time, the US oil industry began spending more funds on exploration and production abroad than it did in the United States (Chevron, 1996, p. 11). There have been few new discoveries of natural gas in the United States. In the late 1990s, no investments in new US refineries were forecast, implying more dependence on imported oil although new oil and gas drillings were allowed according to the US Bureau of Land Management. This included permission for drilling in some 6,000 new areas. In the North Sea, new technology is reported to have improved the efficiency of oil drillings from 17 to more than 40 per cent.

More recent estimates indicate that the reserves of oil and gas may last some 50 years. To meet demands, the EU has projected a doubling of the production of natural gas in 2030 and an increase of oil by 65 per cent (EU, 2003). This is in contrast to estimates by the Association for the Study of Peak Oil, which predicted a declining world production of oil around 2010. Exxon Mobil is expecting a peak around 2010. Many geologists assume that a peak occurred around 2004 to 2008. New oil reserves will be located in areas of political instability or difficult to reach. Reserves may be found in deep-sea beds, requiring significant filtration, nano-tubes, nano-magnetic filters and new nanotech-based materials capable of withstanding strong ocean pressure. Most economists suggest, however, that there will be new alternatives, pointing out that the crises in 1973 and 1980 did not result in a global oil shortage.

Coal reserves may be sufficient for some 200 years, at least. China is the world's largest coal producer and consumer, coal accounting for 76 per cent of its commercial energy consumption. So far, alternatives to fossil energy have not been able to compete. Wind energy supplies less than 1 per cent. Biomass contributes about 14 per cent of energy worldwide and one third in developing countries, equivalent to 25 million barrels of oil (FAO, 1999). The FAO has estimated an increase of global use of bio-energy to 25 per cent during the next five decades, mainly from 100 million ha of tree plantations. Most of it will be in the North, which currently uses only 3-4 per cent bio-energy. The target of the EU is to double the bio-energy consumption by 2010, requiring new plant varieties and new technologies. In the long-term, bio-energy will require so many resources (in production, harvesting and transportation) that it may ultimately be unprofitable.

In the mid-1980s, it was assumed that alcohol could replace petrol in cars as a source of fuel that would require large areas of land. Early production of ethanol started in Brazil and has contributed to 700,000 new jobs in the countryside. Driving a car on ethanol alone would require 10 times the acreage for producing food for a balanced diet. In 2003, one fifth of Swedish cars were driven by gasoline with 5 per cent ethanol. One litre of ethanol requires 6.7 kg of wheat. If all cars were using this mixture, 125,000 ha would be needed (today 190,000 ha are in fallow). In the United States, the mixture contains 10 per cent ethanol and some 50,000 ha of agricultural land have been converted into woody plantations, compared with 16,000 ha of willows planted in Sweden as of 2003. Methanol from bio-energy might be an alternative. In Sweden, it will require some 3 million ha of energy crops for biomass. It may be a more practical step than using ethanol from agricultural production. To meet the energy demand for transport, wheat has to be grown on 10 times the acreage of the currently arable land of 2 million ha (B. Bodin, 2006, personal communication).

In conclusion: Emerging scarcity of fossil energy may become apparent sooner than most consumers of the North would like to imagine. The price of fossil energy will significantly increase because of both emerging scarcity and continued tension over oil resources. Terrorist attacks on pipelines for oil have increased dramatically since 2001. Increased oil prices will directly affect farmers in regions where agriculture is exclusively based on fossil energy. Resource-poor farmers may be less directly hit, although higher prices will affect their use of agrochemicals and plastics. Various kinds of bio-energy may over time provide alternative fuels. This calls for quick political decisions providing solutions in a long-term perspective and giving greater attention to issues of land use.

Climate Change and Agriculture

Climate change is not a new phenomenon. There was a climate change at about the time the Roman Empire collapsed and cereal production declined. Climate change is also believed to have influenced the great cultures of Central and South America. There have been dry and wet periods and the world is now experiencing a 15,000-year-long summer (Fagan, 2004). In Europe, there was a long period with strong climatic variations between 1550 and 1850 (Burroughs, 1997). In the 1870s, dry spells due to an unusual El Nino caused difficulties in many tropical countries. Although El Nino has existed for 130,000 years, it has become both stronger and more frequent over the last decades. Some scientists argue that such changes can be traced to global warming. In 1847, the Directors of the East India Company expressed concern about the danger of artificially induced climatic change (Grove, 1990). The British Association reported on the economic and physical effects of tropical deforestation and produced publications on global climatic desiccation. In 1896, the Swedish scientist Svante Arrhenius warned that the Earth would become warmer from a continuous use of coal.

Since 1700, the content of carbon dioxide in the atmosphere has significantly increased, the use of fossil oil probably accountable for 80 per cent of this increase. Average global temperature has increased by 0.6 degrees over the last 100 years, the highest rise in the last 500 years. Based on meteorological simulation models, the Intergovernmental Panel on Climate Change predicted in 2001 that, with no remedial action, the global average temperature might rise between 1 and 3.5°C by 2100, greenhouse gases being the main cause. More recent findings indicate even higher temperatures (www.climateprediction.net). Most experts believe increased global temperature is an effect of the release of greenhouse gases, mainly carbon dioxide and methane. This led the Second World Climate Conference to focus on the increase of greenhouse gases in 1990. Opponents have, however, argued there might be uncertainties. Complex factors involved in predicting the climate make simulation models too uncertain (Kerr, 1998). An increase in average global temperature of 0.5°C over 140 years might be due to normal variation over time (Christensen Wiin and Wiin-Nielsen, 1996). Danish scientists have proved linkages between average temperature on earth and the period-length of sunspots (Lassen and Friis-Christensen, 1995).

Since the beginning of the 19th century, atmospheric methane has grown two and a half times. About half of the emissions of methane are agricultural in origin, mainly from wet paddy fields, swamps and marshes. A second source of methane comes from bacteria living in the guts of ruminants, producing some 15 per cent of global methane emissions, equivalent to about 80 million tons. Annual emissions from rice fields are about 50 million tons. According to IRRI, flooded rice fields emit annually an estimated 25 per cent of the methane reaching the atmosphere. Reductions of methane emissions from rice fields may involve changes in the rice variety and land use, reduced disturbance of flooded fields and changes in the timing of fertilizer applications and irrigation schedules. The future rice plant must be high-yielding and have high oxidation power to reduce the methane. Deforestation in the tropics produces about one fourth of the world's net annual carbon dioxide emissions and about 10 per cent of global nitrogen oxide emissions.

Assuming global warming, agriculture will be influenced by both increased temperature and rainfall. Thus, overall agricultural production over the next 50 years is estimated to increase at a lower percentage. According to recent simulation models by the International Institute for Applied Sciences Analysis, climate change may create a new Sahara in southern Africa by 2080. Most African countries would feel a negative impact of global warming, Kenya being a possible exception. In a report to the Committee on World Food Security in 2005, the FAO concluded that climate change could lead to the loss of more than 250 million tons of cereal crop production, mainly in countries of Sub-Saharan Africa. India may lose up to 18 per cent of its non-irrigated cereal production, affecting the number of undernourished people. Above all, water availability may worsen at lower latitudes. Heat and water stress at lower latitudes may result in yield reductions and may, for instance, increase spikelet sterility of rice (Rosenzweig and Parry, 1994; IRRI, 1994). More carbon dioxide will also increase the biomass of weeds, though plants may only use a certain maximum of carbon. These considerations are important for global rice production. By 2020, more than 350 million tons additional rice must be produced than in 1990. For maize production in developing countries, CIMMYT has estimated climate change may lead to a 10 per cent drop over the next 50 years. A study on a worst-case scenario in China has predicted a 20-37 per cent drop in yields of rice, wheat and maize over the next 20-80 years. Yields of cotton may increase.

The greatest temperature changes will be at high latitudes. Increased carbon dioxide will encourage growth at higher latitudes, particularly of wheat, potatoes and barley. A doubling of the carbon dioxide will result in a 30 per cent increase of the yields of these crops (Reilly, 1996). In Sweden, the most likely effects of climate change on forestry include an increase of the potential biomass production, possibilities to grow new species and higher risks of several kinds of damage (KSLA, 2004). Another scenario implies a rise in temperature of 4-5°C in 50 to 100 years with an increased forest growth by 10 to 30 per cent, depending on the length of the vegetation period (Orlander, 2000). Malarial mosquitoes will become more common in Europe and even Sweden.

According to international negotiations, industrialized countries should decrease their release of greenhouse gases, mainly carbon dioxide. Committed developed countries have agreed to reduce their greenhouse gas emissions during 2008-2012 by an average of 5.2 per cent relative to 1990 levels. When Russia ratified the Kyoto Protocol in 2004, its enforcement was ensured although the United States has not ratified and Australia has withdrawn. Large producers of carbon dioxide in developing countries are excluded, for instance China. The final agreement, involving the United States, on the follow-up negotiations on the Kyoto Protocol in Montreal in late 2005 only stated that discussions will continue after the year 2012 without any set objectives or a timetable. Industrialized countries may meet part of their commitments by carrying out specified forestation and reforestation activities in developing countries (UNFCCC, 2001). Agroforestry systems might sequester three times as much carbon from atmospheric carbon dioxide than the same areas of crop and grasslands. Systems of agroforestry may sequester 60 per cent as much as newly planted forestry land (ICRAF, 2002).

Efforts to stop global warming will be expensive and probably unpopular. Cars and industry are major sources but also features of "modern" development that are relevant for economic growth. Private cars contribute some 6.3 billion tons of carbon dioxide, a fourfold increase in 50 years. The United States accounts for about one quarter of all releases of carbon dioxide. Drastic cuts in using cars for transportation would be one quick but radical move to reduce emissions of carbon dioxide, in particular with reference to estimates by the EU that road transport in Europe will double in 2010 compared to 2001. Moreover, every third transport of dangerous goods by Swedish rail or road did not pass the regulations according to inspections in 2000. Ethanol is already in use in several countries and may gradually increase. However, recent research by scientists at Cornell

University in the United States has revealed that it takes 29 per cent more fossil energy to make maize-based ethanol than actually is produced. Willow trees may become a more efficient source than maize. In 2004, tests with Citaro buses began in London and nine other European cities. Evobus, a subsidiary of Daimler-Chrysler, has developed this first hydrogen fuel cell bus. The Chinese Government purchased some of these buses to be used in Beijing from 2005 onward. Jointly with the Tekniska Verk, the Swedish railways subsidiary Euro Maint has produced the first biogas train in the world. It is to be marketed in the Baltic and developing countries. Cars with accumulator batteries will require much energy for their production and ultimately cause environmental pollution through heavy metals such as cadmium, lithium, copper and lead.

Another avenue to reduce the release of carbon dioxide is nuclear power, an environmental issue of high political sensitivity. It accounts for about 16 per cent of all electric power in the world, a more or less constant figure since 1987. Today, some 440 nuclear power stations are globally in operation, one fourth of them in the United States. In total, 34 nations use nuclear power. In 2003, eight nuclear power plants in countries of the North were closed down. Owing to increasing demand for energy and higher oil prices, some 25 new nuclear plants are now under construction or planned (for instance in India, Ukraine, China, Russia Iran, Slovenia, the United States and Japan). China is planning to expand its nuclear power stations by 30 new plants by 2020.

In conclusion: Industrial development, a focus on growth, a steady increase in use of automobiles, large-scale agriculture and affluent life-styles are all factors contributing to increased carbon dioxide. They are all characteristics of "modern" development. Increased and excessive use of fossil energy is unsustainable for our life support system. With increased global warming, poor agricultural producers in the tropics will be among the main losers.

Internationally negotiated efforts to reduce greenhouse gases are simply a small step in treating the symptoms in certain countries, delaying full-scale effects within one generation. Bio-energy or other cleaner energy sources can reduce the carbon dioxide level, though they may hardly be a major option in industrial complexes. Substantive reductions of carbon dioxide can be achieved if the traffic in large cities is not allowed to grow freely in parts of the world where two cars per family is a standard. It would be reasonable to introduce a one-car policy per family as a political signal that changes are possible. This is amplified by the fact that some 20 million people die annually in global traffic accidents, most of them in developing countries. Governments should immediately abolish taxes for the use of those cars operated by sources other than fossil energy and emitting low levels of carbon dioxide. Transport by rail rather than by trucks can further reduce the problem.

Water Scarcity and Water Pollution

Over 97 per cent of the world's water is in oceans and seas, being too salty for most productive uses. About 70 per cent of managed water resources relate to agriculture, a share that may fall because of a rapid urbanization. In Asia, agriculture accounts for 86 per cent of total annual water withdrawal. In India, water reservoirs were built over 2000 years ago. Water for irrigation in south India and Sri Lanka was supplied for a fee, a system that soon may reappear. Worldwide, about 1.5 billion people depend on ground water for their drinking water.

Between 1955 and 1990, water availability per capita declined by half. In 2003, a report by the United Nations stated that average access to water might decline further by one third over the next 20 years. According to the World Bank, water scarcity will affect almost half the world population by 2025. Others have argued that anxiety about water scarcity is unwarranted, except in certain countries (Lomborg, 2001). Some 15 countries are water deficient.

Twenty years ago, donor agencies were keen on providing funding for finding ground water through drilled village wells, for instance, funding some 11 million wells in Bangladesh alone. Recent tests have revealed that some of these wells now are polluted with arsenic well above the limits set both by the Bangladeshi Government and the WHO. In 2002, scientists argued that nitrogen leakage in Denmark was 30 per cent below official estimates. In central France, high levels of nitrate and residues from chemical pesticides have been noted, implying that 70 per cent of the water resources in Bretagne would not be safe to drink. Today, nitrates are of less concern in drinking water than pesticide residues. But water pollution is of global concern. For example, 60 per cent of the Chinese water resources are estimated to be seriously polluted.

In 1990, reports from the former Soviet Union called attention to ecological accidents with adverse effects on infant mortality. Drinking water showed signs of pesticides. It has long been known that increased use of pesticides in most Asian countries caused a decline in rice-fish culture. Repeated applications of pesticides prolong the exposure of fish, leading to mortality. In Denmark, pesticides pollute about 16 per cent of the ground water, Roundup being identified as one source. Almost half the water in wells and ground water in the United States is contaminated by pesticide residues. Years ago, traces of triazine were found in 98 per cent of all mid-western surface waters and trichloroethylene (TCE) was estimated to be present in one third of the drinking water (Steingraber, 1997). Although many of these chemicals were introduced some 30 years ago, the risks they pose to food and water are not well documented. In the Netherlands, 16 substances were detected once or more in concentrations exceeding official values set by the Dutch government (Crommentuijn et al., 1997). Swedish investigations on pesticide pollution in water started in the mid-1980s and recently revealed residues of more than 50 different pesticides. Residues are more common in areas with intensive agriculture. Findings in the ground water of southern Sweden have also showed residues of pesticides banned long ago (dinoseb, DDT and fenoxy-acetic acids). The dinoseb content was very high (849 µg/l) compared to a current limit of 0.1 µg/l set by the EU. In 2004, one third to half of all private Swedish wells had polluted water.

In conclusion: Increased consumption of water can lead to scarcity over the next few decades in several parts of the world. Globally, people are becoming more dependent upon ground water. This implies future competition between drinking water and the use of water for agricultural purposes. Clean water will not remain a free natural resource to everybody. Pollution of water should not be a great surprise in view of the large production and use of an ever-growing range of chemicals in general, agrochemicals and pharmaceuticals. They all reappear in water and circulate in the food chain unless strong measures are taken to reduce both their numbers and their use to ensure long-term food security and food safety.

Urbanization and Waste Mountains

Fifty years ago, New York and London together had more than 10 million inhabitants. In 1975, one third of the population of the world lived in cities but two thirds, or 5 billion, will do so within two decades. Today, Tokyo is the largest city with some 35 million inhabitants, followed by Mexico City (19 million). According to the World Bank, the number of mega cities (> 10 million) is estimated to reach 54 in developing countries and five in industrialized countries by 2050. In Dhaka, Bangladesh, there is a daily increase of its population by 3,000 persons. In India, the city of Mumbai is expected to have doubled its population of 1990 to reach more than 24 million inhabitants by 2015. In developing countries, urban poverty levels are generally about 25 per cent but in Latin America, 90 per cent of the poor live in cities. According to Habitat, the UN Human Settlements Programme, about 2 billion people may live in shanty towns and poverty by 2030.

Whereas urban poverty is one issue, the future production, storage and transport of food are equally important. At present, some 200 million urban farmers supply food to 700-800 million people. Supermarkets have gradually appeared also in the large cities of developing countries, making specific demands on the retailing system. Another issue is how to add value to products to increase the profits of the producers. Drying, washing and packaging for retail could add value. Many city dwellers have different diet demands and the urban elite may wish to consume more meat. In Africa, most of the food production still takes place in the countryside but this situation will also change over the next few decades. This scenario is another challenge for planners and agricultural policy-makers in many developing countries.

At present, one fifth of the global population lives in cities where the air is not fit to breathe. In China, some 3 million people died from urban air pollution over 3 years in the mid-1990s. A decade ago, annual releases from Chinese cities and industrial areas equalled 10-18 per cent of the total quantities of the pollutants (sulphur dioxide, wastewater and solid waste) produced in the world (Dazhong, 1995). More than 3 million ha of cultivated areas were polluted by wastewater and 5 million ha was directly degraded by air pollution. This problem is growing in other urban areas as well. About two thirds of all hazardous waste produced in the United States came from chemical manufacturing in the 1990s. More than half of the toxic chemicals released by US industries went into the air, including some 70 different known or suspected human carcinogens. The incidence of asthma in the United States has increased dramatically.

Growing mountains of waste are another global problem. The average inhabitant of the EU produces annually about 500 kg of waste per capita, well above the target of 330 kg set in 1993. This is roughly double the amount produced in developing countries. Industrial wastes, sewage and garbage disposal are already major constraints in the large cities of developing countries. This affects city dwellers' access to clean water. Many of the city authorities have insufficient resources to handle these problems. In the United States, incinerators handle less than half of the nation's trash, releasing dioxin from medical waste and other toxic products from household garbage. In Sweden, waste is dumped in about 38,000 places, about 100 of these being classified as the highest risk. Some recent findings have indicated an increased incidence of chromosome damage to foetuses of pregnant European women living up to 3 km from the site of a waste site. Similar problems are bound to appear in developing countries.

Waste from medical drugs causes damage to both the environment and the water quality, thereby affecting long-term food safety. The US Geological Surveys have found residues of medicine, hormones, steroids and perfumes in river samples from the United States. In 2003, US researchers detected the active substance of Prozac (fluoxetin hydrochloride) in liver, brain and meat of fish from Texan rivers. In the same year, a large number of vultures were found dead in the countryside of both India and Pakistan. They had eaten farm animal carcasses containing Diclofenac, a non-steroid anti-inflammatory drug used in both human and veterinary medicine. An average Swedish hospital usually discards 25-30 per cent of drugs because they have expired. More than one fourth of the drugs prescribed by Swedish medical doctors are never consumed by the patients. Ultimately, this means that one half of all pharmaceuticals are washed down the sewage system or destroyed, a problem that has hitherto been given little attention.

Municipal waste products are rich in nutrients. Stabilized sewage sludge, treated wastewater, compost, landfill leachate and ashes could possibly serve as substitutes for mineral fertilizers as part of a recycling process. Several Swedish municipal facilities with wastewater and landfill leachate irrigation of *Salix* plantations have been established since 1990. These plants can accumulate heavy metals but recycling is a current problem due to high content of cadmium. One fifth of Swedish wastewater treatment plants have metal content in excess of current limits for applications to land (Eriksson, 2001). Swedish farmers refuse to apply sludge on their soils, and farmers anywhere in the world would probably do the same.

Nuclear waste constitutes a very special type of product. The most dangerous radioactive waste comes from spent fuel from nuclear reactors and liquid and solid waste from plutonium production. In the United States, this waste is buried at government sites and in salt caverns in New Mexico. The Environmental Protection Agency (EPA) has ruled that the US Department of Energy must demonstrate that the Yucca Mountain can meet EPA standards for public and environmental health for 10,000 years (Long, 2002). Nevertheless, the peak radiation dose to the environment will occur after 400,000 years. Other nuclear waste comes from remnants from decommissioned power plants, hospitals and research institutions. This problem is emerging in developing countries, if and when they build more nuclear power plants.

In conclusion: Urbanization is a major threat, particularly to people in mega-cities in developing countries. Urban poverty and the number of unemployed might further increase, calling for determined action and planning. Arable land in urban areas will be used for construction rather than for food production. Job opportunities for urban farmers will also be reduced if they cannot adjust to new market conditions, new demands by the food chains and food imports. Growing waste mountains lead to serious long-term health problems and degradation of water and the environment. This problem area is an immediate challenge to policy-makers. To avoid long-term effects from pharmaceutical waste, there is a need for both environmental labelling and screening of the long-term environmental effects of pharmaceuticals before marketing.

Towards Declining Confidence in Food Safety?

The BSE Crisis Led to Panic: What Next?

In the past, people knew the origin of the food they prepared and consumed. To many consumers, food is a commodity on the commercial market. Still, most people take it for granted. "Modern" production methods are less transparent and may lead to misconceptions. In the United Kingdom, farming was some time ago identified as the "site at which contaminated food production is initiated" (Marsden et al., 1994). In fact, access to food ought to be a fundamental right. In principle, it must be safe for consumption. Science is seen to provide a guarantee until evidence appears to the contrary. Control has moved from the individual consumer or the family to outside research laboratories chiefly in the industrialized world. This shift of control has been demonstrated in a number of food scandals in recent years, for instance mad cow disease (bovine spongiform encephalopathy or BSE). The dioxin scandal in Belgium is another example. The Belgian Government knew about the dioxin problem but waited three months to act. Poisonous oil was mixed with animal feed. In 2002, the hormone medroxyprogesteron acetate (MPA) was detected in pig feed and lemonade. In Japan, the Snow Brand Company sold re-marked beef from Hokkaido containing BSE. It was the second food scandal involving the same company, which had earlier poisoned some 2,000 people through its milk products. Incidents such as these and the threat of avian flu add to public concern, leading to a questioning of the scientific establishment and its role in food safety. Altogether, they have led to a breakdown of confidence in both the food production systems and the food safety regulatory policies and agencies in several parts of the industrialized world.

In early 2002, the FAO and WHO jointly organized a meeting on food safety and food quality, justified by an increasing number of people being exposed to and even killed by poisoned or contagious food during the previous decade. About 200 diseases may be transmitted through food and some 5,000 people die annually from them (Mead et al., 1999; Ackerman, 2002). In the United States, some 76 million people each year suffer from food-borne diseases. Food-related diseases are not new to Europe or to developing countries. In 1953, a salmonella epidemic killed 100 Swedes. It resulted in the introduction of strict tests, gradually leading to the lowest rate of infected poultry meat in the world. Since then, salmonella has again been found in eggs in Europe due to contaminated animal feed.

Ten years of misleading reports were published before UK politicians and scientists admitted the existence of BSE in meat and bone meal and its danger to humans. The disease is caused by a prion, an infectious protein, which first occurred in the United Kingdom in the mid-1980s. The danger of BSE is its influence on a new type of Creutzfeldt-Jakob disease (nvCJD). The human form of BSE infected younger people, the first case appearing in the United Kingdom in 1996. In the United Kingdom alone, at least 135 people have died from the disease. Some estimates indicate that 3,800 people may carry the disease without their knowledge. Some experts argue that new cases might appear gradually over the next 30-40 years. Altogether, there have been almost 200,000 cases of BSE in 21 different countries. In 2005 alone, 327 new animals were identified with the BSE. The first BSE case in the United States was reported in 2003 and in Sweden in 2006.

There is much less exact information about the effects of BSE in developing countries. Many of them have imported cattle or meat and bone meal over the last decades, although some countries banned the import of beef and dairy products from European countries affected by the disease. Apart from outbreak of diseases, illegal meat imports, found in inspections of flights at Heathrow Airport, pose a threat to food safety (Edqvist, 2002). This relates to an increased animal trade, for instance, the United States imported 2.4 million head of cattle in the early 2000s. Animals are shipped across the country. On average, about 20,000 animals are transported within the EU every month.

In total, about 2 million cows were killed in the United Kingdom to stop BSE. In November 2000, the European Commission decided that all cows born before 1998 should be tested and certain slaughtering material classified as risk material and burnt. In Sweden, animal feed based on dead cows was forbidden in 1986, a decision taken after intensive public debate initiated by one individual who suspected the animal feed. Five years later, the Swedish Government decided to prevent addition of bone and meat products of ruminants in feed for ruminants. After 1995, it is no longer legal to feed ruminants in Sweden with proteins from pigs, horses and other animals. In 2002, the EU established a European Food Safety Authority to provide scientific facts to consumers on indirect and direct influences on food safety. One task is to predict the next crisis and signal warnings.

The experience of BSE influenced the UK Government to act swiftly, and with some panic, by slaughtering animals when foot and mouth disease appeared. In all, some 2,000 cases were found. However, more than 4 million animals were killed and burnt, thereby producing dangerous dioxins. Since half the killed animals with clinical symptoms later proved in laboratory tests to be uninfected, the principle of slaughtering animals living within 3 km of an infected farm was questionable. One reason for the drastic effects was the large size of the animal herds. The total costs of the whole exercise were estimated at US$ 20 billion, whereas the farmers lost some US$ 4 billion.

Danger from Direct and Indirect Food Contamination by Synthetic Chemicals

Two decades ago, some 5 per cent of all food imported to United States contained residues of banned pesticides. An average Indian meal was reported to include a daily intake of pesticide residues of 0.5 mg, which is above acceptable levels (Alam, 1994). The corresponding figure in the United States is about one third of the marketable food, with 1-3 per cent being above the legally defined tolerance level. In Thailand, more than one third of all vegetables are contaminated with organophosphorus insecticides. Pesticide residues were found in many samples of soil, river sediment, fish and shellfish (CGIAR, 2001). In Sweden, regular control shows that one fourth of cereals and most imported food and fruit contain small residues of chemicals, seldom above the limits set by WHO and the Swedish Government. During the last decade, studies have revealed that the contents of certain chemicals have declined in Swedish bodies (dioxin, PCB, DDT, lead and mercury).

The proportion of food containing illegal residues is contested. Current analytical methods detect only a minority of the 600 pesticides being used in the United States. This may explain why the sales of organic foods have increased in Western societies, including the United States. Simple, effective and cheap analytic methods are lacking. For example, nasarin is an antibiotic added to chicken feed. Since 1996, it is mandatory for all members of the EU to test for this type of chemical. Nevertheless, it is estimated that about 20 laboratories conduct about 100 tests annually for the production of 5 billion chickens. Four times as many chickens are slaughtered worldwide. In many developing countries such tests are few or even non-existent. Imports of frozen chicken to Africa more than tripled between 1995 and 2002, according to the FAO.

Cadmium is a heavy metal that may pose serious threats for long-term food security and food safety. An example is cadmium pollution in a river from a zinc factory at Shaoguan in southern China in late 2005. Cadmium used to be a residual of old phosphorus fertilizers. In Sweden, atmospheric decomposition and commercial phosphorus fertilizers constitute the most important source of cadmium in arable land. The cadmium content of Swedish surface soils has increased by one third between 1900 and 1990. In cultivated wheat crops, the content of cadmium increased, exceeding the limit in certain regions so that bread wheat grown there could not be accepted as food (Stolt, 1998). Phosphorus fertilizers with high cadmium content are now charged with a special fee or even banned in Sweden. They may still be in use in developing countries, although products with low cadmium content are available. Even though the process for accumulation of a heavy metal is prolonged, the problem is not fully resolved, in particular as regards atmospheric decomposition from other sources.

Chemicalization of Food—and Life?

Annual world production of synthetic organic chemicals has skyrocketed, although it is difficult to estimate. A decade ago, it was argued that about 11 million different chemical substances had been produced (van Emden and Peakall, 1995). Global production is probably now in the range of 1 billion tons compared to 500 million tons in 1990 and 1 million tons in 1930. Each year about 1,000 new chemicals are introduced, most of them without adequate testing and review. Altogether, some 60,000 to 100,000 different chemicals are in regular use in the global market. It is not fully known how dangerous they might be or what their effects on later generations may be. Worldwide, existing facilities can only test about 500 substances a year. Furthermore, the time span may be long from manufacturing to noticeable negative consequences, as with synthetic oestrogen (DES). It was believed to be a wonder in the late 1950s but several decades later found to be linked to a rare vaginal cancer in young women. Altogether, some 1,200 to 1,500 chemicals have been tested for their carcinogenic effects. Another issue is to investigate how different chemicals interact with each other in both humans and animals. A third issue relates to chemicals that may not be necessary, such as the addition of silver threads in sportswear and shoes. The intention is to give anti-bacterial effects but silver will cause environmental risks. There is rudimentary knowledge about where and to what extent chemicals accumulate and finally end up in the environment and in human bodies.

In the early 1980s, there were about 100,000 chemicals on the European market. Since then, approximately 3,000 new ones have appeared but less than 5 per cent have been chosen for risk assessment. The EU has approved 325 different food additives, one tenth known to cause allergic reactions in people. The new EU Directive on Chemicals (REACH) proposed obligatory risk assessment by industry on new chemicals. It was much opposed by the chemical industry with its 1.7 million employees, resulting in a compromise when the European Parliament took its decision in late 2005. Risk assessments will be required for only 5-10 per cent of the 30,000 chemicals covered by REACH; the majority will only be registered. Test requirements·already exist on chemicals introduced on the market after 1981. The Stockholm Convention on Persistent Organic Pollutants, which came into force in 2004, is quite a step forward. It was signed by 151 countries, which have agreed to ban or limit the use of the 12 most dangerous and persistent chemicals.[1] In light of the existence of thousands of chemicals, it is a start, albeit slow.

---

[1] Aldrin, chlordane, DDT, dieldrin, endrin, heptachlor, mirex, toxaphene, PCBs, dioxins, furans and hexachlorbenzene

Anybody, almost everywhere, will today find a few hundred chemical contaminants in his or her body fat. In the United States, at least some 40 possible carcinogens appear in drinking water, 60 are released by industry into ambient air and 66 are routinely sprayed on food crops as pesticides. Chemicals may have direct or indirect effects. Examples include caesium residues from the collapsing nuclear reactors at Chernobyl in the 1980s or fatal explosions at Seveso, Italy and Times Beach in Missouri in the United States, all leading to acute human exposure to dioxin. Dioxins are capable of profoundly altering biological processes at a few parts per trillion. The most poisonous dioxin is known as 2,3,7,8-TCDD, dispersed during the Vietnam War to strip away the forest canopy. Agent Orange was a mixture of 2,4-D and 2,4,5-T, the latter contaminated with dioxin during manufacturing. Since then, its effects have been debated in the United States. In 1974, the US EPA decided to suspend the use of 2,4,5-T, a position that was re-evaluated in 1991. The greatest threat of dioxin was not cancer but its power to disrupt natural hormones. Dioxin was not an oestrogen mimic like DES but it somehow disrupted oestrogen responses. In 1995, the US National Academy concluded that dioxin-contaminated herbicides were the cause of three cancers: soft tissue sarcoma, non-Hodgkin's lymphoma and Hodgkin's disease. In early 2001, the US Government again changed its position, since new research had shown that Agent Orange had connections to some leukaemia of Vietnam War veterans. Some 3 million Vietnamese people have been directly harmed and the indirect effects are not yet known. In early 2004, the US manufacturer of Agent Orange was sued in a New York court.

Hormones are important to food safety, being very potent at extremely low concentrations without altering genes or causing mutations. They programme cells, organs and the brain as well as behaviour before birth. They regulate development in basically the same way in a mouse, rat or human. In the United States, antibiotics have been added to animal feed for a long time. Recombinant bovine growth hormone (rBGH) was released in 1994 and approved by the US FDA but not allowed in Canada and Europe. Injected every 14 days, rBGH and antibiotics increase production by some 25 per cent in dairy cows. But declines take place in fertility and dressing. The volume of such antibiotics used in feed equals or exceeds that used in human medicine in the United States. Monsanto is now cutting down its production of rBGH.

As early as the 1930s, studies in the United States indicated that influencing hormone levels during pregnancy might be harmful. Findings from wildlife later showed that a number of synthetic chemicals, including DDT, could somehow act like oestrogen. It is known to be one important risk factor for breast cancer but hormones might also have various effects. Maleness and femaleness are usually seen as mutually exclusive but are not (vom Saal et al., 1992). Even before birth, the female hormone oestrogen increases sexual activity of a male in adult life. Though the sperm delivers a genetic trigger for a male, a developing baby does not commit itself to one course or another. It retains the potential to be either a male or female for more than six weeks. Thereafter, the Y-chromosome directs the unisex sex glands to develop into male testicles.

According to the Community Programme of Research on Environmental Hormones and Endocrine Disrupters, there was a significant spread of oestrogen-like substances in the EU countries in 1999-2001. Estradiol is one of three principal types of oestrogen produced by the ovaries and dispersed in the bloodstream. It is the most active oestrogen. An oestrogen receptor is a special protein, found inside the cells in many parts of the human body. A union of the hormone and the receptor will trigger the production of particular proteins. In general, receptors are said to be highly discriminating about chemical structure but only a tiny fraction of the natural oestrogen in the bloodstream will be free. In contrast, all synthetic oestrogen will be biologically active and will bind to the receptor (Colborn et al., 1997). This has implications for the use of hormones. Prior to 1962, most medical scientists thought a drug was safe unless it caused immediate and obvious malformations. Later it was realized the effect of the drug depended on the timing of the drug use, not on the dose. However, the dose of hormones is a more complicated question. The same dose can have very different effects depending upon age, sex and hormonal status.

Researchers have identified at least 50 synthetic chemicals that disrupt the endocrine system. This may explain why women in the past used a variety of plants to prevent pregnancy and precipitate abortions. It is less known whether such chemicals are beneficial or hazardous. At least half of them are persistent and resist natural degradation processes. Some mimic estrogens, others interfere with testosterone and thyroid metabolism. Plants produce oestrogen substances as a defence against insects and researchers have found oestrogen substances in at least 300 plants. Examples include sage, garlic, wheat, rye, oat, barley, rice, soybean, potato, pea, bean, alfalfa sprout and apple. Like DES, and DDT, the plant compounds can fool the oestrogen receptor. The important difference is that plant estrogens might be eliminated within a day but synthetic hormones might persist in the body for years. Estrogenic chemicals may add to lifelong exposure. In the early 1940s, sheep grazed for an exceptionally long time in Perth, Australia. They were hit by an epidemic of infertility, caused by clover imported from the Mediterranean region. The chemical causing the "clover disease" was later proved to be a substance of plant defence that could mimic the effect of oestrogen. A recent illustration is the hormone MPA. It is used as a drug for women but forbidden by the EU as a hormone for animals.

The food industry is already making use of nanotechnology and almost 200 applications are under development and patenting. In 2006, a Center for Nanoscale Science and Technology was launched for collaborative research between industry, universities and the US Government. The technique may be used for rapid and label-free detection of food-borne pathogens. Since the techniques imply that food will be changed at the molecular level, there are reasons to believe that minute molecules a few millionths of a millimetre may move freely within the human body after consumption of such food products. They penetrate human tissues and organs, thus constituting serious risks not yet identified or investigated. They may be classified as new potentially harmful materials. Nano-ingredients pose risks also in beauty products. This calls for both collection of safety data and strict rules and regulations. So far, there are no labelling requirements for nanofood products. Silver nanoparticles are among the fastest-growing products in the nanotechnology industry. Global sales are estimated at US$ 800 million to 900 million. This expansion may threaten smaller food companies and developing countries since the larger food processors will gain control of the intellectual property rights. Recent estimates predict the total market for nanotechnology-based environmental applications will be US$ 6 billion by 2010.

In conclusion: The steady increasing trend of new chemicals is alarming, constituting both an immediate and long-term threat to human health and food safety. The "precautionary principle" is seldom applied. Chemical screening and risk assessment must consider long-term effects on human beings and animals. This requires the design of cheap methods of risk assessment and testing by the manufacturers, who have the burden of proof prior to the release of any new product. Since this implies higher costs, it will slow down the pace with which new chemical products are marketed. It will be necessary to consider what risks we are willing to take for additional comfort or leisure in the short-term and in the long-term, or at least for the next generation. Developing countries ought to ban import of all chemicals that already show potential risks in industrialized societies without being penalized by the WTO. Governments should introduce extra high taxes on all non-degradable chemicals and additives not absolutely essential for good quality food and feed products. Products today derived from petroleum came initially from vegetation, for instance celluloid, cotton fibre, soybean resins, linseed oil, castor oil. With increasing oil prices, these older products may again attract interest.

The BSE crisis may be the first major food scandal whose full consequences are not yet known. Policy-makers and relevant specialized agencies of the UN must take determined action to guarantee food and food safety for consumers worldwide. No compromises can be tolerated in dealing with long-term food safety and security. This principle will favour local markets rather than one global food market directed from the North. Well-informed consumers will prefer safe food to cheap food, calling for improved education in biology, including food production. There is a great need to identify and apply new methods for the identification, monitoring and assessment of food-related hazards to guarantee food safety in the 21st century.

Obesity—A Growing Problem

According to WHO, about 1.7 billion people were overweight in 2003. In the United States about 61 per cent of adults are overweight and almost one third are extremely overweight. A similar trend is emerging in other industrialized countries. Between 1990 and 2000, the number of overweight Swedes increased from 5 to more than 9 per cent. In consequence, medical costs to society dramatically increase. Excessive amounts of salt, fat or calories are promoters of cancer. The McDonald's fast food chain has already been sued in the United States. The food processing industries are probably next in line. Excess weight is hardly a common problem for a majority in developing countries but does exist in urban areas. With rising incomes and growing urbanization, and with current concepts of development thinking, this trend will expand. To the pharmaceutical industry, it means a growing and profitable market. In contrast, only 1 per cent of more than 1,200 medicines produced between 1975 and 1997 were developed for tropical diseases.

Both the scientific and industrial establishments stress that functional foods are the right avenue for diets of tomorrow. In combination with certain vaccines, they are even expected to be breakthroughs, such as potato with polio vaccine and banana with a malaria vaccine. There are great hopes that food will be tailored to certain target groups. Any successful vaccine campaign will take time and with this orientation the focus may be in highly industrialized countries with large markets for the food processing industry. The crucial issue is whether functional foods constitute a shortcut or simply an illusion. In 2003, there were three approved health products as functional food in Sweden but only Proviva had proven scientific evidence of positive effects. The same effects may also be achieved by using conventional food in more appropriate dietary combinations, improving the knowledge of food preparation and more physical training.

In conclusion: Excess weight is of growing concern to society and to the health of individuals, mainly in industrialized countries. The global market and rapid urbanization over the next few decades will further change food habits. Better education in biology, food and home economics is required. An active life with more physical training in the educational system may help to minimize obesity. The food industry can play a decisive role in leading the movement towards more healthy food.

Migration—A Search for a Better Life

Famine, economic stagnation, population growth, environmental degradation and civil wars have always produced movements of people. Emigration may be motivated by the search for better living conditions. Movements of people also influence land use patterns. But current discussions on immigration and refugees in Europe seem to ignore that Europeans have not only travelled widely but also emigrated in a large scale. The introduction of potatoes and their increased cultivation made it possible to feed the increasing Swedish population during the 19th century. Famine struck the country in 1860, forcing Swedes to emigrate. Over five decades, about 1 million Swedes left the country. Between 1824 and 1924, about 52 million Europeans emigrated. On an annual basis, this is close to current immigration to Europe, reaching 2 million legal and half a million illegal immigrants (Reid, 2002).

At the end of World War II, Sweden had about 200,000 refugees, many of whom returned or went to other countries. During this period, Sweden recruited labour from abroad without which industrial expansion would not have been possible. This was not a new feature of Swedish economic life. Starting in the 15th century, active recruitment by the State took place for 300 years. But the State promoted the "Swedish" to the exclusion of immigrants, including the Finns, Germans and Lapps, all living within the nation-state. This was in contrast to a much stronger change into "Swedish" of the people living in the southern province of Skane, formerly under Danish rule. This implies that national identity is a social construction (Johnson, 1999).

Up to the early 1980s, Sweden received about 100,000 political refugees. For the next 25 years there was a policy of welcoming refugees but the individual capacities of the refugees were mostly ignored. They had a range of qualifications and networks. In the mid-1990s, immigrants numbered 1.4 million, most of whom had great difficulty in finding employment and became dependent on social security support. Several started their own enterprises, contributing to more than 65,000 companies around 2000.

According to the United Nations, there are currently some 150 million migrants worldwide. One million people are applying for asylum and 20 million are refugees. In the 1990s, the number of refugees was some 43 million, a little less than the 48 million during World War II. Today, half of them are internal refugees, displaced within their countries of origin, for example in the Sudan and Mauritania. Some live under very poor and harsh conditions. In fact, more than 12 million people live in some kind of slavery today according to recent data by the International Labour Organization. Most of them live in Asia. Long ago, the Pharaohs, Romans, Arabs and Turks found their slaves in the Nile Valley, including the Sudan and also Darfur. Later, the Atlantic slave trade involved many European countries, profit being the primary concern. The large-scale trade ceased in the late 1880s, when almost one third of the population of the Sudan were slaves. Officially, slavery was abolished in parts of the Sudan in the mid-1930s but is reported to have reappeared. Profit remains the driving force for modern slavery, the trafficking and smuggling of refugees. In 2000, Christian Solidarity International reported that the price of a slave in the Sudan was two or three goats at the local market. Each year about half a million women and young girls from poor countries are brought into prostitution in Western Europe.

In conclusion: As in the past, the challenge of refugees and migration will remain on the agenda as long as civil wars, environmental degradation and poverty exist. Food preferences of immigrants will stimulate food imports from their countries of origin, positively influencing international food trade and production. With political will and determined international action, slavery and trafficking could be stopped.

Agro-terrorism

Agriculture may be a sleeping target for terrorism using biological agents. In the past, such weapons, including poisonous gases, were produced to combat military threats. The United States developed and stockpiled spores of *Puccinia graminis tritici* between 1951 and 1969 (Davis, 2004). Another primary pathogen was *Piricularia oryzae* against rice. Both are causes of important plant diseases. The target was wheat in the former Soviet Union and rice in Asia. A similar programme was developed in the former Soviet Union and officially ended in the early 1990s. Incidents of food contamination have also been recorded. Another threat may come from terrorists with access to nuclear agents. Terrorist groups may also turn to other biological agents. Most observers assume the agent of foot and mouth disease to be the ultimate biological agent to be used against livestock. Anthrax might be another threat. In addition to aspects of food safety, there are great financial costs involved.

Globalization and Agribusiness

The Globalization Process

The term *globalization* first appeared in Webster's Dictionary in 1961. Today, it is a household word but has different meanings for different people. It may mean a tendency to universal application of economic, political and social and legal practices. It can mean cultural homogenization and may lead to economic improvements. Above all, globalization means an international division of labour, in which TNCs constitute one central element. Globalization is certainly a trend but also a process of old origin. The British East India Company coined the term "global reach". After independence, the United States chartered its own corporations and so did Japan some hundred years later. In the 19th century, industrial capital of Western Europe stimulated global expansion. Free trade replaced protectionism. The nation-states became powerful and constituted the basis for negotiations, formulating certain rules of conduct through international agreements. The current globalization process can therefore be seen as an acceleration of historical dynamics with characteristics such as rapid communications, new technologies and free trade ideology with no effective democratic system of global governance. Since capital can move electronically 24 hours a day, the effects of poverty transcend national borders. While poverty has existed for centuries, the rich have not voluntarily adjusted their life-styles in favour of the poor. This is important to remember in discussions relating to the fulfilment of the MDGs.

Even though some TNCs may appear social, their bottom line is profit. If and when problems arise in one location, the TNCs can shift to new production sites. When costs for production increase, they shift their operations from one country to another where the wages are lower. They can sub-contract hazardous activities to smaller companies abroad, thus avoiding direct responsibility. But people in democracies can only turn to their political agencies, namely the nation-state. Since they are losing power as a result of the process of globalization, other groupings may turn out as winners or losers, depending upon whether they are excluded or included in the global production and trade. This is accentuated by the establishment of new regional forms of cooperation such as the North American Free Trade Alliance (NAFTA), Free Trade Area of the Americas,

EU, AU and others. As one result of the NAFTA agreement, General Motors closed down more than 20 factories in the United States and Canada and moved to Mexico.

Transnational Corporations

Transnational corporations are enterprises that control assets, such as factories, mines or sales offices, in two or more countries. About 90 per cent of all TNCs are based in the United States, the EU and Japan. These countries can be seen as three large corporate states. More than half the TNCs come from five countries: the United States, Japan, France, Germany and the Netherlands. In a ranking of the 100 largest economic entities of the world, half of them are corporations. The United States is home to 8 of the 10 largest companies, including the low-price supermarket chain Wal-Mart, General Motors and Exxon Mobile. In terms of revenue, the Japanese Mitsubishi, Mitsui, Itochu and Sumitomo are among the largest corporations in the world. They have played a great role in creating Asia's miracle economies but have also been influential in generating the environmental crisis of that region.

There are about 65,000 TNCs with 850,000 subsidiaries and 54 million employers (UNCTAD, 2002). This number has rapidly increased from 40,000 in 1995 and only 7,000 in 1970. These corporations are estimated to hold 90 per cent of all technology and product patents worldwide. They are involved in 70 per cent of the world trade and the sales by the affiliates exceed the total international trade. Assuming that the world economy is growing by 2-3 per cent annually, the largest TNCs, as a group, have generally a growth rate of 8-10 per cent a year.

The pharmaceutical industry is a global giant, its world sales reaching some US$ 430 billion in 2003, an annual growth of 8 per cent (AstraZeneca, 2004). Roughly 85 per cent of this market is in Japan, Europe and the United States. In recent years, mergers have been common. In the mid-1990s, the merger of GlaxoWellcome and SmithKline Beecham resulted in GlaxoSmithKline GSH. In 1999, Astra merged with Zeneca. Three years later, Pfizer and Warner-Lambert combined their efforts. After two years of negotiations, Pfizer purchased Pharmacia in 2003 at US$ 60 billion. Prior to this, Pharmacia-Upjohn was an exception to the conventional trend of expansion by growth. Its policy was to be the best-managed company of the pharmaceutical industry, not the largest one. Founded in 1847, Pfizer is the largest pharmaceutical company in the world. The Novartis Group has branches in 140 countries but only half as many employees as Bayer, which also works in agriculture and operates in 150 countries. Annually, the pharmaceutical corporations invest 15 per cent of the total turnover in

R&D, equivalent to about US$ 5-7 billion. It usually takes about 12-15 years from the first chemical analysis to the appearance of a drug on the market. About eight years of sales are required to pay for its development.

A Growing Global Agribusiness

The TNCs and large companies are major actors in global agriculture in input supply, commodity trading and agricultural research. They command directly or indirectly some 80 per cent of land worldwide that is cultivated for export crops such as banana, tobacco and cotton. Much of the genetic seed stock is under TNC control, signalling that the viability of public plant breeding might be in question for certain crops. Three companies control the world sales of coffee (Murphy, 2002). Ownership of the five top world food manufacturers is split between Europe and the United States. In 2000, Philip Morris purchased Nabisco to be incorporated in Kraft Food, the second largest food company of the world after Nestlé. They are followed by the US-based ConAgra, General Mills and Cargill. In 2001, two of six food processors were European: Nestlé and Unilever. As to global food retailers, two are US-owned and four are European. They include Wal-Mart, Kroger and Carrefour, Ahold, Tesco and the Metro Group. In forestry, there has not been the same concentration, as yet. Some 40 international corporations control about 115 million ha of the world forests.

The US broiler industry was the first livestock commodity to be rationalized, starting during World War II. Now, globalization of chicken production is well under way; eggs are produced in one region, chickens are raised in another, they are slaughtered in a third and meat is processed in a fourth. In the United States, three major companies handle all cattle slaughter. Contract and corporate production of food in the United States and some Western European countries has increased and contract farming has increased in developing countries, maize being one example. With increasing labour costs for the production of hybrids in the industrialized countries, hybrid multiplication schemes have been relocated to developing countries.

Today, the kiwi fruit is truly global. Likewise, orange juices are reconstituted from concentrates and consumers in other countries add water. Consumers do not think about the seasonality of fruits, fresh products coming all over the world at any time. According to the World Bank, developing countries now export US$ 20 billion of horticultural products a year, the fastest-growing agricultural sector in the developing world. In Kenya, horticultural exports constitute the third largest source of foreign exchange.

Market saturation in industrialized countries has led the global firms to a strategy of market segmentation. In tree crops, tissue culture techniques constitute a major research area for corporations. In animal feed, reduction of costs is just one objective; firms also aim for biological rather than chemical inputs or for the substitution of additives. Instead of bulk production, the corporations focus on functional attributes for specific niches. Wheat may be redesigned to produce wallpaper paste; canola has been redesigned to produce lauric acid, an ingredient in detergent and soap. These few examples illustrate a process very different from past introductions of new crops. National models will become insufficient for the analysis of global and more complex food regimes and supply. In short, the TNCs play a major role in the transformation of agriculture.

The mergers of companies lead to concentration of power. In agribusiness, this process was started in the 1970s by the petroleum industry, Shell becoming the largest seed company. Seeds and agrochemicals were combined with products in both human and animal health. The agrochemical and pharmaceutical corporations purchased some 30 seed companies. Recently, the mergers have accelerated. Novartis was created in 1996 when Ciba-Geigy and Sandoz merged. In 1999, Aventis emerged as a merger of Hoechst and Rhone-Poulence. In 2000, the agricultural activities of Novartis and AstraZeneca were combined into a global Syngenta. In the same year, Pharmacia Upjohn purchased Searle and took over Monsanto. One year later, Monsanto was again sold to make a clear distinction between pharmaceuticals and agrochemicals. Since the agrochemical companies were seen to convey a negative message to the public, the pharmaceutical companies sold off their components of agrochemicals during the last decade. In 2000, BASF purchased the agro-products part of American Homes (Cyanamid). BASF is strong on fungicides and pesticides, whereas insecticides were the comparative strength of Cyanamid. Other examples include the sale of Knoll by BASF and Aventis Crop Science by Aventis to Bayer. The new Bayer, operational in 2002, was estimated to account for about one third of the agrochemicals market as the second largest producer of agrochemicals after Syngenta. In 2002, BASF purchased the fungicide and insecticide components of Bayer. Two years later, Bayer decided to separate its chemical division into a special company, Lanxess. As the biggest chemical maker of the world, BASF has increased its research in genetically modified (GM) plant science, planning its first release of GM high-starch potato in 2007. In early 2006, the AstraZeneca Company turned to more pharmaceutical biotech research by purchasing the UK Cambridge Antibody Technology. In 2004, Norsk Hydro had decided to concentrate on energy and gas by selling out Hydro Agri, its fertilizer component. As the largest fertilizer producer of the world, it has operations in more than 50 countries.

A few years ago, a small number of TNCs controlled some 80 per cent of the market of agrochemicals. Other companies have little or no research of their own. Pesticide development can be quite profitable. In the late 1990s, an investment was expected to yield a return of four times but costs have recently increased, most of them allocated to safety aspects. The largest companies include Syngenta, Pharmacia/Monsanto (previously Monsanto Agro), Aventis and BASF (including Cyanamid). Today, Syngenta has a focus on seed and plant protection for a sustainable agriculture.

Seed and Livestock Corporations

After World War II, large private companies took over seed distribution in the United States. This commercialization of the seed industry was a result of the development of maize hybrids and new wheat. It also led the seed companies to take over much of the R&D of these crops. Some years ago, four companies controlled 69 per cent of the US seed market and the 10 top seed companies controlled about one third of the US seed trade worldwide (Hayenga, 1998; RAFI, 1999). Over the last five years the value of the global proprietary seed market has increased by 30 per cent, reaching more than US$ 17 million in 2005. This change is due to expanded plantings of oilseed crops and maize, mainly by hybrid and transgenic technologies, according to a US firm, the Context Network. This value is almost the same as the total world annual seed sales in the 1970s. Other estimates of the world annual seed sales indicate values up to US$ 30 billion, although lower figures are also reported.

The growing global seed market and the need to handle issues of IPRs have accelerated the trend of reducing the number of seed corporations. Nowadays, the seed industry is under the control of a few large companies (AgrEvo, DuPont, Monsanto, Novartis, DowElanco). In 1999, Pioneer Hi-Bred International Inc. merged with E.I. DuPont de Nemours & Co. The new company operates 75 production plants worldwide (Johnson, 2000). Another factor is the importance of biotechnology research. To be profitable, research investments in biotechnology require quick access to markets where GM crops are allowed. This can be illustrated by the experience of SW Seed, ranked as the 14th largest seed company in the world. Up to 1995, it had 65 per cent of the market for rape in Canada. When GM rape cultivars were introduced there, its market share dropped to only 20 per cent. Within three years, the company got approval of more than 10 GM rape cultivars and regained its position in Canada, where GM rape constituted three quarters of the market.

In some developing countries, the private seed sector has gained in importance during the last decade, mainly for maize. Multinational corporations control about half of the total commercial maize seed market. However, a major part of food production still comes from farmer-saved seed, e.g. 90 per cent in Sub-Saharan Africa and 70 per cent in India. For the future, a mix of public and private sector participation can be expected but it depends upon what role the private sector actually plays in individual countries. Alliances rather than competition will be important for increased efficiency and in reaching large numbers of farmers, in particular the resource-poor ones. This will require more consideration to factors contributing to the use of seed rather than to the techniques of seed production.

Most pharmaceutical companies have been engaged in R&D in and supplying products for veterinary medicine and animal health. During the last decade, more specialized companies have appeared that focus on biotechnology. The US-based ABS Global Company has the impressive motto "Providing Protein for the World". It is represented in more than 70 countries, selling bull semen. Its vision is to serve as the leading biotech centre of the world for agricultural research: the International Center for Agricultural Trade and Technology. An attempt to directly combine efforts in health with agriculture is exemplified by Epicyte, based in San Diego (The Economist, 2001). It is growing crops meant to kill the human sperm. It has produced anti-sperm and anti-herpes antibodies in gel forms. These proteins could stop both sexually transmitted diseases and pregnancies. The company has used GM maize plants, assumed to be more like humans in their cellular structures than usually thought. The maize seed would also be an ideal storage place for proteins. Once harvested, the antibodies are to be extracted and turned into medicines.

Human tissues are an increasing component of global trade. Since pigs may also be used, this area is of certain relevance to future agriculture. In the late 1990s, there were four global companies. The market prospects for 2010 might reach US$ 6 billion. The industry itself is predicting an annual value of US$ 80 billion from products of human tissue culture within a generation. These prospects attract the interest of both pharmaceutical companies and universities not only for production of new knowledge but also for patents and profits. In the United States alone, there are already more than 50,000 patents on organ transplants.

Globally Expanding Agro-biotechnology Companies

The growth of the biotech industry is led from the United States. In the late 1980s, there were about 120 small companies specializing in agricultural biotechnology and annually spending US$ 4-6 million on R&D (Dibner, 1991). Four researchers created AMGEN. Today, it is a global company employing some 13,000 persons. At present, there are some 1,500 biotech companies in the United States compared to about 1,000 in Europe, though the US companies are smaller. With 200 biotech companies, Sweden had the seventh largest biotech sector in the world in 2005. The top transnational bioscience corporations spend approximately US$ 3 billion on agricultural R&D but European spending on biotech research is only one third of that in the United States, a decline over the last five years.

The agricultural biotechnology industry is experiencing a wave of consolidation. One major reason for buying competitors is the desire, and necessity, to get access to IPRs and patents. When Syngenta was formed, the genetic use restriction technology came under its control. In 1998, a CBD conference instituted a de facto moratorium on this "terminator technology". The moratorium remains despite an attempt by the Canadian Government and others to lift it at recent meetings. At the Eighth Conference of the Parties of the CBD in Curitiba in 2006, new efforts are to be made to find a resolution calling for "case by case" assessments of genetic use restriction technology at the national level. Another interesting development is a resolution passed by the Illinois Farm Bureau in late 2005 urging the US Congress to reconsider federal laws that bar farmers from saving patented seeds for replanting from one year to the next year. The resolution focused on GM Roundup Ready Soy developed by the Monsanto Company. It has recently revived its plans to purchase the US-based Delta & Pine Land, a global company for cottonseeds. Such an attempt failed in 1999 because of the controversy over terminator technology, although Monsanto had pledged not to commercialize such crops. In 2005, Monsanto again revised its pledge, confining it to food crops.

In the mid-1990s, Monsanto purchased Holden's Foundation Seeds, Calgene and Ashgrow Seeds and Du Pont's investment in Pioneer Hi-Bred. Other important groupings include Novartis, Dow Eleanco, Monsanto, AgroEvo (Hoechst & Sheriny) and Pioneer-DuPont. In mid-2003, Biogen and Idec Pharmaceuticals announced plans to join forces in a share-swap deal, creating a new company. In late 2004, the US companies Chlorogen and Sigma-Aldrich signed an agreement on chloroplast transformation technology for the joint production of pharmaceutical-producing GM plants. The Syngenta Company purchased the US companies Garst and Golden Harvest. In early 2005, Monsanto acquired Seminis, the world's largest producer of vegetable and fruit seed. In the same year, SW Seed closed down its breeding work on vegetable crops followed by a further downsizing in 2006 and terminated its plant breeding on rye and winter wheat. Syngenta and DuPont formed an alliance, GreenLeaf Genetics, to pool their patents on crop traits for licensing to US seed companies. Syngenta will get the rights to DuPont's GM technology for herbicide resistance and DuPont will have access to Syngenta traits for insect resistance. This joint venture may challenge the dominance of the GM seed business by Monsanto.

At present, the commercial use of agricultural products from biotechnology has grown. GM crops are now grown in some 20 countries. More than 40 other countries conduct research on them. Greenpeace International has claimed that GM crops have been illegally planted in 39 countries during the last decade. Some 50 GM crops have passed the US review process. More than 60 per cent of all processed food on the shelves of the US supermarkets contains genetically engineered soybean, maize or canola (Ackerman, 2002). Projections show that 95 per cent of the US agricultural production would be genetically modified within a decade. In Canada, three quarters of the food is genetically modified; this is attributed to the dominance of neo-liberalism, government support and policy-making by a few influential people (Kuyek, 2002). According to the EU, both GM production and non-GM production will be common in Europe during the coming decade. In 2005, the market was estimated at US$ 90-100 billion. This explains why corporations want field trials allowed also in the South, exemplified by the recent call on the Thai Government by the Charoen Pokphand (CP) Group, the largest agricultural corporation in Thailand dealing with cassava, rubber and maize for animal feed.

Negative public opinion has led European biotech companies to move their research efforts and field trials to the United States, starting in 2002. Syngenta was the last company to leave its crop research in the United Kingdom in 2004. Market prospects for GMOs (genetically modified organisms) were considered bleak over the next four to six years and the EU has problems in deciding upon a GM policy. The move is also a cost-reducing manoeuvre since it is expensive to develop a new GM crop. Estimates in the United States are US$ 300-400 million. To US corporations, the European market for GMOs is quite important. This explains why the United States is actively involved in trying to find political support worldwide, having a centralized approach to biotech within the US Government. India and the United States decided to have joint biotechnology projects in a Knowledge Initiative on Agriculture over three years, starting in 2006 and investing some US$ 100 million. Another example was the joint US/Vatican conference on GM food in mid-2004. At a ministerial conference in Burkina Faso, the USDA provided new technologies and even cottonseed free for possible distribution to African farmers.

The private biotechnology sector in the United States has also been proactive in establishing research links with US universities. In 1998, the Novartis Company signed an agreement with the University of California, Berkeley. During five years the university was to receive an annual research grant of US$ 25 million, giving Novartis the rights to negotiate exclusive licenses on discoveries, irrespective of financial sources. This trend of forging closer links between private and public research was partly a result of declining government funds to research, partly amplified by the results from the merger of the Institute for Genomic Research with Perkin-Elmer, the major manufacturer of DNA-sequencing instruments. That merger challenged the investments by the US Government to the Human Genome Project, involving over a dozen research centres. But efforts to concentrate R&D were common in the private sector. In 2004, a review of the agreement between Novartis and Berkeley concluded it should not be repeated. It had been too problematic and had not turned out as expected; not a single innovation was licensed and no dramatic discoveries were made.

The Private Sector and Environmental Pollution

As early as 1992, it was stressed by the (now defunct) UN Centre on Trans-National Corporations that the influence of TNCs extended over roughly 50 per cent of all emissions of greenhouse gases. But direct foreign investments in developing countries may often be in what has been narrowly defined as "pollution-intensive industries". Investors are attracted by low salaries and less strict environmental regulations. National factories make similar investments and also local factories may contribute significantly to pollution, for example the tanneries in India and Bangladesh that pollute the canals and waterways with large amounts of chromium. To the authorities, these factories provide employment to many citizens, so closing them down is not an easy solution in the medium-term perspective. Other findings indicate that many factories in the South were causing less damage to the environment than they used to do a decade ago (World Bank, 1999). The corporations have created departments of environmental affairs and many have signed the non-binding Business Charter for Sustainable Development. The World Business Council for Sustainable Development has been formed. Globally, 36,000 companies have introduced environmental guidelines according to international standards (Sjoberg et al., 2002).

There are several examples where the private sector has been very quick to respond to new requirements. One example relates to the Montreal Protocol, banning the use of CFCs. McDonald's in the United States eliminated CFCs from its foam food packaging within 18 months. Whirlpool within a year introduced a new model of refrigerator equipped with a cooling system without the CFC refrigerant (WRI, 1996). In 1992, Nissan and Mercedes-Benz became the first car producers to eliminate CFCs from air conditioning systems. Within two years, all their new cars had substitutes for CFCs. These few examples show that the private sector can be quick in responding to new rules and also provide incentives for new markets.

The Union Carbide factory is a different example. It was established for local production of insecticides in Bhopal, India. The production process was complex and the ingredients very poisonous. To save costs, the safety arrangements were gradually diluted. This led to a disaster, killing some 8,000 people and harming many thousands when a toxic gas was released by mistake in 1984 (Lapierre and Moro, 2003). But the agricultural component of Union Carbide was purchased by Rhone-Poulence and taken over by the Dow Chemical Group in 1999. Today, that Group argues that it has no responsibility for an accident caused during the Union Carbide era. In 1991, the Chairman of Bhopal Union Carbide was sued in a court in India. As yet, he has not been traced.

There are several examples where the private sector has failed to take responsibility for its actions. In the past, unwanted synthetic chemicals were simply dumped in the wild. Long ago, the Swedish Astra Company dumped barrels with chemicals in the Baltic Sea or buried them in the factory compound. Frequently, new ingredients are introduced. In Swedish food products some solvents include hexosans, toluene and trichloroethylene. They are seldom registered or traced as residues, as pesticides are. Glyphosate is a systemic herbicide and part of it is carried into the harvested parts of plants, like many other herbicides. Its potential carcinogenic effects may not be detected for three to four decades. Therefore, it is of serious concern that initial success stories and technical breakthroughs may over time turn out to have sometimes severe side effects. A few examples illustrate this dilemma:

- For long, CFCs were seen as the safest substances invented. The inventor, T Midgley Jr, received the Priestly Prize in 1941 and CFCs were on the market for 40 years without any suspicion. Since the first study on CFCs focused on the lower atmosphere, it was not until 1974 that their influence on the stratosphere was reported. In 1995, Midgley was awarded the Nobel Prize.
- In 1938, Paul Müller discovered DDT, quickly seen as a very effective pesticide. At about the same time, oestrogen was synthesized (as diethylstilbestrol, DES). Ten years later, Müller received the Nobel Prize. Two years later, DDT was shown not only to kill insects but also to have oestrogen effects when given to young roosters: it feminized them. The chemical structure of DDT was similar to that of DES (Burlington and Lindeman, 1950). In the 1970s, research showed that DDT had hormone effects (McLachan et al., 1975). Even if it did not look like oestrogen, the body could mistake DDT for oestrogen. Nonetheless, DDT is still one of the cheapest and most effective methods of mosquito control.

- Polychlorinated biphenyls (PCBs) became a commercial success, being produced on a large scale between 1930 and 1970. Monsanto Chemical Works was the first producer in Alabama, USA. They were on the market for 36 years before serious questions surfaced publicly. The synthetic industry has globally produced (exclusive of the former Soviet Union) some 3 to 4 billion tons of PCBs. In 1976, the United States banned their manufacture. But 10 years earlier, it was known to Monsanto that PCBs were poisonous and fish deaths were reported. In early 2002, Monsanto was even accused of having caused human damage, since citizens of Anniston, Alabama, were reported to have levels of PCB 10 times that found in the average person. Statistics of cancer were reported to be alarmingly high from that region. One year later, Monsanto was fined US$ 700 million for poisoning inhabitants with PCB in Anniston. The PCBs are world travellers and quite persistent. Some 250 million tons of PCBs were in use worldwide in 2000, although alternatives exist.
- The thalidomide tragedy broke out in 1962. By the time the tranquillizer was removed from the market it had caused severe deformities of 8,000 children in 46 countries. The human body could mistake a synthetic chemical for a hormone. As with DES, the effect of the drug depended on the timing of the drug use, not on the dose. The Americans largely escaped the tragedy because of one sceptical doctor at the US Food and Drug Administration (Colborn et al., 1997). Just a few years ago, thalidomide was reported as a drug for the treatment of cancer at the Uppsala Academic Hospital.

Towards a Few Large Agricultural Corporations?

Many TNCs have already reached the limit where economies of scale accrue. But access to capital and increased number of strategic options continue to accumulate. Since there are few alternatives, the TNCs may continue to expand while most nation-states play a diminishing role. Most likely, the global corporations will increase their role in influencing global trade and politics in the next few decades. They may stimulate employment in developing countries by moving production there. Since 1995, almost 3 million work places in the United States have been moved to Mexico, China, India, the Philippines, etc. The outsourcing of work by Microsoft, IBM, Oracle, Motorola and Intel is another feature of the same trend. These companies expect to save 25-40 per cent through lower wages and good communications. Estimates indicate that more than 200,000 US jobs will annually be moved to low-income countries over the next decade. The same process has started in Europe. It will include food and fibre production.

One example is the Nordic forestry company Stora Enso, which is seeking to expand in China, South America or Russia rather than in the Nordic region. Low costs of labour and raw material are critical parameters. Moreover, the growing period of the trees will be eight times as short in Brazil as in Finland. It is not surprising then that Finland is now importing paper pulp from its Veracel factory with 165,000 ha in Brazil, designed to produce 1 million tons annually.

Another issue is whether the trend of concentration will ultimately result in one or two global corporations for seed, animal feed, agrochemicals and other products. Some competition may remain with two TNCs, although that is questionable. The choice of products may be severely constrained, for instance with seed, in particular for resource-poor farmers. Another aspect is whether the current quest for complete control will end in stagnation, followed by collapse, due to lack of creativity. Such a conflict between control and creativity is apparent in a historical context. Enlarged institutions face the danger of ending up being petrified, as exemplified by the US Enron, the Swedish LM Ericsson and the Italian Parmalat. Obviously, any global TNC faces similar dangers. On the other hand, corporations are apt to initiate modifications when necessary.

A third issue is whether, and how, TNCs can help alleviate poverty and accomplish the MDGs. Globalization needs market forces so it probably strengthens the rich to get richer rather than help the poor. To avoid this, government institutions must facilitate the equitable spread of benefits. This possibility is doubted by many, who argue that past development has deprived more and more people of direct control over their immediate environment and means of livelihood and that the triumph of the market is a threat to the welfare of people (Martin and Schumann, 1996). An uncontrolled focus on growth has led to environmental destruction, whereas the nation-states have decided to treat only the symptoms. Both the private sector and nation-states advocate maximum growth, for different purposes. Such views explain why movements have begun against globalization. The prevention of global warming requires more than market mechanisms. Resources can have high values without a price tag, sacred forests being an example. What exact role can the TNCs play for society in addition to profit making? Therefore, it is crucial to establish an active dialogue between the CEOs of the TNCs and national policy-makers. They have to explore options for collaboration and conflict resolution. More political vision, and wisdom, is needed on how the TNCs best can serve the people and how the nation-state can be in command to counteract the inequalities.

New thoughts and attitudes are required for capitalism with social concern. Self-regulation of markets is not sustainable when considering human misery, marginalized peoples and ecological degradation. The market is not well equipped to handle labour and land, predominant parameters in agriculture for the use of natural productive resources. Moreover, markets fail where corruption exists and purchasing power does not exist. The private sector usually follows on activities started in the public sector. This will not happen until conditions improve to make such activities practicable. Policy reform is needed and public investments attracting private sector investments and new partnerships must be promoted. Still, it remains to be seen whether corporate globalization as presently manifested can foster a world based on democratic participation, social justice and ecological sustainability. This demands changes in the private sector, possibly initiated through the UN Global Compact and the UN Norms on the Responsibilities of Trans-national Corporations and Business Enterprises with Regard to Human Rights.

Global Food Supply

Past Performance

In terms of global food security, the world as a whole has performed quite well since 1960 (Table 4). World cereal production has tripled over the past 50 years from about 650 to 1,900 million tons. The per capita food supply has increased by almost 20 per cent despite a 70 per cent increase in world population. Food prices have declined by 50 per cent. In the mid-1990s, world grain prices increased substantively, then fell back again. Crop yields grew at about 1 per cent annually in the 1990s versus 3 per cent in the 1970s. Per capita cereal production in all developing countries has increased, except in Africa. Individual developing countries have been quite successful, such as Bangladesh, which now has enough rice for 130 million people. In the past 50 years, India quadrupled the production of both field crops and milk production, agriculture accounting for 25 per cent of its GDP. This resulted in self-sufficiency in many commodities and surplus in some (ICAR, 1999). China is providing food for 22 per cent of the population of the world from 9 per cent of the global arable land. Rice production in China peaked at 500 million tons around 2000, then declined by some 20 per cent in 2004. This is a significant increase compared with an average national yield of some 1,000 kg/ha 50 years ago and an average annual grain yield of about 180 million tons (Borgstrom, 1966). By 2030, food grain demand in China is expected to increase to about 640 million tons–a huge challenge.

In spite of overall positive developments, the proportion of hungry people in the world has declined only from one fifth to one sixth. The absolute number of hungry and undernourished has fallen slightly. Some 850 million people, mainly in the developing countries, are still chronically or acutely malnourished (UN Millennium Project, 2005b). In Sub-Saharan Africa, there was a decline in per capita caloric availability, resulting from high population growth rates and low growth rates in agricultural production. South Asia is still the region with a majority (60%) of undernourished people compared with 25 per cent in Sub-Saharan Africa. To a large extent, grains are mainly used for food in developing countries in contrast to developed countries, where they are used as animal feed. With rising incomes, the use of grains turns toward more livestock

Table 4. World food production 1970-2000.

Commodity	World production (million metric tons)		
	1969/1970	1989	2000
Cereals	1197	1865	2064
Maize	267	470	593
Wheat	311	538	585
Rice	307	506	601
Barley	128	169	135
Sorghum/millet	92	58	84
Roots and tubers			
Potato	299	277	328
Sweet potato/yam	142	133	139
Cassava	91	148	177
Legumes/oilseeds/nuts	167	162	165
Sugarcane and sugar beet (sugar content)	73	105	150
Vegetables and melons	55	433	692
Fruits	144	330	466
Animal products	n.a.	703	992
Milk, meat, eggs	383	509	866
Fish	n.a.	99	126

Source: Cockrill and Marsden, 1970; FAO Production Year Book, 1970, 1989; FAO Yearbook, Fisheries Statistics, 1989; FAOSTAT, 2002

production. Future imports to South Asia and Sub-Saharan Africa will be required for food rather than for feed. Surplus food production is a concern in many developed countries, where aspects of food safety have emerged as major issues during the recent past.

Nation-states are responsible for providing food security for their citizens. This must be solved at the household level. It requires access to food, either through improved income generation, safety net programmes or nutritional feeding programmes. This does not imply that countries must be self-sufficient in producing their own food. But they do require adequate policies in place to make nutritious food available. That requires significant rates of agricultural growth in food-deficit countries, not a common feature of agricultural policy of most governments during the last two decades. Lack of infrastructure and institutional reforms are other concerns and so are wars, political instability and economic uncertainties. Food security can also be achieved by diversified productive activities and by creating employment opportunities in both the rural and urban sectors. Very often, resource-poor farmers practise carpentry, trade and sell charcoal and wood, or earn income from wage labour, as illustrated in the study of farmers in Trinidad, Ethiopia and Sweden. The global community can only marginally contribute to food security, a role confined to the availability of stocks and food aid in case of famines.

International Food Conferences – Why Do Not Policy-makers Act?

When the world food situation emerged as a global concern in the early 1960s, the FAO started its Freedom from Hunger Campaign. The purpose was to establish large international banks for food. Those banks did not materialize. At that time, some 1.4 billion people were starving and an additional 900 million were undernourished. Ten years later, about one third of the population in developing countries was undernourished, the majority living in Asia and only one tenth in Sub-Saharan Africa. These concerns and the population growth led among others to the Second World Food Congress. Even at that time, some of the current issues were highlighted in the FAO's Indicative World Plan for Agricultural Development and were debated in the first plenary session of that Congress (e.g., toxicity of DDT to humans, the effects of discharges of oil, salinity of soils from fertilizer use). Another feature was a call for improved partnerships, nowadays perceived as a novel approach.

In the 1970s, the international community continued to organize a range of international summits, conferences and meetings at which countries, through their policy-makers, agreed to take actions (Annex IV). Implementation has, however, been marginal on many issues. The resolutions of the UN World Food Conference in 1974 emphasized increased food production in developing countries, offers of food aid and more liberal food trade. Rural development and coordinated efforts were to be promoted. At that time, food security was defined as the "availability at all times of adequate food supply of basic foodstuff... to sustain a steady expansion of food consumption ...and to offset fluctuations in production and prices". Within ten years, the Conference concluded boldly "no child, no woman or man should go to bed hungry, no human beings' physical or mental potential should be stunted by malnutrition". The more specific outcomes were the creation of the World Food Council and the International Fund for Agricultural Development. Later, the FAO redefined food security as "physical and economic access to food to all people at all times". The World Bank, like several others, chose its own definition of food security. Since then, considerable time and effort have been given to expose food security to hundreds of definitions.

Increasing concern over desertification led to the UN Conference on Desertification in 1977. However, independent observers stressed that solutions to world hunger were hardly confined to farming techniques and new methods of population control. Rather, hunger was caused by the increasing concentration of control over the food-producing resources. Two years later, policy-makers underlined the need for agrarian reform and rural development at the World Conference on Agrarian Reform and Rural Development (WCARRD). The elimination of poverty required political will at the national level. This was quite well reflected in the resolutions, stating that developmental efforts ought to give due regard to existing power structures to reach the poor and involve them actively in the development process, unusual language from a UN meeting. Less than a decade later, the Secretariat of the World Food Council at the 13th Ministerial Session in Beijing found that "national and international policies have not reduced the number of people suffering from hunger and malnutrition". Growing demands on food systems were harmful to the environment and the need for policy changes was reaffirmed.

The global food situation was thoroughly reviewed in the 1990s. One pessimistic outlook was based on declining production trends of some 10 per cent between the mid-1980s and 1993. This was in contrast to the period 1950 to 1984, when world grain production per person had increased by 40 per cent. It was argued that agricultural technology had little potential to further increase production. Fish production had reached its biological limits and the rangeland carrying capacity had been exceeded. Demand for water was pressing hydrological limits, fertilizer responsiveness was declining and much cropland was being lost to degradation, urbanization and industrialization. Expanded trade and food imports by developing countries were suggested as major solutions. Another extreme position was formulated on the basis that harsh climates in the developing countries offered less potential for tropical agriculture. Modern farming methods were inappropriate to the environmental and social conditions in the tropics. Therefore, the tropical countries ought to produce industrial goods, thereby creating job opportunities and cash to pay for imported food.

Several simulation models were elaborated, all econometric in nature and differing in scope and detail. A study by IFPRI focused on 35 countries, crops and livestock, but ignored possible effects of global warming. An FAO study covered all developing countries and a wide range of food products. The World Bank Grains Model looked at wheat, rice and coarse grains as a group. All the projections for 2010 suggested adequate global supplies and reached similar conclusions. The grain yields were projected to increase 1.5-1.7 per cent per year and the harvested area to increase modestly. The growth of global demand for grain would be slower and the grain trade was to increase. Real grain prices were expected to remain constant or decline. Food problems were expected to persist in South Asia and especially in Sub-Saharan Africa. Among major differences between the models were the impact of environmental degradation, the amount of land added or lost from agricultural production, land subject to intensification through irrigation and/or changed cropping systems and the rate of increased biological yields over three decades.

In the short term, the IFPRI model predicted a shortfall of some 190 million tons of food in the developing countries. It had to be met by imports of primarily wheat, maize and other coarse grains, equivalent to a doubling from some 90 million tons in 1990. In addition, some food aid would be required by 2020. For the long-term view (2020-2025), the challenge was more serious, though there might be a relatively good global food supply in 2020. Trade would have to expand, doubling the imports by developing countries. Food problems would persist in Sub-Saharan Africa, where imports were projected to triple, assuming appropriate policies and expanded investment in research for the development of new technology. During the 1960s and most of the 1970s, developing countries imported only 3 per cent of their total grain consumption.

The World Food Summit (WFS) in 1996 was aimed for heads of state but attended mainly by ministers of agriculture and some ministers of development cooperation. Its target was ambitious, even unrealistic to some observers. The political target was 400 million undernourished by 2015. The FAO projected 680 million by 2010, although most other predictions had indicated that the current number of chronically undernourished people might not change significantly over the next two or three decades. The FAO proposed a global Food for All Campaign and launched a Special Programme on Food Production in Low-Income Food-Deficit Countries. Its aim was to reduce by half the number of undernourished in 15 years. Certain issues were not debated, such as liberalized trade, population stabilization policies and the intensification of farming. Being a UN agency for governments, the FAO gave little attention to the future role of the private sector in agriculture.

The implementation of the Plan of Action of the WFS was a national responsibility, requiring genuine government commitment and actions. Prior to a follow-up meeting in 2002, the FAO concluded that the targets for 2015 were not to be met. There were still some 850 million hungry people in the world. Although the majority were in the developing world, there were 28 million in countries in transition and 9 million in the industrialized countries. There were serious doubts about a reduction of the hungry by half by 2030. The follow-up meeting, attended by some 7,000 participants, simply reaffirmed the goals of the WFS; there was very little subsequent political action.

In 2000, the heads of state and governments of almost all countries of the world agreed the MDGs should be achieved by 2015. The political goals are rather lofty but appealing to the public. The task is huge. During the decade following the baseline period of 1990-1992, the number of undernourished people in developing countries decreased by 9 million. The number of chronically hungry increased at a rate of nearly 4 million per year. This meant that two thirds of the reduction of 27 million during the previous five years was completely wiped out (FAO, 2002a, 2004). But to reach the WFS target, the number of hungry people had to be reduced by 24 million per year, almost ten times the pace achieved since the early 1990s. In 2005, the situation had not changed much. Only South America and the Caribbean will reach the MDG target of cutting the proportion of hungry people by half. No country will reach the more ambitious goal set by the WFS of halving the number of the hungry people (FAO, 2005).

Assisting the poorest countries means a focus on Sub-Saharan Africa, since 34 of the 49 least developed countries are African. In 2001, some 50 per cent of bilateral development assistance went to the least developed and other low-income countries. In contrast, the World Food Program allocated more than 80 per cent of its budget to those countries in 2001 and 2002. In a recent paper on potential global food scenarios for 2015 and 2050, IFPRI suggests progressive policy action, assuming increased investments in rural development, health, education and agricultural R&D (IFPRI, 2005). It stresses yield increases in contrast to a technology and natural management failure scenario that has forced farmers into marginal lands, resulting in yield shortfalls.

Food scarcity and malnutrition remain a problem at the national level in many developing countries, particularly in Sub-Saharan Africa. Increased demand for food will come from both increased population and higher income. Assuming only modest income growth, food needs in developing countries may almost double in the next 30 years. Food demand will also change because of urbanization. The number of resource-poor farmers will continue to decline, but this process will take quite a long time in countries with a large rural population. The market economy, together with globalization, will imply that a larger share of food production will enter the national and international markets. This raises a question about where the required food should or could best be produced. The global trading system must become more effective, in view of the fact that it encompassed only 10 per cent of the world grain production in the mid-1990s.

Putting an End to Mega-conferences

One can seriously question the range of expensive international mega-meetings that are held to repeat the same political messages in thousands of resolutions, often delivered by a new group of people. They seem to have little historical reference to earlier decisions and commitments made and little intention of implementation after the meetings are over. As early as in 1974, the nations of the world pledged to end hunger "within a decade". The notion that Africa was a special case was stressed during the FAO Conference in 1977. Since then, WCARRD and WFS have referred to the need for political actions agreed to by all participating governments.

Almost 100 countries met in early 2006 at the United Nations Conference on Agrarian Reform and Rural Development. They agreed to another declaration on the essential role of agrarian reform and rural development to promote sustainable development to benefit the poor and marginalized groups and also agreed on establishing mechanisms for periodic evaluation of progress in those areas.

Over several decades, the rhetorical target of reducing poverty has led to political commitment on paper but no action for implementation of the agreed resolutions. Although international conferences may stimulate debate, they are very expensive not only in time and costs for all participants but also in view of lack of actions for change. In addition to officials of national delegations and staff of various UN bodies there are also participants from NGOs. For instance, more than 12,000 NGOs registered at the Global Forum at UNCED. When the World Social Forum met in Porto Allegro in Brazil in early 2002, there were 51,000 participants from 151 countries, representing 4,900 organizations. The Johannesburg Summit attracted some 50,000 participants in various activities. In early 2006, the Fourth World Water Forum in Mexico drew 11,000 participants over seven days.

Certain global conventions are needed and will require thorough discussion and time. At the same time, there is a need for a distinction between local and global issues and those that are interconnected. The issue is whether they will turn out to be realistic and in what time perspective. They have to be consistent with a system for easy control. Above all, agreements and resolutions will require national actions. Then, one may consider more substantive meetings for those governments that have shown commitment to change and taken action rather than general conferences that show no progress. It seems timely to introduce a moratorium of five years on international mega-conferences. This calls for constructive improvement and effectiveness of both the United Nations and its specialized agencies and their mandates, mode of operation and collaborative arrangements with all other stakeholders. For the future, it seems more appropriate to make use of existing UN agencies for follow-up mechanisms on the work on MDGs and other global tasks after scrutiny of their mandates, efficiency, accountability and future work plans in addition to the ongoing UN reform process at the political level.

The Challenge

Increased Productivity

By 2020, demand for rice, wheat and maize is expected to grow by 40 per cent, requiring about 50 per cent more water. Urbanization will lead to rising competition for land and water. Fish consumption is projected to increase by almost 60 per cent in developing countries compared to only 4 per cent in developed countries. By 2050, the demand for wood may double. These examples illustrate the need for increased productivity in the use of productive resources for food and fibre. Because of climatic constraints, almost half the earth's land surface is unsuitable for rain-fed agriculture. Altogether there are some 4.7 billion ha available for future land use. So far, irrigated areas have increased by 70 per cent but further expansion might be marginal. Most suitable areas have been exploited and costs for new constructions will be significant. Four decades ago, for each individual on the planet there was 0.5 ha arable land, 0.8 ha grassland and 0.3 ha forest cover. Since then, the global average of cropland has continued to drop. The current available 0.15 ha of cropland per person is expected to fall to 0.09 ha by 2020, a figure that may drop further, depending on future population growth. This trend may not only affect production but also give rise to social unrest due to tension over land.

Growth in agricultural output must come from increased productivity, not from area expansion. The key issue is to produce more food in areas with the most rapid population growth in developing countries. But future food security with food safety must be achieved without further degradation of the environment. This is an enormous challenge to most countries, since almost 40 per cent of arable land in the world is already degraded. Soil fertility has continued to decline and soil degradation is a concern on 65 per cent of the agricultural land. High soil acidity affects more than 40 per cent of the world's arable land and only 10 per cent provides good opportunities for intensive agriculture. Therefore, the need to produce food in an environmentally sound manner is a huge challenge of global relevance. Solutions can be found only if international and domestic policies, institutional frameworks, and public expenditure patterns are conducive to cost-effective agricultural development that meets certain requirements of sustainability. In principle, our future food ought to be safe for consumers without any specific labelling.

In brief, the challenges are the following:

- More than double food production in the next 30 years to keep pace with population growth.
- Increase food production while maintaining the natural resource base, at least with minimum further degradation on approximately the same arable land as today.
- Produce food that is safe for human consumption in a long-term perspective and without detrimental health effects to subsequent generations.
- Reach the hungry, malnourished and poor by accomplishing the MDGs by 2015.

Can the Challenge Be Met?

Some 70 percent of the poor people of the developing world still live in rural areas. In an industrial country such as Sweden it took more than 50 years to reduce the proportion of people living from agriculture from more than 50 per cent to 1-2 per cent. To meet the challenge, the productivity and profitability of millions of small farmers must initially be improved. This calls for sustainable land use systems capable of doubling their output. It requires a systematic attack on all fronts of R&D, for example, ecology, land use, soils, agronomy, breeding, farm management and pest management, to increase the productivity of complex systems including aspects of crops, animals, water, energy, forestry and fish production. One example is the search for higher yields of more water-efficient crops. In dry regions, such as African rangelands, water limits production. This is why they have remained rangelands for centuries. How can they support long-term sustainable land use? In the 1960s, there was solid scientific evidence that selecting or breeding the right dwarf varieties of wheat and rice could substantially increase yields. One question is whether we have a similar solid scientific base from which to infer that progress will take place up to the 2020s. The levels of annual increase of crop yields have been declining. In addition to research per se, the management and organization of future research requires more attention.

First, governments of most developing countries must give political priority to agriculture and rural development. Also, the donor community must reorient itself and accept reality. Domestic and international policies and institutions must provide appropriate incentives to hundreds of millions of farmers in the developing countries to reach the MDGs. Policies that distort farmers' incentives must be abandoned and so must policies that heavily tax agriculture, such as overvalued exchange rates and industrial protection. If countries move away from self-sufficiency, they must be able to use world markets and reasonably stable markets. This calls for freer agricultural trade. International negotiations should attempt to reduce levels of protection in the OECD countries. This may mean improved access for developing countries, but trade alone will not suffice.

The challenges ahead are complex, particularly to the research establishment. Investments in public agricultural research must increase. Decisive and adequate steps must be taken now since changes must start as soon as possible in the farmers' fields. The overall challenge can be met only by a political commitment and action. It will not be easy and cannot rely on business as usual.

Governments must consider agriculture as a major contributor in combating the complex issues of poverty, hunger, undernourishment and poor health. As recently illustrated from China, this challenge cannot be met by a continued conventional sector approach confined to agriculture or forestry or medicine or health. There is need for one lead ministry in a joint process in which all relevant national ministries and institutions participate on equal terms, a ministry in control of any donor involvement. The future task also includes the build-up of effective partnerships with the private sector on sound ethical grounds and with a clear division of responsibilities for the public and private sector.

Some Pros and Cons of the Millennium Development Goals

The MDGs are now the focus of the international agenda, being strongly pushed by the donor community. Directly or indirectly, they relate to poverty, food security and agriculture. They aim for a strengthening of development efforts and policies in each country to meet the MDGs and to be integrated and aligned with the poverty reduction strategy for each country. In that context the MDGs may help governments, international organizations and donor agencies to effectively focus on major areas for development. The MDGs can be used for regular monitoring of major changes at the country level. A more pragmatic, quicker and less bureaucratic approach was demonstrated by the Copenhagen Consensus, identifying priorities such as stopping the spread of HIV/AIDS, improved control of malaria, the addition of micronutrients to food (zinc, Vitamin A, iron, iodine) and freer trade. Its conclusions relate quite well with the MDGs.

The MDGs are viewed by the United Nations as a clear willingness by developing countries to deal with corruption and generate more capacity for economic growth. Based upon the work of ten task forces, conducted by more than 250 of the world's leading development practitioners over two years, the UN Millennium Project has presented a practical plan to achieve the MDGs. There is a call for all countries to put in place MDG-based national development strategies by 2006. The project states that the task to achieve the goals is possible and proposes certain quick "wins". They are based on a market ideology with less regard to cultural, social and political conditions. It also stresses that specific technologies for achieving the MDGs are known. This may partly be true. However, the past has not shown any quick wins in development. The major issue is implementation. Likewise, there are few success stories on partnerships. The New Partnership for Africa's Development (NEPAD) is a partnership of heads of state with a vision that sustainable growth requires the eradication of poverty. Second, national governments, international institutions and bilateral donors agreed to eradicate poverty long ago. Political action is required to provide access to power for the poor rather than access to technology. Rather than a revitalization of the old idea of a technology push for another Green Revolution, it might be more appropriate to refer to an alliance against hunger and poverty for political action towards a real shift of power and focus in line with national MDG strategies.

Overall, the number of poor people has been reduced from 40 per cent of the population of developing countries in 1980 to 21 per cent in 2001. In reality, this is mainly due to good developments in a few large countries, such as China and India. To them, the prospects of reaching the MDGs may be greater than for many other countries. As early as 1980, China developed its own set of goals and indicators ("Xiaokang"). China has been quite successful in increasing food production. It may reach the MDGs by 2015, but there is a need for more balanced economic growth between the rich coastal and the poorer central and western regions. Both China and India are also facing growing problems of the destruction of grasslands, desertification, declining biodiversity, soil and water quality, air pollution and problems in disposal of solid waste. These problems are partly a result of the application of current technologies. Moreover, some 150 million people are short of employment. Nevertheless, the growth in China and India illustrates a dimension less prominent in the discussions of MDGs for a majority of developing countries. The use of natural productive resources to provide growth and thereby income and employment to people is a requirement if the goals are to be achieved. In countries with a large rural population this calls for agriculture and rural development.

To a number of developing countries, the well-intended approach with the MDGs may prove difficult in reality. Forecasts for Africa have shown "the MDGs are not going to be reached for most indicators in most countries" (Sahn and Steifel, 2002). To the UNDP, the MDGs seem unrealistic since the targets to eliminate poverty would not be reached in 2015 but possibly later (UNDP, 2003). The World Bank, projecting that the MDG for poverty will be met globally, takes a more positive position. To halve poverty, the Kenya Government must anticipate the economy will grow by 7 per cent a year in contrast to about 3 per cent during the past decade. Some African countries have increased their food production per capita and others have reduced the number of children who are severely underweight for their age. Reducing some hunger may be possible but to give every child a primary education and reduce child mortality by two thirds may be more difficult. The Ethiopian Ministry of Education recently declared that education for all would be attainable. The Ministry of Education in Trinidad and Tobago has just begun to prepare a National Education for All Action Plan supposed to lead to educational strategies for its Vision 2020. Europe is facing problems in halving the number of poor by 2010. The goal set during the Lisbon Process may no longer be feasible since economic development has been slower than originally predicted in 2000.

In general, there is a great danger that the MDGs will turn out to be another buzzword. Most likely, many recipient countries accepted the concept of MDGs in exchange for more development funds. Exactly the same kind of negotiations led to the CBD in the early 1990s but the promised funds did not materialize. Another problem relates to the fact that a doubling of the funds for development assistance neglects the current problem of low absorptive capacity in many developing countries as highlighted by the EU.

Sustainability in the North is not challenged by the concept of the MDGs. Future development is assumed to be a Western type of societal development, that is, consumerism and exponential growth. This is hardly realistic or sustainable. To reduce poverty in the South, one ought to discuss approaches to minimize the increase of economic wealth in the North, which is chiefly based on unsustainable processes causing environmental degradation. This is also a task for the United Nations and the scientific and political leaders of the developed world. More aid to the South will not solve the problem, as shown over the years by those with the actual power, the governments. Rather than promises of more of the same by those in affluent societies, there is a need for reflection also on other aspects of good life for future generations. This calls for important changes in the educational systems, in particular by the agricultural colleges and universities. Moreover, the debate on development emphasizes one factor at a time, such as technology transfer, capital mobilization or free trade. Although all these factors are important, they do not fully explain growth and technical changes. Social interaction also matters, relating to institutions and social groups. This includes aspects of pride, collective will, trust, social cohesion and even culture. One cannot forget the social contract between citizens of a country. This calls for attention to cultural identity or collective will, in particular if masses of poor people and resource-poor farmers are to benefit.

One requirement for success relates to current deficiencies of the UN development system. Too seldom do the UN agencies operate as a coherent team at the country level. Also, the proposed UN system at country level seems to be too isolated from national activities, assuming permanent external help. In fact, there are already a number of governments in the South with capable institutions that should be given more responsibility than just hosting a UN office. This would be critical if and when the MDGs are used for priority setting and guidance on practical actions at the country level. The UN proposal is also unclear whether bilateral donors will interact with a coordinating UN office or if they simply bypass that entry point with reference to their own political agenda, as in the past. Finally, it is unclear whether there are incentives for the World Bank and IMF, in their urgency for quick actions, to participate in a consultative and time-consuming process at country level, involving a number of UN agencies and bilateral actors. Even in the mid-1970s, the World Bank and IMF overtook the basic need strategy advocated by the International Labour Organization and imposed market reforms. At present, the World Bank is advocating a policy shift in favour of the poor for their access to health, education, jobs, capital and justice, including political power (World Bank, 2005a). It seems to give more recognition to the importance of agriculture in achieving the MDGs (World Bank, 2005b).

The MDGs do not explain whether and how the private sector can or should be involved in the process. This includes the role of TNCs. Neither the UNDP Human Development Index nor the MDGs mention corruption, democracy and good governance as important parameters for development. Lack of democracy is not an obstacle for growth and development, as illustrated by Singapore and China. Democracy has been expanding in countries but is also needed in the IMF and the World Bank to make them more realistic in decision-making. After independence nation-states in many developing countries have been assumed to adopt a European type of democracy. It has even been actively promoted, for instance by the creation of the International Institute for Democracy and Electoral Assistance in 1995. This ethnocentric initiative is based on a traditional concept of the nation-state and that a Western-type democracy is superior. But the nation-state had already started to lose power in an integrating Europe. An early warning was signalled about "the curse of the nation-state" in Africa (Davidson, 1992). Since the African problem has been argued to be a crisis of institutions, one option would be a return to locally based mass democracy and old moral values that prevailed prior to colonialism (Davidson, 1994). Some elements of this were part of the liberalization movement but faded away when focus was given to a more

European model. One interesting example is the institutional framework for democracy in Botswana, which gives adequate weight to traditional aspects of democracy.

In industrialized societies, democracy is changing toward increased centralization and an upward push for decision-making. That strengthens the power of a central group of actors, ultimately leading to bureaucracy. The lobbyists take over from the representatives of democracy. They serve those who pay best, a prominent feature in the EU and also not uncommon in the private sector. Gradually, the AU may experience the same dilemma. These changes promote a further division of the population. The small power elite will try to fill an information gap for the majority of the people. But information without borders is dehumanized, carries no intention and conveys messages to those "below". The elite envisage a global democracy as part of globalization. However, the majority of people have little or no power. They experience the gap as a matter of lack of confidence, leading to contempt for politicians. A shift of power is one critical aspect of the MDGs but governance by people is a threat to the power elite, a major issue in a process of development.

Between 1976 and 2000, some 8 million Kenyans entered the labour market. By 2013 the country must take care of an additional 20 million. If a third of the poor population in a village or region control the animals, they will always benefit initially and gain more. How single households govern their soil and resources is not a matter of equity. Privately owned land is in the hands of a small number of influential people such as ministers, parliamentarians and high officials. Would they be willing to give up their land to help people without resources towards some equity? The majority of people still live in rural areas and are poor. Swedish politicians have recently turned into a special profession rather than having a political task or job with certain duties and representing their constituencies for a limited period of time. The current political breeds control many Swedish institutions as members of the ruling party or are married to each other (Isaksson, 2002). This recalls the old Swedish nobility and times when the Swedish Parliament was controlled by four interest groups: farmers, clergymen, burghers and the nobility. There is a need to nurture all the good aspects of democracy and governance if resource-poor people are also to benefit according to the MDGs.

Towards an Agricultural Vision

Mission Impossible?

Speculations on the future have often proved wrong. The CEO of Digital Equipment argued in 1977 that there would be no reason to have a personal computer at home. There are no experts on the future, only on the present. Even researchers have questioned whether research about the future is possible. Still, the scientific community has a task to tackle problems facing humankind. In a medium-term perspective, the future can be influenced by efforts made or not made today. This calls for a solid knowledge base as the platform in guiding future actions based on a common vision. Visions based on ideas can be powerful, as illustrated by the birth and development of the Tetra Pak Company (Leander, 1995). The question is how realistic visions can be, since many of them simply turn out to be dreams.

In the 1960s, ideas based on a concept of "instrument of expansion" led to some interesting speculations for the year 2000. For example, it was predicted that 80-90 per cent of the global population would live in urban areas (Kahn and Wiener, 1967). Certain regions of great wealth would exist but surrounded by large areas of misery. Research accomplishments on RNA and DNA were expected to guide genetic scientists to define their responsibility regarding how research results should be practically applied prior to the year 2000. In agriculture, increased productivity was to remain important. New improvements would be of interest only to farmers themselves but not to society, as they used to be before. Some of these speculations were almost on target, although those responsible for leadership largely failed to be proactive by offering early guidance on pockets of poverty and misery, the new genetics and future agricultural research.

In the 1970s, most predictions were characterized by doom. One projected large-scale famines to appear around 2050 and early social imbalances by 2020 (Ehrensvard, 1972). Then, agriculture was to regain its importance, being supported by industry on aspects of food production and forestry. That would imply localized food production. Other speculators concluded there was no need to worry, especially about an energy crisis. Since a majority of the US manufacturing companies had turned multinational, the nation-states were losing power (Gerholm, 1972, 1999). When revisiting his early predictions, the same author found further optimism. Per capita food production had increased and food prices had dropped.

More recently, the World Resources Institute used another approach, asking some fifty individuals worldwide to formulate their personal vision for the year 2050 (Foltz, 1995). In brief, the African authors were concerned with basic needs, those from the Middle East with lasting peace and those from Latin America with democracy and civic freedom. The Central American countries would all belong to a free trade system and poverty would be eliminated. New food plant species would be found and the conservation of natural resources would no longer be a sector issue. With a growing population in China, its natural resources would experience further decline, a major change since many governors of ancient China were concerned with long-term development and legislating means to conserve and manage nature. These visionary views by individuals are, in substance, more specific than the four general scenarios that UNEP presented for the next 30 years (UNEP, 2002).

Visions

Laissez-faire

One future scenario, hardly a vision, is simply the continuation of current processes of development. This implies a Western-type society with highest economic growth and consumerism towards affluent life-styles. Through international agreements for the restriction of the use of natural resources and reduction in the use of dangerous chemicals it is expected that environmental degradation can be delayed in the short and medium term. Nonetheless, it would be unrealistic to believe a future with 8 billion people in almost 200 countries reaching the present stage of development of highly industrial countries without repercussions in nature. History has shown that civilizations must prevent the degradation of natural resources. Ecological capacity must not only be preserved but also nurtured, a missing feature of current thinking in "modern" agriculture, which is based on non-biological efficiency and suited to corporate ideas and confined to economic values. Such systems can hardly thrive in fragile ecological regions sensitive to environmental destruction with simultaneous eradication of poverty. We have to estimate what negative consequences we are willing to accept and even afford, and where, and let that understanding influence the future policy on food safety and security. So far, we have assumed the same ultimate vision for all countries, irrespective of their stage of development, time horizons and ecological conditions. But a vision for agriculture based on a Western-type approach is irrelevant for developing countries with a large rural sector, in the short- and medium-term perspective. They must design their own vision. Such a task is, however, also relevant for the long-term food and fibre production in highly industrialized countries that now depend on cheap fossil energy.

Free Trade

At present, free trade has emerged as the approach for development. It is not the first time in history. The primary intention of the Berlin Conference in 1885 was to guarantee free trade in Africa. In reality, bilateral agreements took over and Africa was partitioned from the outside. Except for two, the African states became colonies. Many people were convinced by David Livingston's objectives to fight the new slave trade, a solution seen to come from three C's: commerce, Christianity and civilization (Parkenham, 1991). But to many Africans, a fourth C, as in conquest, was more visible.

The grain trade is much older. During the Roman era, wheat and barley were the main exports. In the 19th century only wheat was traded on a larger scale. By 1800, milling technology had been widely introduced in Western Europe. Over time, milling had become more sophisticated and more wheat was kept for blending purposes. This led to increased consumption of bread, a similar trend gradually taking place in many urban areas in developing regions where wheat was not previously grown. For example, the US Continental Company arrived in Kinshasa in 1967 with a proposal for a modern wheat mill. Just a decade later, wheat imports from the United States to the former Zaire were substantial.

A first attempt to stabilize the global wheat market by international actions was taken in the early 1930s. A plan for an international wheat agreement was approved but collapsed after two years. Gradually, the grain trade was limited to private companies. In the mid-1970s, global food trade accounted for 19 per cent of wheat and 30 per cent of sugar production. The United States exported half of all grains in world trade and five companies controlled most of the global grain trade. Private families were the owners of these grain companies, a difference from other TNCs. Later on, these companies expanded into sugar, meat and tapioca. In the late 1980s, two thirds of global trade comprised industrial goods and was dominated by Western Europe and North America.

Between the mid-1970s and late 1990s, world trade increased more than six times. In 2004, the international trade in goods expanded in volume by 9 per cent, reaching a value of some US$ 11,000 billion. Industrial goods constituted some 60 per cent. The pharmaceutical industry has between 10 and 13 per cent of world trade, compared to 9 per cent for fossil energy. Cereals used to come second to petroleum in monetary terms. The United States remains a major actor in the grain trade, food and fibre, its food industry employing about 20 million workers (Ackerman, 2002). More than one third of the cereal trade is in maize and other coarse grains due to a demand for both grain and feed for livestock in East Asia, parts of West Asia, North Africa and the former Soviet Union. Maize and wheat are the leading export crops, a few companies controlling more than 80 per cent of the maize export. Over the next decade, the FAO is expecting global competition between exporters of wheat, rice, oilseeds, sugar and livestock to intensify in both developing and developed countries.

The share of world exports to developing countries has doubled since the early 1970s. Some developing countries have also increased their exports, for instance China by almost 20 per cent in 2002. Africa accounts for just a small percentage of all exports and imports, a decline since the 1980s. About half of Africa's exports are in the form of mining products, petroleum and manufactured products (Spore, 2001a). African agriculture accounted for one quarter, mainly for Asia and Europe. The issue of GM food complicates some potentials of free trade. As of 2005, it will be mandatory for all foodstuffs entering the EU to be traceable. This means tracking from the field to the table of the final consumers. Food safety and quarantine regulations are becoming increasingly important to developing countries. In a global economy with free trade, there are also lurking dangers with both plant and animal diseases such as BSE and avian flu. Moreover, there are demands on phyto-sanitary standards and maximum residue limits. Products that fail to comply are barred from a market of some 7 billion Euro. This is a great challenge for farmers in the South. Eventually, only the larger producers may be able to sell to Europe's major retail chains to meet their requirements.

In the long term, free trade leads to incentives. Trade liberalization has led to specialization and labour has been more confined. In the mid-1990s, GATT economists estimated that the effects of trade liberalization could annually contribute US$ 510 billion, starting in 2005. Total annual gain for developing countries was estimated at US$ 22 billion (Paroda, 1996). The abolishment of trade barriers is definitely an avenue for more export earnings for the textile and clothing industry. For instance, it constitutes 97 per cent of all exports by Pakistan, 49 per cent by Tunisia and 37 per cent by the Dominican Republic. These figures are different from those of agriculture. The pertinent question is how much a freer trade will benefit farmers of the South, particularly resource-poor farmers. Another issue is what agricultural produce developing countries may export to compete with the major food-exporting countries. Agricultural trade barriers primarily pertain to cereals, meat and dairy products. Prices of grain fluctuate considerably from changes in climatic conditions, market responses and government policies. Current stockpiles of grain have declined from some 580 million tons in 1997 to about 300 million in 2003.

Over the last five years, world cereal production has been flat at almost 1.9 billion tons because of low prices. Only about 9 per cent of dairy products and 6 per cent of meat enter the global market.

Freer trade also opens up new potentials to strong agricultural corporations. In the 1980s, the free market US farm policies lowered the price support to its small producers. As a result, a large number of family farms went into bankruptcy but the US food corporations made large profits. In preparation for NAFTA, Mexico wiped out some protections for its small food-producing farmers. Their land was taken over by US corporations and converted from subsistence farming into intensive agriculture. Strawberries, broccoli, cauliflower and cantaloupes were produced for export back to the United States. In turn, the US corporations began selling US-produced beans and maize to Mexican farmers (Smith, 1990). As a consequence, thousands of Mexican farmers lost their land. This kind of agriculture is efficient in an economic sense but undermined Mexico's production. Such a policy can hardly contribute to the MDGs. If so, resource-poor farmers of the South would benefit little from increased trade. This view has been vividly expressed by the organization Confederation Paysanne, arguing that only 9 per cent of global cereal production reaches the global market. A free global food trade will mainly strengthen countries with large exports such as the United States, Canada, Australia and Argentina. With China as a member of the WTO, the US export of agricultural produce might double according to the US Minister of Agriculture. This would mean an additional 2 billion dollars in export of agricultural products to China, almost equivalent to the actual figure of US$ 1.7 billion in 2002. This would take place in a situation where about half the Chinese population is still employed in agriculture and 75 per cent of the population lives in rural areas, constituting a rural working force of some 400 million labourers (Hunag et al., 2004).

Another concern about free trade relates to agricultural subsidies. Recent studies by the FAO show that cotton subsidies have an impact on the global pattern of cotton production, cotton trade and world market prices. The World Bank has concluded that the number of poor people will be reduced in a 10-year period if all trade barriers are abolished. The refusal by the EU to reduce its Euro 43 billion worth of farm policy support at the Doha Round was a major obstacle to a freer trade of farm products. In contrast, New Zealand has managed to trade in the world without subsidies to agriculture since 1984, except for some minor government contributions for research and pest/disease control.

The United States supports its farmers to sell abroad by paying them for their contributions to food aid and by granting them export credits. The US Farm Bill for 2002 meant increased agricultural support to farmers by nearly US$ 84 billion over the next decade (Bread for the World Institute,

2003). About 80 per cent of the US subsidies go to 7 per cent of 2.2 million farmers, who grow maize, cotton, rice, soybean and wheat. Almost two thirds of US farmers receive no subsidies at all. The favouring of the most affluent US farmers has not changed over the last two decades. As to the EU subsidies, about 80 per cent is provided to 20 per cent of the European companies. As an example, there are some 80,000 companies in the Netherlands that have benefited from EU support between 1999 and 2004. They include Nestlé, Campina, Avbe, Heineken, Shell and Greenary. In Scandinavia, Arla Foods and Danisco Sugar are large recipients. The new EU budget for 2007-2013 will not change much on the agricultural subsidies in Europe. At the recent WTO negotiations in Hong Kong, the member governments agreed to end agricultural export subsidies and give more help to poor nations. Farm export subsidies will be progressively phased out by 2013. However, no date was given for the reduction of domestic agricultural subsidies and there was no agreement on import tariffs. The most critical issue was avoided, that is, of how far to open markets for agricultural goods in developed countries and for manufactured goods and services in developing countries.

A reduction of the farm subsidies in the United States and the EU will have serious implications for large-scale producers. One illustration is the EU decision in late 2005 to reduce by one third the subsidies on sugar production. Still, that decision may not directly favour small-scale producers in developing countries and the fulfilment of the MDGs. Sugarcane is an old export crop well suited to large-scale production. It is under the control of the private sector, frequently a TNC or its affiliates. Meanwhile, the IMF has recommended that industrialized countries abolish their agricultural subsidies, since society would gain by it, though two thirds of the gains would be for the industrialized countries. Because of large imports, it is assumed the Middle East and North African regions will lose but Latin America and Sub-Saharan Africa would make some gains. In mid-2005, the G8 countries took a step forward, stating their wish to deal with this issue, although no timetable was set.

Global trade in agriculture put high demands on fossil energy but it also means transport of large amounts of water, an issue that has received much less attention. It is of concern since it can be managed by more local production, in particular since water is becoming a constraint. According to the FAO, agricultural water use at the global level is expected to grow by 14 per cent by 2030, leading to a bleak outlook for water in countries with two thirds of the world population in 2025. With increasing energy costs, this aspect of agricultural trade must be given more serious thought in parallel to expectations of profitable expansion of food exports. On balance, nations with a free trade policy have generally shown a doubling of annual growth compared to nations with trade barriers. But freer trade may only help agriculture in the South for products in demand in the North. It may easily add to the exploitation of natural resources, not just water.

Towards an Agricultural Vision

A vision of agriculture must relate to both political and geographical factors, calling for specifications. One global vision will be complicated, although it might envisage peace, health and food for all with an absolute minimum of poor and starving people by 2050. However, it must also relate to the national level, as exemplified by India: "By 2020, India will be free of poverty, hunger and malnutrition, and become an environmentally safe country. This we believe will be possible through accelerated social and economic development–by harnessing the advances in science, and blending them with our indigenous knowledge, wisdom, and unique socio-cultural ethos" (ICAR, 2000). This means 4 per cent growth of a sustainable agriculture per annum with equity and based on the efficient use of resources and conserving soil, water and biodiversity.

With reference to Trinidad and Tobago, Ethiopia and Sweden, one may give some indications of major issues for a visionary approach. Unless population growth in Ethiopia is reduced, the country may have to feed some 100 million people over the next 15 years. This would be an insurmountable task, even challenging its expressed policy of poverty eradication in line with the MDGs. It may be inescapable for any Ethiopian Government not to expand its recent policy of moving parts of the population from the productive highlands to the lowland areas. Their infertile soils and shortage of water may hardly sustain increased population pressure. Whichever step is taken, there are reasons to expect growing tension over land and food availability, leading to social unrest. It is urgent to formulate a political approach that mobilizes the population and the farmers for birth control efforts and the government to provide necessary incentives for improved land use with increased food production, raising incomes and creating jobs.

Most likely, the Government of Trinidad and Tobago will continue to rely on its petroleum products, importing most of its food, including GM food products. If so, few farmers will remain active by 2020. Urban farming may create some new jobs for the production of vegetables and fruits. The issue of land use will increase in importance and soon require decisions on how best to make use of the most land under cocoa and the large areas of fallow land previously planted with sugarcane. With the abolishment of the EU subsidies on European sugar production, sugarcane may reappear as an economic option for a commercial plantation industry. It may provide some employment but will not help resource-poor farmers. On the whole, unemployment will further increase, a ground for further segregation and social and political tension. This will require actions for social security, also including the poor segments of the population and those living in the countryside. A policy of agriculture that continues to be discriminatory may prove disastrous in the long term. Imported food may not be cheap and locally produced food can probably be competitive with adequate government incentives on agricultural commodities for specific markets, nationally and internationally. Such a policy will provide jobs, slowing down the growth of unemployment. Ultimately, the oil reserves will be depleted, calling for alternative sources of government revenue in the long term. In fact, this is a major issue for all oil-producing countries currently dependent on food imports.

In the mid-1990s, the Swedish Prime Minister considered agriculture a feature of a sustainable Swedish society. In spite of the rhetoric, the political ideas never materialized. To some observers, smart food and tailored pharmaceuticals were thought to provide good health and save medical costs in the future. By 2020, GM crops will probably be commonly grown by large-scale farmers within a strictly regulated system. These crops will have been well tested and thus be safe for both consumption and the environment. There might be one or two perennial crops, tolerating mild winters. Renewed efforts in plant breeding may have led to suitable soybean varieties for locally produced feed. Crops will be produced with agricultural inputs less dependent on fossil energy. Much increased oil prices will have greatly curtailed the effectiveness of the large-scale agricultural producers and made ecosystem research the main avenue. Bio-energy crops will cover large areas, resulting from a new policy of land use, under the responsibility of one ministry. Small-scale producers will profitably grow medicinal plants. Diseases transmitted by humans and animals may have increased. Pigs will probably be used for xeno-transplantation and then be produced by special farms. The food industry will have re-examined its policy of supplying additives to most food products.

Food safety will have become even more important to people, leading to a situation where many consumers wish to put on their tables what was produced in the neighbourhood and with transparency. This will have led to a "marriage" between organic or ecological farming and conventional farming, favouring the one that is transparent, provides superior quality with no chemical side effects, uses a minimum of external resources and offers a competitive price in relation to quality. The consumers will partly guide this development in case the agricultural research establishment fails to provide direction.

Within Sweden, these developments may even lead to recolonization of regions with fallows, resulting in an integration of immigrants into agriculture and forestry with available jobs, rather than importing guest workers. By 2020, the role of the EU in agriculture will have been finally tested. More likely, agriculture and land use will again be an item of the national budgets of all EU members. Probably, most of the agricultural subsidies will have been abolished.

In spite of quite different sets of problems faced by various countries, it seems possible to indicate elements of a future vision of agriculture. National policy-makers in close cooperation with the agricultural expertise need to refine them to accommodate national peculiarities, ethnic features and value systems.

By 2025, every person should have access to sufficient food to sustain a healthy and productive life, where malnutrition is absent and where food originates from efficient, effective and low-cost food systems that are compatible with the sustainable use of natural resources. This implies that the government should ensure the following:

- Nationally produced food and animal feed and all imports are safe for immediate consumption and do not contain chemical products harmful to human health in a long-term perspective.
- Food produced within a country uses natural resources without their long-term degradation. This means a balanced land use, less land degradation, less pollution of water and soils and no use of dangerous agrochemicals of high persistence but only those degrading into products harmless to both humans and the environment.
- The policy for land use and agriculture is closely integrated with employment creation both in the food industry and outside the agrarian sector for increased incomes.
- The private sector is encouraged to participate in the development of national agriculture under mutually agreed terms, initially designed by the government towards social development and not only for profit to agribusiness.
- Increased human and financial resources are allocated for the development of the agrarian sector, in particular funds for agricultural and forestry research that benefits society over both the short and long term.

Agricultural Technology and Trends in Agricultural Research

Agricultural Technology and Innovations

For centuries, farmers themselves and/or artisans developed new technology. It was produced without any specific theory (Barzun, 2000, p. 205). Technological developments were made so that the individual could have a better life. They were tested in reality before they spread widely, a process usually taking a long time. Diffusion took place as a by-product of, for instance, travelling and contacts between farmers.

Several agricultural innovations emanate from old ideas, such as the "vallus" in Roman Gallia. When population growth was high, this harvesting machine for cereals was developed about 70 BC (James and Thorpe, 1994). It was a wheel-based trough with sharp knives at the front; the ears were cut and fell into the trough. It was not much used since the Romans did not wish to cause social unrest by replacing slaves with such machines. For the same reason, officials in northern China in the early 14th century did not allow a mechanical scythe to be commonly used. In the mid-1960s, the Government of Trinidad and Tobago took a similar decision by not allowing the use of a new mechanical harvester for sugarcane. The jobs of the sugarcane cutters were threatened. The vallus reappeared in the 5th century. In 1825, its design was printed in the Encyclopaedia of Agriculture. Later, John Ridley used the idea when, as an immigrant in Australia, he developed his harvesting machine in 1843. In the United States, the Mac McCormick reaper appeared in 1831 and was introduced in Sweden in the 1890s.

The British inventor Jethro Tull, considered a father of modern agriculture, created the seed drill in 1701. Seed drilling in rows allowed for more efficient weeding using a horse-drawn hoe. It took a long time for it to spread among farmers. Ploughs were known from North Africa during Roman times but the horse-drawn plough spread much later. The Rotherham mould-board plough was patented in 1730 in Yorkshire (UK). The John Deere Company started making iron ploughs in 1837. They were introduced in Sweden in the early to mid-19th century. In southern Sweden, a wooden beam was in common use in the 1850s. In Uganda, a ploughing school was set up in 1910. The milking machine for dairy cows was invented in 1903 in Australia. By 1930, it had been adopted by 2 per cent of Swedish milk producers (Danielsson, 1981). After 50 years it was in use by some 95 per cent of Swedish dairy farmers. Appearing in 2000, the milking robot has today been adopted by less than 2 per cent of Swedish dairy farmers. Worldwide, some 400 robots have been sold.

These few examples of agricultural innovations illustrate the time span between the appearance of a new product or innovation and its wide adoption. Technology is not adopted until people are prepared to do so in a suitable social culture or with the help of institutions. This must be realized when the UN Millennium Project argues that Africa could triple food production by 2015. The type of innovation influences whether existing economic differentiation continues unaltered or the economic structure is changed. Technical changes also have an impact on traditional religious or ritual roles and relations, which often determine social status. Greater economic mobility undermines traditional principles of political organization. It may lead to loss of power of the leaders. It usually takes years before the old values are displaced by economic values, resulting in a cultural change. Composition, size and livelihood strategy of households are important parameters that determine whether farmers innovate, since farming is only one of many other activities in a large rural system, not simply technology as such.

Basic Concepts behind Emerging Agricultural Research

Initially, religion set the rules for science. The faculty of theology was superior in academia and the power structure of the medieval universities. The idea of the Western world was deterministic and mechanistic. The growth of technical innovations formed the idea of the world, for instance by technological developments in the application of power of wind and water.

During the Renaissance, nature and cosmos were to be read accurately to guide people in the Western world. Galileo Galilee refined the deterministic view of nature. Anything truly real was measurable. Secondary qualities were simply a matter of subjective impressions, feelings and/or opinions. Therefore, knowledge had to be objective. This led to a component of independence in classical science, a notion that remains very strong. But intellectual breakthroughs came after an upsurge in the fine arts, leading to the emergence of two new ideas. One was about nature and man, nature being the object and man the subject. René Descartes viewed both animals and the human body as machines, their workings to be understood in terms of mathematics. This gave rise to reductionism.

Nature was separated from humans and they were superior. The second idea implied a need to analyse the parts of the whole. By making analytical observations of nature, splitting them into pieces and reducing everything to its basic components, manipulation was possible. The use of experiments, which can be repeated, was an expression of the fact that nature could be manipulated. Human nature was separate, leading to objectivity. This dualism set a paradigm for understanding most of Western culture. Innovations were seen as human choices of the future.

During the Enlightenment, improved life was promised to all since science and technology was fundamental to progress in the Western world. Scientists should have a leading role in decision-making. The elite was given a prime role in research and education. This process began with the appearance of the steam engine, followed by the mechanical loom, electricity and the identification of electromagnetic properties, all leading to a strengthened role for science. This took place together with a growing capitalism. But the control of nature by man was also considered natural to the magicians of the Arab culture. In contrast to the magicians, the new emerging science gradually became more open but only to those interpreting the scientific language. In this context, it is interesting to realize that the gap in science between the South and the North is of relatively recent origin. The dominance of science by the Western world did not start until after 1450 AD (Salam, 1989). Prior to that period, the growth of science was dominated by non-Western societies. After 1100 AD, they shared the lead with the present North. One may speculate on what this transition has meant for scientific attitudes and development thinking.

The Malthusian scientific contribution stressed that agricultural production could not be increased indefinitely. Marginal lands had to be used with a growing population. But Thomas Malthus also argued that, since the aristocracy were educated, they would not convert their wealth into more children. Therefore, the solution to poverty lay in the hands of the poor, not with the government. At about the same time, Ernst Haeckel coined the term *ecology*, referring to our biological ecology. But life is more than biology. A holistic sense of ecology includes culture and the presence or absence of justice, hope, beauty and joy. Later, there was a distinction between deep and shallow ecology (Naess, 1973).

Even today, many observers believe in a myth of a balance of nature. Nature is not harmonious but always changing and difficult to predict. People living close to their environment often know far more than visiting "experts", although such indigenous knowledge has been much neglected in past agricultural research. Charles Darwin found it natural for animals to behave in competitive ways, struggling for resources, although it was Herbert Spencer who initiated the phrase "survival of the fittest". In contrast to competition between species, James Lovelock developed the

Gaia concept that the sum of all life on earth is locked in a symbiotic relationship with the total environment. That idea was further developed, while others have ridiculed Gaia as pseudo-science (Dawkins, 1986; Gould, 1989). When life is becoming more and more complex, an ordering principle, sometimes named anti-chaos, might have been more important than natural selection in guiding the evolution of life (Kauffman, 1993).

With reference to the recent buzzword of *sustainability*, mainstream scientific ecology today offers a paradigm that nature is something to be rationally manipulated, rather than exploited, in order to maximize its returns to humans over time. A *paradigm* is the period in scientific work during which scientists take for granted and are committed to a particular view of the world. This determines what problems are to be studied and represents a consensus among researchers in a discipline. But paradigms change as our cultures change. There are changes within disciplines and paradigm shifts at various levels. Scientists do not easily challenge a current paradigm, since it would question their own research approach. This implies conservatism in agricultural science. Certainly, classical science has produced progress to many people but also wars, unemployment, nuclear dangers and global environmental threats. At present, some two billion people struggle with poverty and our global ecosystem is threatened. We are thus justified in re-examining past agricultural research and indicating new avenues for future agricultural research to benefit those living in misery and poverty.

Past Trends in Agricultural Research

The Colonial Period

With Christopher Columbus, the West embarked upon agricultural research in economic botany and plantation crops and the export of raw materials. Tropical experience was gained from colonists, missionaries, governors and botanists. Individual settlers began field experiments, for instance on tobacco and sugarcane in Barbados. Missionaries introduced coffee to Kenya in the 1890s. A decade later, mission supporters introduced seeds of American cotton in Uganda. Above all, the missionaries brought medicines, hospitals and schools in addition to an idea of both progress and salvation. They also started sending local people abroad for training.

The attention to plants in early British colonial agriculture led to a spread of the idea of botanic gardens. The first one was established in Mauritius in 1735, followed by one in St Vincent in 1764. Over the next hundred years, the Royal Kew Gardens (founded in 1759) played an impressive role in this global development. The botanic gardens were the precursors of the Agricultural Departments in the British colonial service.

France established botanic gardens in Africa in the beginning of the 20th century and tropical medicine at a laboratory of biological medicine in Saigon in 1871. In the early 1940s several French organizations were created, focusing on the tropics for the development of science. As late as the early 1990s, some 3,000 French scientists were still actively involved in Third World research on cash crops (Gallaird and Busch, 1993). Founded in 1992, the European Consortium for Agricultural Research in the Tropics was seen as a new tool for the promotion and facilitation of institution building for research in tropical countries. Even today, it brings together some 4,000 technical experts to serve the South under an umbrella called the European Economic Interest Grouping.

The plantation industries played a proactive role early on. Commercial funds were invested by the Dutch and British plantation industry in commodity research stations. As exceptions, some British Government funds went to commodity stations in Barbados (sugarcane), the Gold Coast (cocoa) and the former Nyasaland, now Malawi (tung). Spices were the first truly global commodities, followed by beverages. In contrast to a strong involvement in tropical agriculture by the private sector, the British Colonial Research Service was not officially established until 1949. It was to support research in agriculture, forestry, fisheries and social sciences. It reached its peak in 1958 with over 200 positions but was halved in 1961 (Jeffries, 1964). As to policy, it was not until after World War II that a seven-page document on colonial agricultural policy was produced, having little practical influence (Masefield, 1972, p. 63).

In contrast to the missionaries, the colonial powers gave little attention to the training of local people for higher education, particularly in agriculture. By 1960, there was only one college of agriculture in French-speaking Africa. Between 1952 and 1963 only four university graduates in agriculture were trained in Francophone Africa and 150 in Anglophone Africa (Eicher, 1986). After independence, Congo Brazzaville had four graduates (Ohadoma, 2002). Since then, national agricultural universities have taken over higher education. Donor agencies and other organizations have offered scholarships. After completing their higher degree training in Europe and North America, only a small percentage of students with higher degrees in agriculture abroad have returned to their home countries. Lack of funds for research and satisfactory research environments are still major bottlenecks. Or they may prefer to work for international organizations offering well-paid jobs or feel that their continued research on a specific subject might be better pursued by being employed in the country in which they received their research training. Even today, there are examples of PhD students trained abroad on topics hardly of immediate relevance to their own country.

Topical Research Based on Ideas of Western-type Agriculture

Over the years, plant introductions have made significant contributions to tropical agriculture. Tobacco and vegetable seeds were brought to Somalia in 1905. A settler introduced a variety of pyrethrum to Kenya. From the 1940s onwards, improved cultivars replaced species in breeding work throughout most colonial territories. Targeted plant collections were introduced to find new genetic material, for instance of cocoa in Latin America and South-East Asia and banana in the Pacific. Many plant exchanges took place to improve production during a period labelled as the Golden Age (Simmonds, 1991). The international work of seed and plant exchange has been of great importance.

Although there was no British Veterinary Department in Jamaica, the Agriculture Department broke administrative rules by starting to interbreed Jersey and zebu cattle in 1910. The outcome was the Jamaica Hope breed of dairy cattle, which became a highly productive milking breed in the tropics. In 1934, West African shorthorn cattle were introduced from the Gold Coast to Nigeria. A few years later, Ndama cattle were purchased from French Guyana, another example of an exchange of animal genetic resources. Regrettably, there is little information about the impact of these endeavours.

In the 1920s, mixed farming was preached in Europe. This led to cattle experiments in Nigeria, followed by campaigns on crop-farming systems. When a Kenyan settler published a book on this topic in 1935, the research establishment had already shifted its focus to alternate husbandry, e.g. the alternation of cropping periods with a grass cover on the same piece of land. As a result of the effects of the Dust Bowl in the United States in the 1930s, soil erosion had emerged as the new topic. The British Colonial Service began to recruit Soil Conservation Officers to conduct studies on mulching and grass fallows. Bush fallows were no longer believed to be sufficient and early land preparation and early planting was recommended. Although agricultural extension was distinguished as a speciality of its own, initially called "propaganda" or "educational work", the term *agricultural extension* appeared much later. Practical extension activities did not appear until after World War II. With the arrival of the first ecologist in the British Colonial Service, vegetation mapping became topical in the 1940s. At that time, there was also an early insight that malnutrition was more widespread than starvation in colonial populations (Masefield, 1972). Six decades later, the UN Task Force on Hunger again discussed this issue.

Contour farming was one approach to integrate soil conservation in Africa with efforts to increase soil fertility by crop rotation and the use of manure. These aspects became prominent in the 1950s. Fertilizers, pesticides and credit were advocated together with ploughing with oxen and improved transport by ox-carts. In turn, this led to a focus on farmer associations, cooperative marketing and development. Community development turned into an agricultural extension service. Great expectations were generated from the introduction of mechanical equipment for land clearing and cultivation. In the late 1950s, agricultural economics became central, one theme being farm management, where linear programming was applied even to tropical peasant agriculture. Later, settlement schemes became part of development objectives in Africa, combining health with agriculture. This is again relevant as demonstrated by the HIV/AIDS epidemic.

As developing countries gained political independence, there was a great deal of optimism. Development assistance expanded. The transfer of Western-type technology became a major theme of agricultural development with a focus on agricultural extension, assuming technology was available, a view still prevailing and also advocated by the UN Millennium Project 45 years later. But tropical agro-ecosystems are characterized by greater uncertainty and variability than temperate systems. Certain ecological differences may restrict a successful transfer of technological practices to tropical regions. Technical knowledge gained by the colonial powers was ignored or simply perceived as another feature of colonialism. There were a few opposing voices and their proposed ideas were not welcome, for instance the need to focus on agriculture using existing resources. The attitudes of the elite, their indifference to the peasantry and adoption of various colonial attitudes were condemned (Dumont, 1967). Corruption and despotism were highlighted as other reasons for the underdevelopment that prevails even today.

During the 1960s, more thought was also given to the theoretical approach to development. In the "dual-economy" model with growth over time, the small farmers were to disappear. Only those adopting new technology were to make their way out of poverty. A decade later, another model referred to a theory of unequal exchange or exploitation between the "centre" and the "periphery" of the world economy. The marginalizing of the peasantry was a prerequisite for the transfer of surplus value from a less developed periphery to a developed centre at local, regional, national or international levels. A third model for development saw the solution of the peasant problem through the collectivization of agriculture. The alternative technology movement appeared somewhat later, an idea of back-to-the-land with technological determinism. That concept had already been defined in Mahatma Gandhi's social doctrine but fallen into oblivion. In reality, the alternative technology served merely as a supplement to other technologies rather than replacing large-scale and capital-intensive developments funded by external sources.

Towards Food Crops

Triggered by famine situations, increased research attention was devoted to food production. This led to the establishment of the Consultative Group on International Agricultural Research (CGIAR) in 1971 (see Chapter on the CGIAR, and Annex V). Major emphasis was given to crops in favourable ecological regions with the greatest potential. The CGIAR became closely involved in the Green Revolution; the term was first applied by William S Goud of the US Department of Agriculture in 1968. Its origin goes back to 1951, when plant breeding started in Mexico. By 1960, new high-yielding varieties covered one third of the area under maize. Five years later, wheat imports were no longer needed in Mexico, since the average yields of the new varieties were in the range of 7 tons/ha. After 25 years, the average national wheat yields reached 3 tons/ha. Then, virtually all of Mexico's wheat land was under improved varieties.

From its very start, there was a debate whether research through the CGIAR could help poor farmers, the UN Research Institute for Social Development being a major critic of the Green Revolution in the 1970s. In turn, this led to considerations by the agricultural research community on the theory of technical change. Several theories were developed, such as the "induced innovation theory", "the structural theory," and the role of institutions (Hayami and Ruttan, 1971; Griffin, 1974; Binswanger and Ruttan, 1978). Although the CGIAR maintained that technology was neutral to scale, several studies gradually concluded that scale-neutrality was belied by the greater access the larger producers would have to necessary agricultural inputs. It was also argued that genetic diversity was reduced and pesticide use expanded. On balance, it is true that the effects of the Green Revolution bought time to handle food security, particularly in Asia. This owed much to the foresight of agricultural policy-makers in India, Pakistan and other countries and the action taken by them.

In the mid-1970s, the oil crisis led to high costs for agrochemicals. Farming systems research now emerged as the major topic, mainly confined to crops, at least within the CGIAR. However, animals are a vital component of the food system. Mixed cropping reappeared as a research theme and biological control became more feasible with the escalating costs of pesticides. Gradually, the focus on farming systems research faded away. But the notion that crops were more important than animals was recently reaffirmed as having the greatest potential for poverty and hunger reduction and economic growth in the next few decades (Dixon et al., 2001).

With the appearance of CGIAR, its priority setting of international agricultural research also influenced national research agendas. In the 1980s, cover crops emerged as a new theme in Latin America. No-tillage was seen as a new avenue for R&D. Although most observers agreed to the basics of the definition of sustainability by the Brundtland Commission, there was little policy guidance on how that specific goal could be achieved. Overall, the agricultural research establishment showed little enthusiasm for further investigations. In the 1980s, organic or ecological agriculture was emerging as a new topic for research, particularly in the United States and Western Europe. Since its appearance it has been debated. Gradually, the market for ecological products has grown, although it is still small. In Sweden, it now accounts for a small percentage of the total food market. Also in the 1980s, policy issues of agricultural research became important. A new discipline of policy analysis was suggested. Policy relates to both environmental and agricultural concerns. In fact, "where the environment is at risk, there is no clear-cut boundary between science and policy" (Wynne and Mayer, 1993). To most observers, such a linkage was perceived as too political. Nonetheless, a course of action or policy was quite common within the donor community. In the late 1970s, policy on support to international research was fundamental in the early operation of the former SAREC. Around 1990, there was an explicit focus on policy matters in forestry and agricultural research (in the World Bank, US National Research Council, SAREC and other bodies.). The need for policy research on international forestry led to the creation of the Center for International Forest Research (CIFOR). Over time, IFPRI turned more in that direction. Policy research has, however, been marginal at most agricultural universities.

As briefly illustrated above, the orientation of agricultural research has often been based more on trends and topical features and even ideology than on a precise definition of the problems of farmers. In general, agricultural research has continued to focus on technical problems. The colonial heritage of developing countries has influenced the role of their agricultural institutions and may still do so. National agricultural research systems operate on a historical basis in management and approach confined to disciplines rather than complex systems of land use and natural resource management. This is of equal concern in highly industrialized countries, where the success of agricultural research has gradually undermined its legitimacy. With a much reduced farm population, public support has lost most of its former political influence.

For long, significant R&D has been carried out by the private sector in industrialized countries, aiming for commercial markets. Privatization of research is not as recent as is sometimes believed. Many technical and chemical processes were the private property of individual craftsman in the Middle Ages of Europe. Until recently, there has been agreement that technology was private and subject to the rules of the market. Agricultural research was public and agricultural technology focused on practical implications. This view was first challenged with the development of hybrid maize in the United States, gradually leading to a Plant Variety Act of 1970. Later on, patenting was extended to plants and animals. A recent feature is the increasing interest by the private sector in more scientific work for the demands of a "modern" agricultural sector. In its past work on technology, the private sector was less exposed to errors of judgement. This is now changing dramatically. Also, the power of the corporate industries has increased, partly through their involvement in biotechnology. Between 50 and 70 per cent of global research is paid for by the private sector.

The state is no longer the single actor in agricultural research in industrial countries. Between 1976 and 1995, public funding for agricultural R&D had almost doubled to US$ 22 billion, the greater share (56 per cent) conducted by developing countries (Pardey and Beintema, 2001). The private sector spent 94 per cent of its research in developed countries, where it accounted for just over half the total agricultural research. Globally, it spent about US$ 11.5 billion, one third of total global investments to agricultural R&D. In most developing countries, the private sector generally accounts for much less of agricultural R&D. These figures show interesting changes compared to 1977, when the US National Academy of Science estimated global expenditure on agricultural research at almost US$ 4 billion, out of which some 15 per cent was spent in developing countries. Public expenditure for agricultural research in the United States amounted to almost one third of the world total.

Agricultural Research and Society

Some time ago, it was said that there is an end to "science at its purest and grandest, the primordial human quest to understand the universe and our place in it" (Horgan, 1996, p. 6). If so, science tends to reinforce rather than challenge prevailing paradigms. When a field of science begins to yield diminishing returns, scientists may lose their incentives to pursue further research. New research questions will become more trivial as they will concern details but will not affect our basic understanding of nature. Then, there are few surprises. Biotechnology will, however, not only uncover surprises but also blur the previous distinction between basic and applied science. On complex research issues of societal relevance, both science per se and applied science are required. The last five decades have clearly increased the visibility of the global dimensions of environmental problems, all resulting in even more complex problem areas. A significant question is whether, and how, science can contribute to the eradication of poverty in line with the MDGs. If not, society may be prepared to pay for not more but rather less agricultural research.

150

Successful cooperation between scientists of different disciplines will require knowledge of each other, confidence and joint ideas and approaches. There is a strong belief that the evaluation of good or bad science can only be done within the scientific community itself. This view neglects external demands. It is true that the scientific establishment has accepted certain ideas that are seemingly logical but have not been entirely scientifically proved. Truth is used in a selective way to reinforce opposite views in polarized areas of public debate. Lay people believe science is exact and provides the truth. Agricultural research also needs a philosophical dimension that may not be required by the journalist or TV reporter. With the possible exception of medicine, science journalists report new scientific discoveries frequently as if they are new sport records, all leading to progress. Seldom are discoveries presented in a contextual framework with reference to utility and the possible long-term implications for human beings.

At many agricultural universities, the intellectual environment is fading away. Time for reflection is being reduced to a minimum. When industrialists and businesspeople joined scientific societies they acquired social status and political power. Generations later, they still advocate both scientific activity and values but their intention is now much less the breaking of any social order. In an alliance with science they actually preserve an existing order from which they originally benefited.

Although societal problems have gained in complexity, reductionism prevails in most research. Providing good control over a small number of variables is codified as "good science". In principle, agricultural universities and colleges are characterized by professionalism and institutionalization. By defending their own research sectors or territories, they promote reductionism and fragmentation. This leads to further splits of disciplines, thereby threatening intellectual overview and synthesis. Any new research project will give rise to more unanswered questions. It certainly provides many more details but lesser overview in a social context and greater specialization leads to diminished personal responsibility. If one's area of specialization is confined to one particular enzyme, it is very difficult to take responsibility for the overview. Still, agricultural researchers have to be accountable in a societal context, in particular when the challenge is to tackle major problems of humankind. The questions and answers of scientists cannot be fully comprehended without an understanding of the social context in which they operate.

By discovering new facts, scientists make the world. But discovery also means to uncover and reveal. Therefore, anybody who studies the natural world is a social scientist, although those who are seen as social scientists rarely change society (Busch, 1994). Many social scientists criticizing the effects of the Green Revolution seldom provided constructive proposals for improvements. Being employed by somebody, the researcher is not free to say what he or she wants, particularly in the private sector. Public research funds are not tied to specific interest groups, implying there is material dependence for agricultural researchers. Also, agricultural science has been considered "value free". But in creating "objective" knowledge, scientists emphasize certain values such as objectivity, quantification and replication. Agricultural researchers also argue for productivity, efficiency, speed, effectiveness and standardization, all key values of the capitalist world. Not so long ago, all research applications were actually read. Nowadays, it is too often sufficient to look at the publication profile and whether the reports of the scientists have been published in acceptable referred journals and are being quoted. This may not always be reliable, as shown recently in medical research both from South Korea and Norway. A Norwegian cancer researcher was recently found by external examiners to have cheated not only some 60 co-authors but also got a large number of articles below scientific standards published in well-respected journals such as *The Lancet, New England Journal of Medicine* and *Journal of Clinical Oncology*.

Good science flourishes on criticism and toleration of opposing views. Principles such as openness, impartiality and self-criticism represent a high ideal, not always reflected in reality and sometimes not even recognized by the scientific community itself. When the colonial masters left Sri Lanka, a new elite corps of the Civil Service used suppression to consolidate their own positions (Seneviratne, 1993, pp. 2-3). Science was not given an adequate chance to assist in national development. Instead, there was a need to cleanse the science constituency of corruption and intrigue. Earlier examples show the power of the scientific community. Nicolaus Copernicus, born in 1473, did not publish his major works until immediately before his death in 1543. Since his ideas were very different from those of the scientific establishment, the timing of publication became critical. Charles Darwin used the same trick. In the mid-1960s, the Swedish research establishment forced George Borgstrom to resign from his professorship mainly because of his very negative perspective on the future and his prediction of a total disaster for the planet and its natural resources. Likewise, Rachael Carson experienced difficulties in finding a magazine or periodical willing to publish her findings later to be presented in her book *Silent Spring*. Even scientific colleagues refused to give her details they might have provided. More recently, the Mexican ecologist Ignacio Chapela rose to international attention when reporting that transgenics had flowed into local varieties in southern Mexico in 2001. Subjected to further review, *Nature* later issued a statement that because of technical flaws the paper should not have been published, although Chapela stands by his findings. In 2003, he was denied tenure at the University of

California, Berkeley, where he had opposed the 1998 deal between the University and Novartis giving the biotechnology firm privileged access to research findings by scientists. This conflict ended in compromise in early 2005. In 2003, Bjorn Lomborg was accused by a Danish scientific committee of scientific dishonesty by manipulating his findings on environmental issues. One year later, its ruling was overturned by the Danish science ministry.

Instead of intensive, transparent scientific dialogue on major issues, there is more often just silence by the opponents, for instance on climate change, ecological agriculture, biotechnology or IPRs. A good feature of biotechnology research is that value questions become visible. Still, serious debate seems difficult. Different applications of biotechnology call for ethical assessments. They should specify what is distinctively objectionable about their targets, that is, not interfere more with nature than others. Scientists should accept a responsibility to recognize undesired consequences of their research to society. There is a danger if they are reluctant to re-examine current paradigms and they, sometimes for good reasons, are unwilling to accept other points of view. Science then becomes a form of over-belief or may turn into political correctness. Both agricultural students and the public ought to be taught also about these complications.

Long ago, Bertrand Russell warned that science could make politics too dangerous, since the research establishment tends to believe that it can reach its objectives. For instance, the plant breeder George Stapledon saw Benito Mussolini as a new teacher. He combined a celebration of rural life and localized non-alienating production with a eugenic programme that would draw upon the pure country stock and improve the race (Pepper, 1996, p. 228). To solve Swedish problems of declining population in the 1930s, both economic and social reforms were suggested and so was the weeding out of unfit individuals by sterilization (Myrdal and Myrdal, 1938). The US magazine *Time* declared Adolf Hitler Man of the Year in 1938. Paul Feyerabend objected to the claim that science is superior to other modes of knowledge. There was no one scientific method but a box of different tools. He objected to scientific certainty for moral and political rather than epistemological reasons. Although not condemning Nazism, he was critical of the moral self-righteousness and certitude that made Nazism possible (Feyerabend, 1987, p 313).

"I say that Auschwitz is an extreme manifestation of an attitude that still thrives in our midst. It shows itself in its treatment of minorities in industrial democracies …becomes manifest in the nuclear threats, the constant increase in deadly weapons…. It shows itself in its killing of nature and of "primitive" cultures with never a thought spent on those thus deprived of meaning for their lives, in the colossal conceit of our intellectuals, their belief that they know precisely what humanity needs and their relentless efforts to recreate people in their own sorry image... in the lack of feeling of many so-called searchers for truth who systematically torture animals, study their discomfort and receive prizes for their cruelty. As far as I am concerned there exists no difference between the henchmen of Auschwitz and these "benefactors of mankind"."

Agricultural Research at the Crossroads— Reflections on Future R&D

The Changing Framework for Agricultural Research

The CBD changed significantly the context for research on access to and use of plant genetic resources. National sovereignty rules instead of free access and exchange. This has led many governments to establish national authorities on bio-safety. Benefit sharing and sustainable use are concepts that still require further clarification. For example, the use of agricultural biotechnology opens up great expectations of mapping plant genomes. Considering on-going debates one may think biotechnology is of recent origin. In fact, the retiring Chairman of the American Association for the Advancement of Science argued as far back as in 1970 for a scientific programme to change human beings by controlling their own genes. Molecular biology and biotechnology research were early initiated at universities on human health problems. The Rockefeller Foundation began with financial support to biotech research on rice in 1984. At about the same time, biotech research was initiated in China. Possible threats of a gene revolution led to both a call for regulation in the 1980s and the conclusion that a bio-revolution contained both pro- and anti-poor elements. The Agricultural Biotechnology Support Project funded by the US Agency for International Development (USAID) was launched in 1991 with a vision of making transgenic crops available to developing country partners. In the late 1990s, the World Bank was recommended to network with the world scientific community by supporting high-quality research programmes to exploit the potential of genetic engineering to improve the lot of the developing world.

Biotechnology research carries both a political and corporate dimension. Since trade in GM products was expected to pose major issues for US policy-makers, the US Under-Secretary of State for Global Affairs presented a plan at a symposium of the American Association for the Advancement of Science in 1999 on how to handle scientific and technical matters. The plan called for more scientifically trained staff, greater efforts to use "science as a tool for diplomacy" and a request to the US Academy of Science to organize a first discussion on GMOs. This led to a joint meeting of the US Academy of Science and the CGIAR, which concluded that financial investments in biotechnology at the CGIAR Centres should be increased. One year later, more than 1,000 scientists joined in a Declaration of Scientists in Support of Agricultural Biotechnology. The academies of science from different countries issued a White Paper on transgenic plants, calling for global efforts to feed the hungry of the world. Since then, some developing countries have greatly expanded their cultivation of GM crops, while others have slowed down the commercial release of GM crops for reasons of bio-safety. Other countries do not even allow their cultivation.

The corporate dimension of biotechnology research is connected to the TNCs, demonstrating a growing integration of global food research and the privatization of agricultural research. From 1984 to the mid-1990s, US companies increased threefold their use of exotic germplasm (Goodman, 1998). The use of temperate exotic germplasm increased from 0.8 per cent in 1984 to 2.6 per cent. For tropical germplasm, the increase was smaller, from 0.1 per cent to 0.3 per cent. The corporations have also launched public relations efforts, for instance on "Golden Rice". So far, only the United States has allowed field tests of this rice. The Rockefeller Foundation has continued its promotion of biotechnology research, which has resulted, among other things, in the African Agricultural Technology Foundation, officially launched in 2004. Its objectives are to facilitate the transfer of biotechnologies, bio-safety regulation and the use of proprietary technology. It is registered as a charity, incorporated in the United Kingdom, based in Kenya and financially sponsored by the governments of the United States and United Kingdom and some global corporations. Four corporations have agreed to donate patent rights, seeds and laboratory know-how (Monsanto Co. of St Louis, Dupont Co. of Wilmington and Dow AgroSciences LLC of Indianapolis, all based in the United States, together with Syngenta AG of Basel in Switzerland). Potential crops include sorghum, millet and cassava, all staple crops for resource-poor farmers.

Increased use of biotechnology adds more weight to IPRs and proprietary science. Intellectual property temporarily transforms public good into private good. This transfer of power is a fifth factor that will dramatically change conditions for future agricultural research. Patents are a long-standing institution providing another incentive to profit-makers. In turn, public research institutions may be forced to use IPR to get research funds since governments in industrialized countries are decreasing their funding of agricultural research. Certainly, private companies may step in to provide some of this funding, on their terms. In addition, agricultural university scientists may not provide all information free but may increasingly request payment, as do other professionals such as lawyers or doctors. Moreover, there are more private universities not only in the United States but also in

Indonesia, Russia, Jordan and other countries. Nonetheless, public agricultural researchers must satisfy the needs of the public as well as work with the private sector. This signals a new paradigm regarding private-public relations in future agricultural research.

If governments stimulate further devolution of R&D to the private sector, farmers may have to pay a higher proportion of the costs for the development of new crop varieties and animal breeds. Between 1971 and 1995, investments and number of plant breeding scientists increased substantially in the public sector but not in the private sector in the United States (Alston and Venner, 1998). Out of 373 applications filed on new wheat, one third was public and two thirds private sector varieties. However, the private sector employed two thirds of all US plant breeders in the mid-1990s. If other industrialized countries follow US developments, as is common, this means that they have to adopt strong patent rights for gene constructs or they will have little access to findings from biotechnology. This trend is of much relevance also to developing countries.

Farmers will be affected by the way in which institutions are shaped to issue patents. In member countries of the International Union for the Protection of New Varieties of Plants (UPOV, 1978 or 1991), a plant breeder can obtain protection through plant breeders' rights for a new plant variety if it is novel, distinct, uniform and stable. Such plant variety protection was originally assumed to be adequate for the private sector. This changed when patenting of genes became common. A landmark was the 1980 judicial decision by the US Supreme Court extending patent protection to microorganisms. It was the first patent on life, the bacterium *Pseudomonas*, a process that took 10 years. In 1985, the US Patent Office began utility patents for plants. In 1997, the EU approved patents on gene-manipulated plants and animals. A current issue is how broadly life forms can be patented.

Following the EU approval of patenting of genes, critics have argued this is going to have serious implications with reference to the US company Myriad Genetics, which has a patent on a diagnostic method for breast and ovary cancer. Another example is from Iceland, which has a homogeneous population and records of relatives dating long back in history. In 1997, the deCode Genetics Company was established with the objective to trace the genetic background of 35 diseases, partly using a financial grant from Hoffman LaRoche. The Icelandic Parliament approved a law that offered the company exclusive rights to operate and own central data of all health clinics for 12 years. The National Institute of Genomic Medicine in Mexico now plans a similar mapping of the human genome. In Norway, foreign companies have been granted patents to hundreds of products or production methods requiring genetic material or human genes. The Swedish Parliament approved a government bill on patenting of genes in 2004. But the patenting of gene sequences and biotechnology processes can block the development of new plant varieties. Broad utility patents can monopolize a biotechnological research area rather than an innovation. GM crops may contain many patented genes, making negotiations with patent holders increasingly difficult. These developments amplify the crucial role of the TNCs and their growing control of global food production.

A sixth aspect of future agricultural research relates to the creation of global norms for agricultural produces by WTO, in particular. Streamlining of norms is an important task for all organizations and it has serious implications for future agricultural research. While stipulating free trade, TRIPS Agreement requires all countries to allow for patenting of all inventions (processes and products) in all fields. This could be accomplished through patents, by an effective *sui generis* system or by a combination of both, a "UPOV system". For years, the World Intellectual Property Organization (WIPO) has aimed for global patents, a task it resumed after TRIPS came into force in 1995. If the Substantive Patent Law Treaty does sort out differences between the United States, the EU and Japan, there may be a further concentration of power over the patent system. Its importance is reflected in the on-going conflict between the United States and the EU in WTO. Its preliminary verdict in 2006 may imply there is a de facto EU moratorium on agricultural GM products, which is inconsistent with the WTO rules. To make the interests of developing countries more central to WIPO, the organization created a new committee in 2005 and agreed to set up a fund to help indigenous people to attend WIPO meetings in Geneva. Global patents according to WIPO would be detrimental to plant genetic research work by the CGIAR. This is well illustrated by Syngenta's patent application with "claims to have identified the biological pathways controlling plant meristems". This and similar steps call for strict independence of the CGIAR Genetic Resources Policy Committee (GRPC) rather than serving as a committee with a diverse agenda coming close to WIPO.

Future Agricultural Research—For Whom?

It is usually claimed that science shall only express itself about what it is. Culture and governance should decide what science should be. In considering the overall problems of humanity, this becomes questionable, since the North is directing most R&D. With moral and ethics split from science, a certain value vacuum has appeared, illustrated by the emergence of BSE when dairy cows turned into cannibals, with scientific support. But a value vacuum may even lead to fundamentalism, another reason to re-examine the agricultural sciences. Agriculture will continue to provide food, jobs and services. It deals with common goods and public concerns for food security and food safety.

Future agricultural research should not only address the long-term productivity of natural resources but also contribute to the elimination of poverty. This will require concepts for meeting complexities rather than merely disciplines to reach resource-poor farmers, not just large-scale business farmers. It may focus also on sub-optimal conditions for farming rather than ideal conditions. Future agricultural research cannot neglect more than 600 million subsistence farmers, some 190 million herders and more than 100 million landless people. In addition, more attention must be given to approximately 450 million waged agricultural workers.

Food and agriculture will remain important for human survival. The experiences of revisited resource-poor farmers in Trinidad, Ethiopia and even Sweden show that the basic concepts of research approaches were more or less the same, over several decades. In the North, agriculture is business to some farmers and a social service to others. Food safety is emerging as a new issue whereas the use of hormones and increasing environmental pollution of water and soils have long been recognized. In developed societies an increasing lack of meaning in life is filled by leisure activities. Human stress is tolerated with a belief in a better future. In the past, religion offered certain answers and trust and people were part of a larger family, even over generations. These aspects remain important in many households in developing countries. To them, food shortage, malnutrition, diseases and lack of income and jobs are serious problems. Besides, all questions in life do not have a scientific value. Altogether, this requires adequate problem definition for different categories of farmers not only within countries but also for agricultural research at the global, national or local levels. There is a dimension of great urgency to fulfil the goals set by the MDGs.

Poverty eradication can only be combated through increased incomes at the household level. In a global context, agriculture fully dependent on cheap fossil energy is illusory. How then can production be intensified and what environmental side effects are acceptable in a short-term or long-term perspective? Long ago, a concept of development research was designed that called for changes both in the culture and organization of research. It never caught on because of certain limitations. These limitations remain, such as academic imperialism, inappropriateness, bias of concepts and models and theories, research in service of exploitation and domination through a superior and self-reinforcing research infrastructure and illegitimacy (Streeten, 1974). An orientation towards the problems of poor farmers was implicit to many donor agencies when the CGIAR's initial focus on increased food production in developing countries was abandoned in the mid-1980s. Those in command of its research orientation did not adhere to the rhetoric towards resource-poor farmers. The need to shift the focus of agricultural research towards the poor was even proposed by an international network at WCARRD (Bengtsson, 1979). Recently, a new type of research has been advocated for the CGIAR and on general sustainability to meet challenges of poverty and environmental stability (Sayer and Campbell, 2001, 2004).

Agricultural production systems in most developing countries vary greatly. Except for those in "modern" agriculture, farmers do not confine their work to planting crops or rearing livestock. They may also engage in fishing, charcoal production, off-farm opportunities, beekeeping or carpentry. A concept of sustainability would imply management of agro-ecological systems in ways that ensure their sustenance. This requires a systems perspective rather than reductionism, for example, emergy analysis, evaluating resources and services in both ecological and economic systems (Odum, 1996; Brown and Herendeen, 1996). They are both energetic systems, exhibiting characteristic designs that reinforce energy use. Sustainable farm management is site-specific, dynamic, productive and managed for diversity. This is in contrast to a current notion of agricultural research that to a large extent has become related to aspects of Western-type industrial management, characterized by specialization, standardization and consolidation of control for the highest economic output.

Although most agricultural research is location-specific, some research problems also carry a global dimension, such as biodiversity and the ozone layer. To many researchers in developing countries, global research problems may be attractive, although lack of funding and staff resources may restrict their active involvement for the time being. National agricultural research systems of developing countries must make their own priorities in problem solving and by participating in international networks. Such networks offer opportunities to get quicker access to and better use of internationally available research results. Research for sustainable production calls for methodologies better suited to resource constraints and to cope with micro-ecological variations, as illustrated by the revisited farmers in Ethiopia, Trinidad and Sweden. New technologies for increased productivity must be decentralized and also consider . traditional cultures and environments. The researchers must be accountable and realize the pros and cons of the possible long-term consequences of their research as part of an overall agricultural policy. Most of the future agricultural R&D ought to work toward the MDGs, probably requiring primarily public research funds. In addition to this criterion of relevance there is a need for reflexive objectivity, whereby values are taken into consideration (Alroe, 2000). The following discussion is based on these considerations with reference to MDGs and the earlier suggested agricultural vision.

Land Use

Some 30 years ago, arable land worldwide was estimated at 1.5 billion ha, half of which is located in developing countries. Since then, area expansion for agricultural production has been some 10 per cent, expanding cultivation into more marginal soils. Possibly, an additional 5 per cent of potential land may be cultivated by 2010. Area expansion is not possible in South Asia, West Asia and North Africa. The potential for expansion is unevenly distributed in Sub-Saharan Africa and Latin America. One third is in two countries, Brazil and Zaire, but three quarters of that area is subject to soil and terrain constraints, has no deep soil fertility and is deficient in micronutrients. As an example, Tanzania has scope for expansion since only 7 million ha is cultivated out of a total of some 40 million ha arable land. China has launched a new approach. It started to rent arable land for food production in Cuba in the mid-1990s, followed by Mexico in 1998. In early 2004, China also rented 5,000 ha in Laos. Would this trend imply that current areas of arable land under conservation schemes in the United States and Western Europe can again be used or rented by large and powerful food-deficit countries?

Common ownership of land can cause natural resource degradation. Each user will maximize his or her share of the resource without regard to other users, or even to his or her own future. A good example is the rangelands, occupying some 2 billion ha. They support large populations that are dependent upon cattle, sheep and goats and they are often an important source of fuel wood. People have a variety of sources of income. In Sub-Saharan Africa, there is a general reliance of 30-50 per cent on non-farm income sources. Honey, fruits, fuel wood and charcoal production may contribute half of the cash income of an East African farmer. In India, the poor obtain some income from common property resources. In Brazil, the northeastern region was originally perceived as homogeneous and much emphasis was given to large projects with centralization and supporting activities rather than a search for solutions to real problems in a heterogeneous environment. In such semi-arid areas, consideration should be given to the natural and human resources existing in each different ecological location, as well as to water, soils and R&D (Prisco, 2000). A mean annual rainfall below average is not as important for agriculture as originally believed. Therefore, pastoralism should be advocated where there is no adequate water or soils and no R&D. This calls for a different approach to R&D.

The management of an ecosystem must meet conflicting goals and take into account the linkages among production and environmental problems, sometimes including both food and tree products. In Sri Lanka, mango and tamarind fruit trees provide wood. In some countries, furniture for export is made from rubber trees and coconut palms. Farmers keep cattle for grazing in forests. When the issue is land use, local knowledge becomes much more critical, as illustrated by CIFOR research in Indonesia. In the shifting cultivation system of East Kalimantan, people considered black soils most fertile and productive for farming. Black soils used to be covered by forest, in contrast to the infertile swamp and steep land areas. Local knowledge of soils was also common among revisited aroid farmers in Trinidad and at Alelto Silosa in Ethiopia. Such detailed information was not as relevant to Swedish farmers nowadays, a marked difference from the 1940s.

Agroforestry is well known from Roman and Greek agriculture. Defined in a variety of ways, several agroforestry systems can meet the nitrogen requirements of moderate yields of grain crops but not the phosphorus requirement. Fast-growing nitrogen-fixing species (*Gliricidia* and *Leucana*) were tested in Africa by IITA. Another example is the combination of trees for nitrogen fixation (*Sesbania* and *Tephrosia*) interplanted with maize crop and allowed to grow as fallows during dry seasons. Yields of maize under this system increased by a factor of two to four (Sanchez, 2004). The wood is harvested and pods and leaves are hoed as green manure before maize is planted for the coming rainy season. Phosphorus is added from indigenous phosphate rock deposits. This system provides both increased productivity and a more efficient land use. In India, winter crops have been planted in association with multipurpose trees, including fruit trees. When considering the total production from that piece of land, yield reductions of some 10 per cent for wheat and chickpea were considered to be of little significance (Gill, 2002). This confirms that farmers' old practices with a traditional mixture of plants are quite often superior to monocropping. This kind of ecosystem is resilient to external pressure. At present, coffee is usually produced in the full sun with high yields but beans of poorer quality. Recent research from Brazil in both coffee and rubber plantations shows that trees among the coffee plants provide shade, thereby reducing the temperature and water loss through evapotranspiration. Shading coffee trees has long been controversial but decades ago it was common wisdom that the shading of coffee trees usually gave beans of higher quality.

According to the FAO, the forests cover almost 3.5 billion ha, including plantations and natural forests. Forested areas in developing countries (land with a minimum of 10 per cent crown coverage of trees) constitute some 2 billion ha. Forests cover some 30 to 40 per cent of the land area of the world and their products provide income, food, fuel and employment. More than 300 million people are directly dependent on forests for their survival. Latin America and Europe are the most forested areas of the world, each area accounting for a little more than 900 million ha. Russia has one fifth of the world's forests. About half of the tropical forest lies in South America, much of the rest in Central and West Africa and the islands of South East Asia. Protected forests, including national parks and reserves cover some 5 per cent of the tropical land areas.

Since 1990, the forested area has decreased in developing countries, often turned into agricultural land and infrastructural development. In developed countries, the forested area has grown, mainly on abandoned agricultural land. Constraints in food production explain why farmers have cleared little land of the humid-Amazon for crops and livestock compared to drier areas. When agriculture becomes less profitable, farmers tend to forest products, specifically exemplified by some revisited Ethiopian farmers. Some 15 million Chinese farmers have now converted more than 7 million ha of cropland into forests or grasslands. This trend underlines a need for research on future land use. One research issue is the role of tree planting in dry areas and its influence on water availability. Trees compete with other plants and their expansion may even result in crop failure and drying land areas, as has recently been observed in Western Kenya and elsewhere.

In 2000, the global forestry industry generated over US$ 430 billion in value added, furniture included (Lebedys, 2004). Global forestry exports doubled in a decade, reaching US$ 144 billion. Developing countries accounted for only one quarter of this value, although they had 70 per cent of all forestry jobs of the formal sector. In the late 1990s, the International Labour Organization estimated that forestry and forest-product industries provided 47 million jobs worldwide. The pulp and paper industry is global and recycled paper supplies nearly 40 per cent of global fibre demand for paper. Old forests provide almost 10 per cent and other natural forests one third. The remainder comes from plantations and non-wood fibres. According to Pulp and Paper International, world production of paper and pulp reached 318 million tons in 2001 compared to 80 million tons in the early 1960s. Projections of overall commercial timber harvest show an increasing annual demand of 1.7 per cent for industrial round wood up to 2010. Developed regions of the world dominate production and consumption of industrial wood products. This is an area for further research. African governments spend little funds on their forests, particularly for research. Swedish policy on forestry is of interest with its two goals of equal importance, one environmental and the other timber production. Annual forest regrowth is approximately 95 million tons, whereas net logging is a little more than 60 million tons. Such a policy, with possible modifications, could possibly assist in combating deforestation in tropical regions.

In tropical forests, most companies practise good forest management in plantations but less so in the wild forests. Long-term investments in slow-growing natural forests are not attractive to investors, a serious obstacle to promotion of sustainable forest management. This is also a problem for the donor community. In operation since 1993, the Forest Stewardship Council certifies that forests are managed according to certain environmental and social standards. About 30 million ha of forests have been certified and 70 million ha under other schemes. By mid-2002, some 50 communities with about 1.1 million ha of forested land were also certified under the council. Only a small proportion of tropical timber is marketed in areas demanding certified products.

Today, fuel wood accounts for 25 per cent of the energy consumption in the developing countries; in Africa it accounts for 50 per cent. More than 2 billion people depend on fuel wood, although it represents only about 1 per cent of global energy consumption. A major policy issue is the increasing price of fossil energy. Wood for both fuel and construction work needs more research attention. Another theme for research is bio-energy. For the last 5 years, the demand for ethanol has led to an increase in the price of sugarcane. New sugar and alcohol factories are planned over the next 5 years; for instance, 20 in Brazil and 16 in the United States. Plantations for bio-energy purposes can be expected shortly. Commercial plantations are well suited for research on transgenic trees with short maturity. Transgenic trees may add positive effects from improved status of soil organic matter and biological activity (Bernhard-Reversat and Huttel, 2001). On the other hand, they may experience constraints through contamination, as already demonstrated in China.

At present, forest plantations on 100 million ha supply one quarter of global industrial round wood (Brown, 2000). About two thirds of the forest plantations are located in China, India, Japan, Russia and the United States; Africa accounts for less than 4 per cent. Each year about 4 million ha of forest plantations are added. Plantations of tree crops of rubber, oil palm and coconuts cover some 14 million ha in Asia. An expected expansion of forest plantations may soon give rise to problems similar to those encountered with monocropping in agriculture, thus calling for further research with this orientation. For instance, China had tree plantations on more than 33 million ha, ranking first in the world in the mid-1990s. Paper has been made from straw and crop residues for centuries but pulp is gaining in importance. After the great flooding in 1998, China imposed a ban on logging. Small factories have been closed down, partly to avoid pollution of rivers. This has led to imports of pulp, timber and lumber, mainly from Russia, Canada and Indonesia. Ten years ago, China was the seventh largest importer of forest products but it is now the world's leading importer of tropical wood. Through its Sloping Land Conversion Program, China is planning to increase its total forest area by 10 per cent by 2010. Competition for land use between food agriculture and forestry will continue to rise both globally and nationally.

Agricultural Biodiversity

Recent research confirms traditional knowledge that the more species there are on a piece of land, the better they utilize space, water and soil depth. This insight is well established among farmers in the tropics, who have practised multiple cropping systems over the centuries. Certain species may be a prerequisite for others to co-exist in a specific environment, such as tannia under the shade of trees and banana plants. Diversity, it is argued, may imply that climate changes can be more easily tolerated. Although there has been an overall decline in agricultural biodiversity, past plant breeding efforts have also added to the gene pools and not simply reduced the number of landraces. More than 80 per cent of rice varieties have been produced by the national agricultural research systems (NARS). Only 6 per cent crossed international boundaries (Evenson, 1998).

Most plant breeding has usually been directed towards new cultivars of cereals with shorter and stiffer straw, leaves growing upright to avoid lodging. They are less competitive against weeds than are plants with long and loosely hanging leaves, for instance landraces. By changing the plant type of a crop, breeders can improve its competitive ability. A standard practice in "modern" agriculture has been to plant one pure crop variety with maximum disease resistance. A different approach would be to use a mixture of cultivars, each of which is resistant to a certain disease. Alternatively, it would be better for a farmer to grow a few different varieties than focus on one or two major ones. In Ethiopia, revisited farmers used this approach when testing improved barley seed. A similar experience is illustrated by the recent introduction of NERICA (The New Rice for Africa) in West Africa, where adopting farmers planted it on 15 per cent of their area under rice. The West Africa Rice Development Association has advocated, however, that all rice farmers grow the same kind of rice continuously rather than the mosaic that farmers used to grow. For instance, "Sahel 108", occupies 75 per cent of the irrigated rice grown in the dry season in the Senegal River Valley (de Grassi and Rosset, 2003). This implies less diversity for farmers. In cooperation with IRRI, Chinese scientists have recently reported that rice grown in trials with a mixture of different varieties gave higher yields and was not as susceptible to diseases as monocultures. This can serve as one future option to expand diversity and calls for dialogue with the processing industry to find joint solutions.

Agriculture can co-exist with biodiversity. Monocropping and single varieties prevail on large commercial farms. Small farmers use their land in a more intricate way as regards both crops and species, though it may not be done on the basis of current scientific rationale (Kaihura and Stocking, 2003). Rather than maximum sustainable yield, a concept of resilience analysis has been advocated. It would preserve biodiversity and reduce the risks of uncertainty. A major issue would be how to integrate this analysis in on-going practical research processes. Moreover, work towards sustainability demands measurements other than just kilograms per hectare per year. Such measurements and evaluations ought to include basic facts on soil fertility, the extent of polluted soils, changes in agricultural biodiversity, the level and content of microbial activity and changes in the content of organic matter over longer time horizons.

Worldwide, the FAO has estimated the total number of mammalian and avian livestock breeds to be roughly 7,000. In all, some 40 animal species have been domesticated and thousands of breeds have been developed. In animal science, a small number of breeding companies provide the genetic material of the Holstein-Friesian race. Already millions of cows are, and will be, relatives. The breeding of chicken is concentrated in two US-based companies, Ross and CobbVantress, which control about two thirds of the global market. In 2001, the Swedish Meats Company decided to import and make use of 400 Danish boars per year on a preliminary basis. This decision was based on a belief that Sweden is too small to conduct its own genetic research on pigs. These trends do not promote diversity in animal breeding.

The growth of the number of supermarkets in many developing countries is influencing the future of agricultural biodiversity. The supermarkets will transform urban food markets also in developing countries. The fastest growth is in Latin America but it is also starting up in Africa, for instance the Pick'N Pay and Shoprite/Checkers chains. In Brazil, the share of sales by supermarkets has grown from 30 per cent in 1990 to 75 per cent in 2000. The current 3,000 Chinese supermarkets are expected to be at least five times as many over the next five years. With more centralized procurement systems, the insistence on standardized products, quality and even imports, the supermarkets may discard locally and nationally produced food. They require large quantities of food of consistent quality and quantity. These demands will most likely discriminate against resource-poor farmers. Supermarkets will also stimulate changes in consumer habits. As shown by the IITA, the traditional leafy vegetables are being replaced in Africa by introduced *Brassica* species, including cabbage, kale and mustard greens. More than 100 different leafy vegetables, so-called orphan crops, are used both as food and fodder in Asia. Together with fish and tubers these vegetables provide a major source of nutrition to rural people. So far, they are given little research attention. It is up to plant and animal breeders to decide on the future research orientation to avoid further erosion of the food culture and the present concentration of research efforts on a small number of plants.

Although diets are becoming more diversified in Western societies, food consumption is based on a small number of crops and animal species, implying a cultural decline. There are also potentials in a search for new crops and farm animals, a neglected area for research. A recent attempt was made by IPGRI in a project focusing on millets in India and Nepal, medicinal plants in Egypt and Yemen and leafy vegetables in Africa. Other examples of hitherto neglected plants include coriander (*Coriandrum sativum*) for medicinal use as a spice and green herb; the sago palm (*Metroxylon* spp.) for food, paper, adhesive gels and house building; sea buckhorn (*Hippophae rhamnoides*) for medicinal use and as a multipurpose berry; and the Japanese quince (*Chaenomeles japonica*) as a fruit crop and ornamental. Sweet grandilla (*Passiflora ligularis*) is another fruit crop and so is the pomegranate (*Punica granatum*), the latter also used for treating ailments such as sore throats and rheumatism. Food crops with future higher potential may include taro/dasheen (*Colocasia* spp.), oca (*Oxalis tuberosa*), fonio (*Digitaria spp*), breadfruit (*Artocarpus altilis*) and choyote (*Sechium edule*). Crambe (*Crambe abyssinica*) has potential as a possible lubricant. Other possible crops are the protein-rich leaves of *Moringa olifiera*, ackee (*Blighia sapida*), mitoo (*Crotalaria brevidens*) and mulukhiya (*Corchorus olitorius*), whereas *Solanum villosum* provides a highly sought-after nightshade. The industrial hemp (*Cannabis sativa*) may become a fibre crop if the hallucinogenic effects are eliminated. It can be used for paper, textiles and plastics. It provides thick shade and stops weed growth. The plant is frost-tolerant and can be grown in Europe, yielding four to five times the yield of linseed. *Urtica diocia* can be grown for fibre. As a perennial it can yield for up to 15 years. Modified starch from plants is another area of growing importance to replace plastic products as the price of plastic increases. In short, there are interesting alternatives, often more important at the regional or local level than at a global level. This will also guarantee and even expand dietary diversity without industrial fortification efforts. This calls for more research investments in crops of regional importance and relevant to the needs of the most needy people. In this perspective, it is also crucial to increase the financial support to the International Center for Under-utilized Crops, which was recently relocated to Sri Lanka and hosted by the International Water Management Institute.

The basic idea of collaborative protected area management is to manage environments to achieve the twin goals of economic growth for local people and long-term sustainability of both natural and cultural resources. The local people must handle the management and make decisions. According to the United Nations List of Protected Areas the total number of such sites has increased to 102,000 from 1,000 in 1962. This is almost 13 per cent of the global land surface. Paying cash for biodiversity has been suggested as an alternative to development projects promoting sustainable forestry and agriculture. It pays off quickly and provides clear incentives. But practical experience shows this approach rarely works (Ferraro, 2001).

Direct payments also face obstacles such as the conflict over property rights and knowing whom to pay.

Bio-piracy is a specific aspect of agro-biodiversity, supposed to be partly controlled by the political and administrative framework of the CBD. Its requirements place more responsibility on individual researchers and institutions to accept concepts of prior informed consent and benefit sharing of gained results from research collaboration. Several recent reports indicate an increase of bio-piracy. One example is the patenting of a local herb (*Swarzier madagascariensis*), reported to treat malaria, oral thrush and foot rot. Lausanne University took a US patent in spite of its collaboration with the University of Zimbabwe. Ultimately, it withdrew the patent. Another example is Oxford University, which took steps to patent its own findings of an AIDS vaccine developed jointly with Nairobi University. A third example refers to a US patent giving exclusive rights to market a yellow bean of Mexican origin. The variety, grown for years in several countries, was brought into the United States without an export permit from the Mexican Government, a violation of the CBD according to CIAT.

Piracy is an issue in bio-prospecting. Conservation International is conceived as an environmental NGO but is also financially supported by several US corporations. Operating in about 40 countries, it may help pharmaceuticals and corporations get access to the biodiversity of indigenous people. Their representatives often stress the failure by Conservation International to consult with them. Hawaii has a unique biological diversity with about 8,800 endemic species. Nonetheless, the US Diversa Company was given exclusive rights in 2002 to discoveries on genes from all existing collections. In early 2004, the Hawaiian House of Representatives passed the first bill on bio-prospecting, which had a negative impact on sales of GM crops. The only crop that could be affected would be taro or eddo, numbering some 300 varieties. This would be highly significant for its future worldwide breeding work. In response to local protesters, the US University of Hawaii has recently decided to give back three of its patents on taro hybrids to the native Hawaiians. The Diversa Company has also an estimated gene collection of more than 3 million micro-organisms with daily DNA screenings of million of genes. With some 100 patents and many more pending, its power over microbial genes is well illustrated. Recently, more than 50 Syngenta researchers were transferred to Diversa to strengthen this type of research, another illustration of the power of genes in biodiversity.

Soil Fertility and Replacement of Nutrients

Estimates by the Global Land Assessment of Degradation have indicated a global degradation of some 2 billion ha since World War II. Degradation also occurs in industrialized countries, often by pollution. In developing countries, some 400 million ha of cropland are degraded by various causes. In Africa, losses due to erosion are reported to have caused annual yield reductions of 2-40 per cent (Lal, 1995). Since the topsoil is lost, this affects soil quality regarding both the content of organic matter and plant nutrients. In China, the total eroded area is about 160 million ha with a total soil loss of 5.5 billion tons/year, estimated to represent about one fifth of the total soil loss in the world (Dazhong, 1995). During recent years, sand storms have carried away the topsoil, partly caused by the Chinese policy of compensating for a decline of arable land. In southeast China, urban and industrial development has expanded. Instead, new land areas were opened up for cultivation in areas such as Mongolia, Gansu, Ningxia and Quinghai. Since the use of non-agricultural land is a major source of this erosion, it is a mistake to think only about agriculture (Swallow et al., 2001). Although the figures may be exaggerated they point to a long-standing, increasing trend of soil erosion calling for soil conservation measures. A classical remedy for erosion is terracing, used for centuries in Asia, which requires much labour and organized collective action. On the other hand, such work may provide job opportunities for many unemployed. In southern China, trees have also been used successfully as windbreaks to control erosion of coastal land, leading to rising yields of wheat and rice.

Areas with the worst degradation are generally not those where yields are currently declining (Dyson, 1996). If soil fertility is reduced, productivity gains will be hard to achieve. Soil fertility is the capacity to supply adequate nutrients to produce a desired crop, not synonymous with agricultural productivity. At harvest, nutrients are taken away and they are also lost in water runoff. In rice fields, some 70 per cent of nutrients are lost. Without replacement there is no agricultural sustainability and an application of more nitrogen fertilizer will fail without adequate soil fertility. There must be sufficient organic material to make full use of the phosphorus. Its availability to plants is critical, irrespective of how much is applied. In "modern" agriculture, nutrient leakage is caused by excessive application of fertilizers, illustrating a need for balancing various parameters. Also, micronutrients are important for such a balance. This is illustrated in a recent study by the UK Medical Research Council. During 1940-1991, soils were observed to be impoverished with respect to sodium, copper, magnesium, iron and zinc. Over time, much less attention has been given to trace elements (Ames, 1998).

A key question is how to best supply an adequate amount of plant nutrients, since their use must double both in Africa and Asia. In Africa, average annual mineral fertilizer use is less than 10 kg/ha compared with a global average of 98 kg/ha. Some 60 per cent of all fertilizers are applied to cereals, half of them to rice, followed by wheat, sugarcane and cotton.

Even under good agronomical practices, the recovery rates are some 50 per cent in dry-land crops but seldom more than 40 per cent for rice. Mineral fertilizers are not the only option for nutrient replacements. Plants can provide the nitrogen, for instance in agroforestry systems. During Roman times, lupine was incorporated in crop rotation systems. Much later, clover was introduced to Western Europe, reaching Denmark in 1730 and Sweden some years later. Danish forage production increased dramatically, leading to an increased nitrogen balance (Kjaergaard, 1995). The growth of the potato was totally dependent upon clover and "clover has had greater influence on civilization than the potato" (Taylor et al., 1985). The increased European agricultural production between 1750 and 1880 amounted to some 175 per cent, two thirds having been attributed to the effect of increased cultivation of clover (Chorley, 1981). Like certain agroforestry trees, green manure and cover crops can supply from 50 to 100 kg of nitrogen per ha. In rice, blue-green algae (*Anabaena azollae*) can produce up to 400 kg N/ha per year under experimental conditions if fern is incorporated in the soil.

Livestock manure is another source of nutrients. In the past, it was spread on almost every farm in developed countries. Half a century ago, organic manure provided more than 98 per cent of the nutrients applied to Chinese soils, a proportion now reduced by half. Its major disadvantage is the high labour demand. In areas with large-scale production, organic manure has become a major polluter, being heavily concentrated in fewer places. Although mineral fertilizers can replace nutrients, their exclusive use cannot prevent a gradual degrading of the organic material of soils, which is crucial for water holding capacity, microbial activity and other qualities. This calls for specific research efforts in regions with relatively low fertilizer applications. In zero N treatments in southern Sweden over 24 years, losses of organic matter were nearly 10 tons/ha in a non-livestock rotation and about half as much in the livestock rotations. With 100 kg of N per ha, the losses were negligible (Carlgren and Mattson, 2001). Apart from certain exceptions, such rates of nitrogen application in developing countries can hardly be expected in the next decade. Soil fertility can be enhanced by certain conservation measures. Among the best practices are cover plants, contour strips and ridges, reduced tillage and control of water runoff.

For many years, cheap energy made industrially produced fertilizers abundant in most parts of the world so nitrogen-assimilating crops were not able to compete. This situation is changing, calling for a shift in research attention to include crop rotation systems, nitrogen-fixing plants and more dynamic on-farm research by integrating organic and inorganic nutrient sources to suit local ecological variations. Researchers at Cornell University (USA) have recently discovered genes that allow certain plants to form symbiotic relationships with soil-dwelling fungi that absorb phosphate. This may lead to GM plants growing effectively with less phosphate-based fertilizer.

To meet the MDGs, mineral fertilizers are most likely the most simple and effective way for replacement of nutrients and increased food production. They would be environmentally sound provided they are not applied in excess and provided fertilizer mixtures do not contain harmful chemicals. More research attention must be given to the design of tailored mixtures of macro- and micronutrients for different crops, in particular those cultivated by resource-poor farmers. Well-balanced nutrient mixtures are required for new potential crops including those for bio-energy production. Such a mixture may also be required for forest plants, although they are seldom put into a forest stand nowadays compared to the mid-1960s. Nitrogen fertilizers are not even allowed in Swedish forests, although it is a major growth-restricting nutrient in Scandinavian forests (Ingerslev et al., 2001). The shorter the plant rotations, the greater need there would be for adding nutrients in developing countries.

One feature of on-farm research relates to the need for field demonstrations to show positive effects of the use of various fertilizers. This is more crucial than a number of trials to find out the highest yields for certain fertilizer mixtures. Such demonstrations are now carried out under SG 2000 in some 15 African countries. In West Africa, the Soil Fertility Management Project is assisted by the International Fertilizer Development Center (IFDC). This approach should be supplemented by a research design in which farmers are responsible for operating these plots. Their annual findings and observations should be collected and analysed by the research establishment. This would require the acceptance of a concept of sub-optimal field approach to assist farmers rather than conventional rigorous scientific testing on a few research stations, which takes too long and still does not take ecological variations into full account. A countrywide network of participating farmers ought to supplement the system of a few scattered agricultural research stations.

By the late 1960s the FAO Fertilizer Programme, financed by the world fertilizer industry and donor governments, was operating in a number of developing countries. Since individual donors had their own specific preferences for the kind of fertilizers they wanted to support, this made it difficult for the recipient government to make its own choice. The experience gained was not used and funds were wasted. Had farmers been actively involved in managing some field activities of those early demonstrations, they would have acquired the new knowledge. Then, there would have been no need for SG 2000 to repeat demonstrations three decades later as is now the case, for instance, in certain regions of Ethiopia.

During the past two decades, the international community agreed to eliminate fertilizer subsidies. It was hoped that this would stimulate the development of a private sector that would deliver more efficiently and at more reasonable prices. This proved wrong and subsidies per se are not the real issue. Farmers will make use of mineral fertilizers, if available at reasonable prices in relation to the price they can get for their produce. Revisited Ethiopian farmers and the aroid farmers made this argument explicit. They purchase fertilizers only if they can make a decent profit by using them. This would be in some contrast to the message from the UN Millennium Project, which signals a return to subsidies. However, the issue is about government policy or some kind of guaranteed price for the growing season, avoiding surplus production. Such financing might come from an extra fee on all future use of pesticides harmful to humans and the environment. Meanwhile, governments should initiate a substantial dialogue with the private sector on how and when it can take over the future distribution of different fertilizer mixtures and also provide packages of not only 50 kg but also 5 and 10 kg. Resource-poor farmers in many countries buy small quantities. To make mineral fertilizers available in a timely manner, new delivery systems should be tested to include small agro-dealers, the old middlemen, shops, NGO offices, private salespeople in a village, proactive farmers and even priests. They could probably be more effective than the current reliance on a rigid government system.

Even today, mineral fertilizers are relatively expensive in many developing countries. About one third of mineral fertilizers are imported. Delivery systems are inefficient; the costs of transportation are high both within countries and internationally. The price a Sub-Saharan African farmer must pay for fertilizer is often two or three times the world market price. If so, its use becomes often uneconomical, in particular on food crops. Aroid farmers in isolated areas of Trinidad expressed the same concern. Reduction of costs can be achieved from procurement to shipping, unloading and bagging and to transportation to the farm gate. These are all part of government policy and not significant issues for research.

For the future, it is realistic to assume increasing prices of mineral fertilizers, nitrogen being the most serious problem. Natural gas is the preferred source of hydrocarbons in the fertilizer industry. By 2020, estimates indicate natural gas will account for one third of the global energy use compared to one fifth in the mid-1990s (IFA, 1998). Phosphorus and potassium reserves are mainly deposited in the Middle East and South Africa. In 1998, the US Geological Survey estimated that the world phosphate rock reserves, at present rate of mining, would last for about 80 years and the reserve base for about 240 years. Morocco has most of the reserves. The known reserves of potash are much larger and may last a

172

little more than 300 years. One single country, Canada, has over half the reserves and half the reserve base.

Natural Resource Management

Even today, some 300 million people practise shifting cultivation. Often, they lack a title to land and move as squatters. To them, the principle of natural resource management is hardly a new concept. They are forced to consider the whole ecosystem, not only single factors at a time. If they fail, the lives of their families will be jeopardized. If opportunities arise, these farmers also act for commercial purposes, clearly illustrated by the revisited aroid farmers. With low population pressure, the traditional slash-and-burn agriculture was once a sustainable system, albeit with low outputs. Now, it is impossible due to declining soil fertility and increasing population growth.

Also in developed countries, a research focus on natural resource management is becoming a necessity. In terms of sustainability, the overall trend in modern agriculture shows problems and decline. Many consumers in the Western world are increasingly questioning facets of "modern" agriculture. It favours large-scale operations of both production and the food processing industry. Certain negative consequences of large-scale agriculture led to discussions of alternative approaches in a study by the US National Research Council by the late 1980s. The International Movement for Ecological Agriculture requested a change. To some, the basic idea of ecological agriculture has been related to overall natural resource management. Other observers have called for sustainable agriculture without a precise definition. In general, research on natural resource management has mainly been directed to an awareness of environmental damage. Following UNCED, a CGIAR Task Force on sustainable agriculture recommended consolidation of activities regarding soil, water and nutrient management, strengthening the system-wide programme on integrated pest management and the formation of a Consortium on Integrated Pest Management. The eco-regional approach and the system-wide initiatives were considered as sustainability issues. In the mid-1990s, CIMMYT proposed a five-year training project on principles and practice of sustainable cropping systems since "no institution in the world" offered a "practical focused training in sustainable systems research and development". It is less clear what these changes implied in reality. The FAO made organic farming a priority area for 2002-2007 and alternative research methodologies have been suggested (Madeley, 2002). A recent handbook provides both theoretical aspects of sustainability and more practical approaches on how issues of sustainability are intertwined with the environment and society (Leal, 2005). On the whole, much less research has been directed to methods of increased productivity combined with a sustainable management of natural resources.

In spite of the new challenges to modern agriculture, skeptics of organic agriculture have refrained from formulating any alternative agricultural research. There is much misconception about ecological agriculture, which has a range of definitions. In the Western world, the concept was initially derived from biodynamic agriculture. In 2005, the International Federation of Organic Agriculture Movements introduced a revised concept. Organic agriculture is based on the principles of health, ecology, fairness and care. Among others, it should avoid the use of agrochemicals, account for the real environmental costs and consider solutions tested over time. In current EU terminology, ecological agriculture is intended to describe a self-sufficient, sustainable agro-ecosystem with people integrated as a supporting factor. This principle is relevant to many resource-poor farmers. To them, one major problem is low productivity, their soil fertility requiring reasonable amounts of plant nutrients. The basic issue is an adequate problem definition. Norman Borlaug finds it impossible to turn back to so-called organic farming, urging African policy-makers not to be "misled by the current prophesies of doom coming from extremist environmental groups whose elitist leaders neither understand agricultural science, nor the need to raise farmers' incomes substantially as a precursor to increased investments in environmental conservation" (SG-2000, 1999). Results of a long-term experiment in southern Sweden show it would hardly be possible for "an individual farmer to survive the transition to organic farming on a soil that has not previously been given better plant nutrient status with farmyard manure and/or fertilizers in accordance with conventional methods" (Gesslein, 2001). This illustrates the role of soil fertility.

In Zambia, some 40,000 farmers have adopted conservation farming. It means no tillage, sowing at the right time, use of organic manure and frequent weeding. One may dispute whether or not this is classified as organic or ecological agriculture but it illustrates one alternative approach, both effective and focusing on resource constraints. It emanates from experiences supported by SIDA/Sida for the last 15 years. Zero tillage or conservation agriculture is also applied to 20 million ha in the United States and 13 million ha in Brazil and other Latin American countries. Some 220,000 Brazilian farmers are using cover crops and green manure, integrated with livestock, resulting in a doubling of their maize and wheat yields to 4-5 tons per ha. During the past decade, Brazil has preserved arable land and saved US$ 11 billion and more than 1 billion litre of fuel. One major problem remains, that is, keeping the weeds at a minimum. A few large farmers have started to use a similar technology in Kenya, Zimbabwe and South Africa.

The world market on ecological products has recently grown by some 15-20 per cent annually, in the United States and elsewhere. The Natural Marketing Institute and the Organic Trade Association have projected an annual turnover of some US$ 26 billion by 2005. Studies by the Caribbean Agriculture and Research Institute in the West Indies indicate expanding regional markets for organic products, leading to the recent launch of a Regional Organic Association. In the mid-1990s, the Danish Ministry of Agriculture and Forestry formulated a plan on organic food production but that process came to a halt after a change of government. In the early 2000s, ecological production had doubled in the United Kingdom, whereas only 0.2 per cent of the food in Greece was produced on ecological farms. The Swedish ecological food market in the early 2000s had a turnover equivalent to the value stolen by shoplifters in the country: about 3 per cent of the total food consumption. On the other hand, ecological production has expanded since the late 1980s, when 600 farmers belonging to the Swedish Control Association for Ecological Production cultivated some 8,000 ha. The association's label certifies, among other things, that the product has been manufactured without the use of chemical pesticides and fertilizers, that nutrients are provided through clover and organic manure, that farm animals are treated well and allowed to spend their lives in open air and that no GMOs are allowed to enter the production chain. According to the Swedish Parliament, ecological agriculture in Sweden should constitute 20 per cent of the cultivated area by 2010 as compared to 13 per cent in 2000. Also, the proportion of milch cows and beef cattle in ecological production has grown.

In the State of Rio Grande do Sul in Brazil, more than 12 per cent of the production comes from ecological agriculture. The local government has taken specific steps to promote small-scale agriculture through its Programa Agroindustria Familiar, serving a population of 11 million people. Egypt is one African country with local "certifiers" accredited by the International Organic Accreditation Services Inc. Such recognition is important, since it may require some 10 years to become a standing exporter in organic products (Spore, 2001b). If developing countries are entering new markets as a result of free trade, they will face new requirements of standards. Such standards may emerge as new barriers in place of those supposed to be dismantled by the WTO.

In short, there are some important features of ecological or organic agriculture of relevance in designing alternative approaches to future agricultural research. Others require much more scientific attention and possible modifications. The major problem in the debate is the lack of any other conceptual alternative. A quite conservative agricultural research establishment has been by-passed by the political establishment, still in favour of the old alternative. Some of the undisputed positive effects of ecological agriculture include the requirements of a balance between the number of animals and fodder production and the reduction of nutrient leakage from large animal factories. Biodiversity is increased and there is a reduction in the use of fossil energy and chemical pesticides. The use of nitrogen fertilizers may be reduced. The rejection of mineral fertilizers would be unrealistic in developing countries. At this stage, there is an urgent need to present an up-to-date summary of current, conspicuous research findings in ecological agriculture worldwide, including its pros and cons. Experiences of older forms of Chinese agriculture may provide useful data. It would be useful for an international network to collect and analyse available information regarding research on natural resource management in agriculture through the FAO or the CGIAR. Moreover, such an analysis could also provide scientific justification how the current concept of ecological agriculture should be further promoted, modified or rejected as an avenue towards sustainable agriculture. If rejected, alternatives to the current concept of "business as usual" are still required.

As previously indicated, the introduction of emergy analysis may be one alternative. Our life-supporting system must be viewed as the foundation for a global future sustainable society in relation to the use of natural resources and the use of energy. Up to now, energy analysis has primarily aimed at the stocks of fossil fuels. It does not account for natural renewable resources, human labour and economic services, all features of both agriculture and an ecosystem. In such a context, the current thinking on cheap food is illusory since environmental costs and those of renewable natural resources are neglected. In the United States, the value of the food and food industry is three times that of the car market and that figure is even greater in the United Kingdom. Short- and long-term environmental costs for the production of food are rarely accounted for. Other aspects include transportation costs, water constraints and the fact that low wages of people in developing countries allow for a trade that mainly benefits the North. These accumulated costs are currently not paid for by food consumers. For example, shrimps are produced in cleared mangrove ponds in the South with much lower labour costs to serve European consumers a cheap product. Originally, the mangrove forests acted not only as breeding grounds for various types of fish but also as natural barriers to storms and even tsunamis. Thus, there are consequences to nature in transferring those areas to commercial activities. Current lifestyles in the so-called developed world are also influencing our life-support system. A major challenge for future research would be to try to capture all these parameters within the whole ecosystem. This is hardly possible with current research approaches in ecological agriculture or natural resource management. Emergy theory may provide a step forward (Odum, 1996; Brown and Herendeen, 1996). By synthesizing different forms of energy and materials, human labour and economic services, and evaluating them on a common basis into equivalents of solar energy, all costs may be taken into account (Franzese et al., 2006). The emergy theory comprises also ideas and proposals for new thermodynamics laws in addition to the three classical ones. Altogether, this is a major challenge for future research. It calls for the urgent design of a conceptual framework as well as the development of new methodologies with particular reference to agricultural research and land use issues.

The new approach will include some concepts from ecological agriculture, energy analysis and natural resources management for a future global society with sustainable food security and food safety. The theoretical approach deals mainly with the quantitative aspects of the natural resources but omits qualitative parameters such as water quality or soil fertility. Still, fossil energy is important, since scarcity will give rise to increasing prices. For various reasons, the current focus on alternative energy sources would not suffice. In 2006, some 20 per cent of all maize production in the United States was for ethanol. Fuels from food crops hold little promise, however, according to the US National Academy of Science. The entire US maize crop of 2005 would have supplied only one tenth of US demand for gasoline. Plant fibre cellulose of synthetic hydrocarbon fuels may be the alternative research orientation. In South Africa, the Ethanol Africa Company is planning eight bio-ethanol factories. Since white maize is a staple crop the company will use yellow maize to avoid competition for food. Small-scale farmers are given a price set before the planting season and they are assumed to produce one third of that maize. Probably, this approach will give improved incomes to involved farmers in the short-term. It will change the agricultural system in food-deficit regions. It will not take land degradation into account.

Another example where land use and the natural resources are given less consideration applies to new efforts by the Ericsson Company. Together with mobile operators, it has started a test plant for bio-energy in Nigeria. Only one fourth of the country has access to electricity. By using crops such as sugarcane, groundnut and others for energy instead of diesel to operate the radio-based stations, electricity can be locally produced in regions without electricity and where roads do not allow regular transport of diesel. If successful, this model is anticipated to lead to a more rapid spread of mobile phones in remote regions of Africa and Asia. The trial in Nigeria may create local employment and be profitable to participants. However, it may also lead to land use competition and may not easily cope with issues of natural resources management.

Water Resources and Their Agricultural Use

Any human being can live one day without food but always requires about

2 l of water. A rice crop will require 2,000 kg of water for 2 kg of organic material, compared to 350-500 kg for wheat. A maize crop yielding some 8,000 kg/ha of grain consumes water during its growing season equivalent to 1,000 mm of rainfall or 10 million l of irrigation water (Pimentel et al., 1997). Animal production requires even more water and so does the food industry, for instance in sugar production. These conditions also restrict the type of plants that can be grown under different ecological conditions. With increasing water scarcity, the balance between livestock and crop production becomes sensitive, requiring coordinated research attention. Another issue for research is the balance between irrigated and rain-fed agriculture. Now, irrigated land accounts for about 16 per cent of the global cropland but produces 40 per cent of global food. Half the cereal production of developing countries comes from irrigated lands, a significant increase since the early 1960s. In India, the irrigation potential has grown from almost 23 million ha in the early 1950s to currently 90 million ha of a total arable land area of 142 million ha. A 10 per cent increase of the productivity of rain-fed agriculture would have twice the impact of the same increase in irrigated agriculture (Seckler and Amarasinghe, 2004). According to the Indian Council of Agricultural Research it may be extended to 132 million ha. These examples illustrate the need for policy research rather than an exclusive focus on technical matters.

In India, agriculture uses some 80 per cent of the national water consumption and ground water accounts for 60 per cent of the irrigated areas. This calls for both improved management of water and methods to increase water productivity. Irrigation is now using 70 per cent of the world's supplies of developed water. However, costs for irrigation have increased, slowing down irrigation. The efficiency of water management has become a major difficulty in the control of large-scale irrigation schemes, noted by aroid farmers in Trinidad several decades ago. This calls for less government control but more delegation of responsibility to the farmers for the operation of such schemes. In fact, the tank systems in Sri Lanka and southern India had such a feature many hundreds of years ago. The tanks were managed by farmers' committees. Moreover, there is a need for technologies to suit different target groups of producers. One example is the foot-operated treadle pump made from bamboo and a PVC or flexible pipe for suction to pump water from shallow aquifers or surface water bodies. It has proven to be a powerful tool for poverty reduction in India and Nepal on farms of less than 1 ha. Through an investment of US\$ 12-15 with one year as the payback period, the pump may raise annual net household income by US\$ 100 (Shah et al., 2000).

In view of looming water scarcity, two scenarios on the world water supply and demand assumed the per capita irrigated areas will be the same in 2025 as in 1990 (Seckler et al., 1998). By increasing irrigation effectiveness, water requirements in 2025 for all sectors were to be reduced by one half. But global data are less relevant than country data. Groundwater is depleted for instance in the most pump-intensive areas of India and Pakistan, where water tables are falling at rates of 2-3 m a year. The number of tube wells has increased significantly over four decades. This will lead to deeper wells and increasing costs for water extraction. At the same time, the water tables are rising, causing water logging and salinization, a well-known feature of the past. Excessive use of water with poor drainage has lead to annual degradation of about 10 to 15 per cent of all irrigated land. Globally, saline land is currently estimated at more than 25 million ha. The effects in individual countries can be quite pronounced, as seen in China, Egypt, Iran, Pakistan and Syria.

For years, tribal conflicts over access to water have been common in dry regions. Water is also a reason for larger political conflicts since some 300 major river basins and many groundwater aquifers cross national barriers, for instance the Nile and the Mekong River. The delta of the Nile no longer grows and the water quantity has diminished. Political negotiations are now creating difficulties for other nations wishing to make better use of the water of the Nile and questioning the old agreement reached in 1929 that gives the final say to Egypt. In certain countries, Sweden being one of them, the implementation of laws regarding water and water pollution emerges as a problem. Under one law, Swedish farmers are granted freedom to use water on their lands provided it does not cause damage to others. Under another law, however, municipalities can ban farmers from applying agrochemicals on lands close to intakes of ground water. The issue is what law should provide the final ruling. Since different ministries are involved, this is another illustration that the old sector approach does not allow for consistency.

Improved Productivity

Towards Higher Crop Yields

Since 1960, food production has more than doubled on almost the same acreage. During the last decade, the increase of yield levels has been slowing down, probably because plant breeding efforts were focused on disease and pest resistance rather than the yields. New crop varieties have not always been well adapted to marginal lands or stress conditions. To meet demand, productivity must double by 2020. For instance, average rice yields in irrigated areas must increase to about 8 tons/ha within 30 years.

Most countries are well below their attainable practical yields. The critical factor is the current yield level in many developing countries. Nevertheless, the raising of yields in the field is a great challenge. Over six centuries, the average wheat yield in the United Kingdom increased by less than 2 kg/ha per year (Plucknett, 1993). According to data supplied by Tanzania to the FAO, national maize yields have increased at an average rate of 0.3 per cent per year during the last 20 years, relating fairly well to the long-term trend of wheat indicated above. At the same time, one should recall that plant breeders argue that 1 per cent annual yield increase may serve as a general rule of thumb, a figure applicable to the wheat yields of revisited Swedish farmers. Estimates indicate that yields generally take off at a transition point for productivity growth, about 1,700 kg/ha for wheat (de Wit et al., 1979). Below that transition figure, the annual increase is low. Above it, the annual increase was about 4-5 per cent. This gives an indication about the contributions that plant breeding and other development efforts can achieve in considering current, low national average yields. Among revisited Ethiopian farmers, the wheat yields started from 1,100 kg/ha in 1966 and had reached 2,300 kg/ha after some 35 years. That was much higher than the national average. According to India's Protection of Plant Varieties and Farmers' Rights Authority the recent drop-off in quality germplasm from CIMMYT and cumbersome procedures under the ITPGR have negatively affected its wheat breeding programmes. Among Swedish farmers, the initial yield level of wheat in 1925 was almost equivalent to the national average figure of 1,700 kg/ha, and the transition point for change. Thus, the yields of wheat increased more quickly, reaching about 6,200 kg/ha after 80 years.

Ever since the Green Revolution, there has been emphasis on a genetic fix in agricultural research. Although important, it is not the only way to increase production. Research on plant genetics must be combined with soil and water conservation, pest management and improved cultural practices. Yields can be doubled just by an increased use of fertilizers, better weed control, integrated pest management and the use of already developed and available plant varieties. Clean seed alone could boost yields by 10 per cent. Rice-duck farming reduces weeding and insecticide requirements, increasing grain yields and incomes of farmers (Ahmed et al., 2004). It is vital to conduct research on resources with reasonably good and guaranteed access to the farmers.

Not only research for higher productivity but also maintenance research is vital to ensure yield stability. It is of equal importance in both conventional breeding and through GMOs. Maintenance research in breeding for resistance to diseases and pests is a continuous process and must be given much increased allocation of funds. This is particularly so if the application rates of pesticides will continue at the same level, producing more and more resistance. A focus on methods of biological control would allow efforts in plant breeding to concentrate on higher yields. An issue is to find the right balance between higher yields and maintenance research.

In the 1960s, the focus of rice research was on the reduction of plant height, leading to short, sturdy stems, high tillering, and dark green erect leaves. A large number of unproductive tillers were seen as major constraint so breeding concentrated on new types with low tillering capacity, few unproductive tillers and a vigorous root system (Virk et al., 2004). Most of those lines lacked resistance to tropical diseases and insects. Exploitation of hybrid vigour became the next step. Rice hybrids have high vegetative vigour and stronger root systems. China is the largest producer of hybrid rice, although the area under hybrids has somewhat declined since 1997. Other major producers are Vietnam, India, Bangladesh and the Philippines. The FAO estimated rice hybrids were grown on some 15 million ha in 2003. Some 60 Asian seed companies are involved in seed production (Virmani and Kumar, 2004). The Chinese companies are linked to the state and their international activities are limited to licensing and joint ventures. Rice hybrids yield 1-1.5 ton more per hectare but seeds would be more costly for resource-poor farmers. Also, this research work takes a long time, as shown in China. Subsidies are required to speed up the adoption process. Since hybrids cannot be replanted, this type of seed production is of particular interest to the private sector. If the intention is to reach resource-poor farmers in accordance with the MDGs, an exclusive research focus on hybrids crops by IRRI and other Future Harvest Centers seems hardly justified.

To reach resource-poor farmers and those using marginal lands, more attention must be devoted to plants growing under stress. About 20 million ha of Asian rice lands are subject to uncontrolled and unpredictable flooding; there, new varieties are required instead of traditional ones, yielding about 1 ton/ha. In areas with adverse conditions, plants need to be bred for drought and salt tolerance. Drought-tolerant cowpea is one option in dry areas. Research at the M S Swaminathan Foundation in India has identified and isolated the gene responsible for salt tolerance in mangrove plants. Eventually, it may be transferred to crops. In Egypt, a drought-tolerant GM wheat variety has recently been developed, requiring irrigation once instead of the normal practice of eight times. Australian researchers have recently grown GM wheat in laboratories, using water that has one third the saltiness of seawater. In 2004, a specific gene was identified by American scientists reporting on the development of prototype tomato and rice plants able to thrive in salt-rich soils. In view of increased degradation of soils, more research should be devoted to a search for crops tolerating modestly polluted soils and new crops that absorb heavy metals, for example *Salix*. Another area for research concerns the identification of leguminous cover crops and grain legumes tolerant of low soil pH and high levels of aluminium.

Cassava and potato have increased both in acreage and in yield. The same is true of sorghum and millet in India, but progress has been slower in Africa. Sorghum and millet cover more than 90 per cent of the total area planted with cereals in the Sahelian region. In West Africa, sorghum yields are now reaching about 800 kg/ha but yields in Tanzania have not changed very much in the last 15 years (Rohrbach and Kiriwaggulu, 2001). With a decreasing trend in the global per capita consumption of these crops, there might be less need for international research efforts by the CGIAR. In areas where sorghum and millet are still common, research attention must also be given to higher value crops and/or vegetables. Diversification is needed, as illustrated by Chinese farmers who are now turning to fruits, vegetables and non-traditional crop production in addition to their rice crop.

A strategic issue in crop production research is to find new perennial crops, either by introducing genes for perennial growth or by domesticating new wild perennial species. One example is perennial linseed (*Linum perenne*), previously grown in Ukraine. Several years ago, the IRRI initiated research work on perennial rice. Another example, without public agricultural research, was shown by the ingenuity of an eddo farmer in Trinidad using dasheen as a perennial for leaf production for callalo soup. The identification of trees may provide new opportunities in farming systems to produce nitrogen, wood or fruits. One example is a possible domestication of the African tree *Allanblackia floribunda*. It is medicinal but produces a food oil of good quality. In Burkina Faso, a Swiss NGO, the Albert Schweitzer Ecological Centre, attempts to make better use of surplus mango during the growing season. The fruits are fermented but processed according to a local method that retains the flavour of the fruit.

Improved seed is one of the most important innovations for resource-poor farmers. For long, seed supply has generally been a constraint. One major problem is the difficulty of building a sustainable seed market. The implementation of national regulations for seed quality control has proved expensive. To reach resource-poor farmers quickly will require immediate attention by governments to the quantity and quality of new seeds. To demonstrate new seeds, many more simple demonstration plots should be set up, combining fertilizers and improved seed. Well planned and simple in design, they may reduce the role of the currently unsatisfactory agricultural extension services. Selected, qualified farmers should be allowed and supported to carry out local seed multiplication schemes. Such an approach may ensure easier access to improved seed and may also cope better with location-specific aspects of profitable crops.

Seed production has not been very attractive to the donor community, SIDA being one exception. As early as in 1977, support was given to seed production in Mozambique. The state farms were responsible for seed multiplication and a 2,500 ha irrigated rice farm was developed. SIDA contracted the private Swedish Seed Association/Svalöf Seed Company to manage the seed development programme through an independent company, SEMOC. A similar approach was used in Zambia, where large-scale farmers constituted the target group. With a monopoly, and by using aid funds, the company made profits. But the principal question remains, namely how well a private company can serve a family sector in developing countries such as Mozambique. For the immediate future, governments may have to take the responsibility of providing seed of major staple crops for resource-poor farmers. Most likely, the private sector will focus on hybrids and the larger commercial farmers in developing countries, some of whom may adopt GM crops in the future. Over time, and when market prospects and other conditions seem more favourable, the private sector may become more active also in the least developed countries. An interesting initiative is a new investment company, African Agricultural Capital, which will promote the production and marketing of high-quality seed varieties and other agricultural improvements in East Africa. It will provide loans to small companies with a potential to deliver products and services in rural areas. Another new approach is the Alliance for a Green Revolution in Africa, expected to spread some 200 improved crop varieties over a five-year period.

Livestock Research

More than half of the world's people are dependent on keeping livestock for their livelihoods. In addition to providing protein, cattle, sheep and goats play important roles for the environment and in the cultural and social lives of people. Per capita meat consumption in developing countries has increased from 10 kg in the 1960s and is expected to reach 37 kg in 2030 (FAO, 2002b). Two thirds of the increase will be in pork and poultry meat. The volume of meat consumption in developing countries grew nearly three times as much as it did in developed countries during the last three decades (Delgado et al., 1999). In view of managing the natural resources, this trend is interesting. A Western diet based on meat requires some 5,000 litres of water a day, twice that required for a vegetarian diet.

The world must produce over 200 million tons more milk by 2030 than the current levels of consumption. This would be a doubling of the consumption in the 1990s. Globally, more than half of all meat and more than 90 per cent of the milk come from mixed crop-livestock systems (CAST, 1999). This pattern is, however, changing since livestock is getting more based on grain-based industrial systems, now consuming one third of global cereal production. Significant gains in livestock productivity are required and cannot be based on an increase of the number of livestock.

The introduction of artificial insemination in the 1960s led to expectations for a gradual breeding in countries such as Cuba, Kenya and Ethiopia. Local Zebu cattle were crossbreed with Holstein stock. But crossings between local breeds and exotic breeds have had less overall success in developing countries, partly because they also require more feed, concentrates and intensive management. This led to production methods dependent on extra feed, even using imported animal protein in the commercial dairies. Rather, there ought to be research on animal feed that can be produced locally, both in developed and developing countries. Strong mechanisms for surveillance and regulations are lacking for animal diseases and hazards such as BSE in many developing countries. More than 10 of the world's most important animal diseases occur in Africa. Losses due to animal disease in Sub-Saharan Africa are significant, estimated at more than US$ 4 billion annually. The tsetse fly is an old problem, infesting 37 African countries. About 300 million people in Africa, more than one third of the African population, live in tsetse-infested areas. According to the International Livestock Research Institute (ILRI), the disease annually costs livestock producers and consumers about US$ 1.3 billion. So far, research by ILRI and its predecessor, the International Laboratory for Research on Animal Diseases (ILRAD), has not shown progress in finding a vaccine. Recent outbreaks of avian flu illustrate another threat. There are increased risks of infection by avian flu and other diseases in large-scale animal production due to global transport, growth of population and industrial rearing of animals.

For quite some time, policy-makers in developing countries have chosen a seemingly quick option by simply importing cattle from Europe or North America for a rapid change of the genotypes. Most attempts on genetic improvement have centred on increased production and much less on adaptation. A few years ago, purebred dairy cows were imported for milk production to several Asian countries without risk assessment, infrastructure development or even training of farmers (M Khalili, 2003 personal communication). Holstein-Friesian animals for breeding were exported from Sweden by boat to the Middle East in the early 2000s, in total some 2000 heifers. Such an approach has also been used to assist small-scale dairy farmers, for instance Jersey heifers in Byumba, northern Rwanda, or through the Heifer International Livestock Project in South Nyanza, Kenya, to help AIDS-afflicted widows and orphans. This concept neglects the fact that cattle in developing countries face climatic stress, disease, parasites and other endemic features of the local environment. Most farmers have few resources to modernize their farms; indigenous cattle have an inherited ability to withstand these challenges, unlike cattle from temperate regions.

Another research issue relates to the current focus on Holstein-Friesian cattle, now making up about 60 per cent of the European and North American cattle. It has been argued that the genetic diversity within this breed corresponds to that of only 66 animals (Gura and Lpp, 2003). It would be more realistic to advocate an animal breeding policy, comprising both new technology and methods adapted to existing socio-economic systems and cultural values of the people who own the cattle (Payne and Hodges, 1997). An approach based upon adaptation will take much longer but shortcuts may prove hazardous. Thus, this work must be strengthened soon. There are a multitude of local breeds of cattle, sheep and goats whose biological values have to be studied and used in the development of a sustainable livestock sector. In agro-pastoral production systems, the herders are skilled in efficiently using seasonal feed resources in remote areas in harmony between humans, animals and nature. This knowledge must be tapped and used.

Global markets also affect future livestock development in developing countries. As an example, Jamaican dairy production fell during the 1990s since it was cheaper to import powdered milk from Europe to collect local fresh milk (Spore, 2002). Other developing countries are facing similar problems when large companies get involved in strict market prospects. This is counter-productive to domestic milk and livestock production in developing countries. It requires strong government action by lobbying by the developing countries. Its importance was illustrated in the United States during the BSE crisis. When WHO issued a worldwide ban on feeding animal tissues to dairy cows, aiming at minimizing risks to public health, meat producers in the United States, which has the highest per capita beef consumption in the world, were worried. The US Government agency concerned considered that that ban posed major problems for the livestock and rendering industries. The National Cattlemen's Beef Association argued that although the beef industry could find feasible alternatives to feeding rendered animal protein it did not wish to set a precedent of being ruled by "activists".

Biotechnology for Food Security and Food Safety

The Overall Picture

Biotechnology involves the manipulation of biological systems. It is not only a matter of molecular techniques and GMOs, since there are ancient practices of fermentation in brewing or baking. In agriculture, the non-GM technologies include the use of tissue and cell cultures, molecular marker techniques and DNA and immune-diagnostic techniques. However, most recent debates have focused on GMOs. Biotechnology includes the use of naturally occurring microorganisms, such as *Bacillus thuringiensis* (Bt), viruses or endophytes, selected for their activity against certain plant pests and diseases. Genetic engineering refers to the creation of new plant types through genetic modification of the gene pool of an organism by the introduction of non-species-specific genes, often from other taxa or phyla. Genetic modification is a process of improving an organism, for example, enhancing the salt tolerance of a plant, by adding gene complexes from within the species or between species (transgenic engineering). Another type of genetic modification might be the engineering of extinction, for instance in the case of the malarial mosquito (Judson, 2003). The Monsanto Company is now reported to focus more on health aspects. It is trying to develop a GM soybean with reduced levels of saturated fat and other GM crops containing omega-3 poly-unsaturated fatty acid, improving heart health. The Danish company Aresa Biodetection has developed GM plants of *Arabidopsis* sp. that detect explosives in the soil from land mines and unexploded ordnance. Within four to five weeks of growth, the GM plants turn from red to green in the presence of explosives.

As the first cereal, rice was transformed in 1988. Attempts to produce transgenic fish commenced in the mid-1980s. Cloning is seen as a new avenue for agricultural research, although it has failed in nature, except for the nine-banded armadillo. The introduction of genes to cows and sheep for the production of pharmaceuticals in their milk has been slow. Somatic cloning of farm animals has had low efficiency, usually less than 10 per cent. Nonetheless, the situation in the United States in mid-2005 showed that commercial farms may start large-scale production of cloned cows, pigs and sheep. As an example, the Viagen Biotech Company has signed an agreement with the Smithfield food company on cloned pigs. Thus, cloned animals may soon appear on the agenda of the international meat trade, specifically meat exported from the United States. Transgenic research in forestry has developed trees that flower after some years instead of about a hundred years. In 1999, the Worldwide Fund for Nature argued for a temporary global moratorium on GM trees. The standards of the Forest Stewardship Council do not allow the use of GMOs. In mid-2005, the FAO reported that some 35 countries worldwide were involved in genetic modification activities in forestry. Major actors are the United States, France, Canada, India and China. Greenpeace has warned that gene-modified fish was a growing threat to wild species in parts of the world.

The first plant to be fully DNA sequenced was *Arabidopsis*, followed by rice. The whole rice genome of 37,500 genes is now publicly available. In 2004, an international research team reported the complete mapping of the DNA sequence of poplar. The full genetic sequence of the tomato is about to be available from the International *Solanaceae* Genome Project. The full sequencing of maize and potato is to start and there is also a proposal to make the cassava genome a sequencing priority. Efforts to sequence wheat have been initiated by the Wheat Genome Sequencing Consortium in the United States. Several scientific groups recently presented findings about most of the DNA of the hen, with some 20,000 active genes. The Joint Genome Institute of the US Department of Energy plans to sequence sorghum, a work that may provide information also about related crops such as millet and sugarcane. Almost US$ 60 million have recently been awarded in 19 grants by the US National Science Foundation for research on the genomes of economically important crop plants, such as wheat and soybean. In early 2005, the International Rice Blast Genome Consortium reported the complete mapping of the genome of the fungus causing rice blast (*Magnaporthe grisea*).

Biotechnology R&D has been directed to a great variety of plants such as melons, papaya, pepper, sugarcane, sweet potato, alfalfa, tomato and cabbage in public sector trials. About 140 tree genera have been used in biotechnology, although six genera were most common in some 60 per cent of the research activities (*Pinus, Eucalyptus, Picea, Populus, Quercus* and *Acacia*). In India, work on the eggplant has now reached the stage of field trials. On some perennial crops (oil palm, cocoa, coffee, tea and rubber) a marker gene has been inserted as a test of transformation methods. Trees of walnut, citrus, cherry, peach and grape have been genetically modified. Papaya is currently the only fruit tree to have been approved for commercialization in the world. Scientists at the University of California have recently developed GM tobacco plants with elevated levels of vitamins, thus able to tolerate increased levels of ozone in the atmosphere. Research in an African context includes work on the resistance to pests and fungi in papaya, black Sigatoka disease in banana, reduced cyanide content in cassava and increased iron content in maize and rice. In Kenya, GM sweet potato in Kenya was reported to yield smaller tubers than the non-GM varieties and even showed susceptibility to the feathery mottle virus it was designed to avoid. That transgenic variety was imported from the United States. Moreover, weevils were a more serious problem on sweet potato than the feathery mottle virus. The first transgenic groundnut from ICRISAT is carrying genes for resistance to peanut clump virus, ready for approval of large-scale trials in 2005. Drought-resistant GM chickpea is being developed. The ICRISAT pearl millet with high content of beta-carotene aims at improving both crop productivity and helping to fight nutrition-related child blindness.

There is an interesting development with herbicide-resistant sugar beet. In the late 1990s, Syngenta launched a tropical sugar beet in India. Early results indicate approximately the same yield levels in 5 months compared to 12 months for sugarcane. The tropical sugar beet has also been tested in several African countries. It grows in saline soils and requires only half the quantity of water that sugarcane requires. On the other hand, it involves more sophisticated equipment for its cultivation. Because of low labour costs, sugarcane is cheaper to produce. It is doubtful whether a tropical sugar beet crop, requiring mechanical harvesting, can succeed on heavy tropical soils where a sugarcane crop can be easily harvested by mechanical harvesters.

According to the International Service for the Acquisition of Agricultural Biotechonology Applications, there were 90 million ha planted with GM crops in 21 countries in 2005. The growth rate was 11 per cent worldwide, which is slower than in both 2004 (20%) and 2003 (15%). More than one third of this acreage was in developing countries. It was claimed that some 8 million poor subsistence farmers planted GM crops in 2005. At the same time, an activist group, the Friends of the Earth, doubted this figure, making reference to the lack of sources for this information at national level in certain developing countries. In all, there is a significant expansion from about 28 million ha in 1998. The largest producers include the United States, Argentina, Canada and Brazil. In the United States alone, the acreage under GM crops expanded by 11 per cent between 2003 and 2004, reaching some 48 million ha. However, the growth rate was only 5 per cent in 2005. In South Africa, GM crops have been grown on 200,000 ha, reportedly the result of the interest of individual scientists and some donor agencies (Dazie, 2001). In China, transgenic rice varieties have been grown on 200,000 ha. Since 1998, the EU has not approved commercial cultivation of GM crops. After a moratorium of six years, the EU approved import of tinned GM maize food (Bt 11) in 2004. The European Commission has asked some EU governments to end their bans on certain GMOs (such as maize T25, MON810, Bt176 and the oilseed Topas 19/29); otherwise, legal actions may be pursued. The Commission is also drafting stricter rules to prevent the import of US animal feed containing Bt 10 maize.

Soybean, cotton, oil crops and maize are GM crops in large-scale cultivation. In the United States, 85 per cent of all soybean and 76 per cent of cotton were GM crops in 2004. Monsanto has reported that its "Roundup Ready" GM maize represented almost one third of all maize planted in 2005. Maize is by far the crop that has been most extensively tested in field trials. Between 1987 and 2004, the USDA authorized some 47,000 field tests of GM crops. In 2004, Monsanto withdrew its herbicide-resistant GM wheat variety from regulatory tests. It may be released to the US market only when there is a better domestic platform and more public support. Since more than two thirds of the GM crops in the US field tests conducted in 2005 contain secret genes, classified as "confidential business information", they may carry contamination risks, in particular an economic threat to certified organic farms. Field tests are much less frequent in developing countries, where there are often difficulties in testing.

Globally, soybean accounts for two thirds of all protein-rich feeds. Oil crops rank second, followed by cotton, sunflower, fish, peanut and palm kernel. Global soybean production is dominated by Brazil and the United States but production is also common in Argentina. Recent reports from Argentina indicate, however, that overall productivity has fallen by some 10 per cent. It has been argued that when the area under GM soybean tripled, thousands of farmers were forced out of business. In the State of Rio Grande do Sul in Brazil, GM soybean is reported to have responded poorly to the last drought compared to conventional types. Not only has the modified soybean received an alien gene but the genetic system of the plant has been altered; this is acknowledged by Monsanto but without new proteins being observed. It was later found that the DNA of the plant was abnormal and sufficiently large to produce a new protein (Windels et al., 2001). Some 80 per cent of soybean worldwide is used as animal feed, representing about 70 per cent of the total economic value of the crop. Partly as a result of the ban on meat and bone meal, the proportion of soybean in animal feed has increased in many countries.

Most common transgenic traits are resistant to insects and viruses. Genes conferring resistance to insects have been inserted into maize, rice, cotton, potato and soybean. In 1994, bromoxynil cotton was approved as the first herbicide-tolerant crop in the United States, followed by Roundup-tolerant soybean two years later. The first successful insertion of a Bt toxin gene into a plant took place in the United States in 1987 and four years later in rice in Japan. Bt has been widely used in Cuba since the country shifted to chemical-free pest control and lost its privileged access to subsidized markets in Eastern Europe. In 1996, US farmers were the first ones to grow transgenic insect-resistant cultivars containing Bt genes, encoding proteins called delta-endotoxins (Cohen et al., 2000). They are highly toxic to certain insect pests but generally safe to humans. In 1997, the EPA approved the first full registration of Bt maize in the United States. At the same time, the first field tests of Bt cotton were initiated in India. In China, a Bt toxin gene in cotton is reported to have 80 per cent insect resistance.

In West Africa, GM cotton field trials were started in Mali in 2004 subsequent to experiments in Burkina Faso and Senegal. Field tests of two Bt cotton hybrids in the state of Andhra Pradesh, India, have given ambiguous results (Qaim and Zilberman, 2003; Qayam and Sakkhari, 2003). The Bt cotton cultivation was more expensive and gave lower yields. The Bt plants produced more cotton bolls but suffered from premature drying and boll shedding. Since the market value of Bt cotton was lower, the Indian farmers mixed their cotton for the market. In India, cotton production rose during 2004-2005, mainly as a result of the cultivation of Bt cotton according to the authorities and even Monsanto. A survey conducted by a market research company (IMRB International) concluded that Bt cotton yields were 58 per cent higher than those of conventional cotton. Monsanto claims that the Bt cotton farmers can expect increased profits ranging from 30 to 60 per cent. Greenpeace International has argued that Bt cottonseeds are four times as expensive as the conventional seeds and increase yields by just a small percentage. So far, there are few assessments of the socio-economic impact of Bt cotton crops in developing countries.

The basic idea of the first GMOs was to benefit food producers and to enhance foods with vitamins and proteins. GM crops may also be more environmentally friendly and the time required to breed a new variety may become shorter. Nonetheless, genetic engineering research will probably face obstacles. Another factor is the need for the private sector to fund research for the next generation of GM crops. It has also been argued that GM crops may reduce production costs, although the seeds are more expensive. In addition, farmers must buy new seeds every year. Unofficial EU reports indicate that costs for cultivating GM oilseed rape (or canola) might increase by 15-20 per cent and for potato and maize by 5-10 per cent. This trend is confirmed in India. Farmers planting GM seeds would also be at a greater risk from crop losses, since India's emerging Farmers Income Insurance Programme would not cover the risks involved in growing GM crops. In the Philippines, the cost of Bt maize seed is currently twice that of conventional hybrid seed, amounting to about US$ 90 for a bag of 50 kg.

In late 2004, the FAO affirmed for the first time that transgenic crops are promising for poor farmers. It stated that GM products are as safe as conventional ones but require more testing for long-term safety and effects on the environment. However, more than 650 NGOs and about 800 non-government individuals from some 80 countries directly denounced that point of view. Australian scientists reported in late 2005 that insect-resistant GM pea had negative effects on mice lungs. A transferred bean protein, initially a harmless insect poison, had changed after the transformation to the pea plant. The US Academy of Science has concluded there would be no unique risk from genetic engineering. Still, recent reports from India state that some 1,500 sheep have been killed after grazing on harvested Bt cotton fields. According to WHO there have been no negative effects from the consumption of GM foods. Since there is no "zero biological risk" in growing transgenic crops, the precautionary principle is considered simply a "ruse" by opponents (Borlaug, 2004). Sceptics have, however, recommended the development of a database to track the presence in genetically altered plants of new compounds or of natural compounds occurring above healthy levels. Some impacts of GM crops at the global level between 1996 and 2004 have been presented (Brookes and Barfoot, 2005). Farm income benefits from GM crops have been some US$ 5 billion, particularly high in the United States and Argentina. Pesticide sprayings have been reduced by 170 million kg. The cultivation of GM crops reduced greenhouse gases equivalent to that released from 5 million cars. The overall "environmental footprint" of GM crops was found to be 14 per cent smaller than for conventional crops. The per-hectare environmental gains were highest for insect-resistant cotton.

The new tool for research should be used but certain scientific issues or uncertainties have to be seriously considered before common use. Some of them are of a policy nature and require decisions by policy-makers. An introduction of monitoring systems would allow for early detection of long-term problems due to GMOs. Risks can be assessed only through field trials, requiring a national bio-safety regulatory system and research in bio-safety in most countries. Public opinion is vital for a general acceptance of GM food products, calling for a transparent and constructive dialogue between the researchers, policy-makers and the public. Although South Africa is one of the major leaders in biotechnology, the term means very little to four fifths of the general public. Likewise, most people were indifferent to biotechnology according to findings in a South African government survey. Better science communication is a prerequisite, keeping in mind that 24 per cent of the interviewed felt that universities were telling the truth about biotechnology. According to a study from Cornell University in the United States, Americans have become slightly more sceptical over the period 2003 to 2005.

Some Issues

A Complex Modification Process
There are several ways of introducing selected genes into the chromosomes of the target organism, and vectors bring along their own DNA as well as the added genes. The DNA acts only in collaboration with a multitude of protein-based processes. A single gene may produce a number of different proteins but gene complexes also regulate the production of amino acids. Moreover, the proteins transmit genetic information and there are many more proteins than genes in species. This is accentuated by the fact that an introduction of new exotic genes in the chromosomes will be done on unspecified sites. The introduction of a promoter or vector may pose some uncertainty by causing unpredicted effects. Concerns have also been expressed about the new science (The Royal Society, 2003). The view of molecular biologists that DNA is the secret of life is doubtful since life created DNA (Commoner, 1984, 2003). Agents other than DNA contribute

to genetic complexity. The 95 per cent of the human genetic code that is not part of the genes is necessary for mutations. Genes only account for 5 per cent; transposable elements account for 35 per cent of the genome and can protect mechanisms of introduced changes. Since some of the genes used for manufacturing GM foods have not been in the food chain before, WHO has stressed that their introduction may cause changes in the existing genetic make-up of a crop (WHO, 2005). Moreover, there is a need to understand what happens at the level of RNA. Very early in life, genes may be influenced by other external factors modifying their future behaviour. It has even been argued that food-related effects might influence the genetics in succeeding generations. The diet of a grandfather can hit his descendants. The findings of the Human Genome Project have shown there are too few human genes to account for the complexity of our inherited traits by referring only to single genes. Thus, genetic modification of crops may not be specific or precise or fully predictable. Sceptics have argued that genetically engineered crops may represent an uncontrolled experiment with unpredictable outcomes. If this is so, it would be a serious challenge to the legitimacy of the agricultural biotechnology industry.

Precision breeding is a new crop technology recently developed by New Zealand researchers. It is argued to be a very precise transfer of gene sequences responsible for a particular characteristic into a new plant. A similar concept has been applied in Swedish plant breeding by TD Foradling Ltd. since the late 1980s, leading to new varieties entering the Swedish national variety list with Plant Breeders' Rights (T Denward, 2004, personal communication). They include winter wheat ("Henrietta"), spring barley ("Malin") and winter rapeseed ("Hilda"). The method saves time and money. It is not transgenic since it does not contain DNA from unrelated species. In consequence, tests to detect GM crops would no longer be necessary for plants developed under this technique. This would eliminate some concerns of the sceptics of GMOs and might be a future approach for the CGIAR. Another approach is the use of the chloroplast transformation technology for pharmaceutical-producing GM plants. Instead of a conventional introduction of a new gene into plant cell nucleus, the new method involves the introduction of a gene into each of approximately 100 chloroplasts within a plant cell. It is expected to increase output and reduce costs and will not affect plant pollen.

Contamination
Genetically modified organisms and their DNA are self-replicating. A small contamination will therefore quickly spread, giving rise to concern. Genes move from one population of plants to another. If undesirable, exotic genes may be difficult to remove, since crop domestication is based on wild plants. Thus, the main safety issue with GM crops concerns the

environment more than consumers. Perceived risks in the field are the transfer of allergens, introduction of new toxins, and cross-fertilization through pollen passing from one plant to another, that is, gene flow. So far, none of the present biotech products have been implicated in allergic reactions in people. In 1994, US Maxygen produced an *E. coli* bacterium by manipulation that turned out to be almost resistant to antibiotics. Australian researchers have created a deadly version of an otherwise harmless virus by introducing a special gene. In Britain, super weeds have been observed. Most of the hybrids were sterile but about half of them were able to breed. Scientists in the United Kingdom have reported a greater frequency of genetic exchanges between conventional oilseed rape and one of its wild relatives. Gene transfer from GMOs to bacteria is considered highly unlikely.

Smuggling of GM seeds has facilitated their spread in South America and in Punjab, India, and they have spread through food aid, as in Romania. Recent reports by Greenpeace International indicate smuggling of transgenic rice in China. The first report of possible contamination by GM maize of other maize fields caused both concern and debate (Quist and Chapela, 2001). Its DNA was found in wild species in Mexico. CIMMYT scientists denied this was their material. But the private sector had been aware that commercial varieties could cross with landraces. The developments in Mexico had consequences for GM wheat in the United States. In 2001, GM wheat trials were laid out with the intention to grow GM spring wheat commercially in 2003. The GM wheat was to be kept separate from ordinary wheat, an approach that had not worked with the modified maize variety "Starlink". Aimed at animal feed, it appeared in tortilla chips in American supermarkets, the genes moving from one population to another. This experience led both Japan and Egypt to reject the offer of GM wheat from the United States. Ultimately, this kept GM wheat from appearing commercially, at least for the time being. Contamination in the field has generally occurred with maize, cotton and canola seeds. One example from 2005 is reported from Western Australia: low levels of the Monsanto GM canola were detected in two non-GM canola varieties. So far, there are 88 cases of contaminations in 27 countries, according to the online GM Contamination Register by Gene Watch UK and Greenpeace International (www.gmwatch.org). Recent research has shown that fertile GM tobacco plants can produce GM-free pollen, thereby preventing the escape of undesired genes to neighbouring plants through cross-pollination.

Findings by the EPA in the United States show that GM "Roundup Ready" bent grass can pollinate test plants of the same species some 20 km downwind. Wild grasses of the same species were pollinated 15 km away. This was different from previous findings with a distance of only 1.6 km. More pollen reached greater distances and herbicide resistance could spread to wild relatives. The new findings call for more analysis, although the representatives of Monsanto and Scotts have argued that no weed problems had occurred in the past, a similar view maintained by the Weed Science Society of America. In China, more than 1 million GM trees with insect resistance have been planted over the past five years. Experiments in Xinjiang province have revealed that GM genes appeared in non-GM varieties. In contrast to a control system by the Chinese Ministry of Agriculture, there is no GM licensing system in the Chinese State Forestry Bureau, making tracing difficult in these plantations.

In 2005, the EU Agriculture Commissioner requested a study of existing national regulations on GM crops to assist in issuing regulations for acceptable distances between GM and non-GM crops. A Spanish Government decree goes into effect in 2006 requiring farmers to establish a 50 m buffer zone between GM and non-GM crops. Spanish seed producers have found this too strict. In Sweden, official plans of mid-2005 suggest, however, much smaller distances between GM and non-GM crops, e.g. 3 m in potato and 15-50 m for maize varieties.

Crops continuously making Bt may hasten the evolution of insects impervious to the pesticide. As in products of conventional breeding, resistance will ultimately develop also in GM crops. Insect evolution may be of greater concern, so plans to grow non-GM together with GM crops require a large regulatory intervention. Regulations should be similar to those for the introduction and release of other exotic plants and animals. In 2003, UNEP and CropLife International drew up guidelines to address contamination in research and field trials. But the question is how guidelines and draft regulations are applied in practice. When Bt maize was passing the regulatory systems in the United States and Europe it was understood that some 98 per cent was used as animal feed. In Africa, the situation is the reverse since maize is primarily produced for human consumption. In such a context, Bt maize is rarely approved.

Need for a Streamlined International Regulatory System
There is a lack of coordination between the international frameworks on GMOs. International guidance is both inconsistent and conflicting. Some countries are not members of all the international organizations or have not ratified relevant conventions. So far, the United States, Canada, Argentina and Australia are not members of the Cartagena Protocol on Biosafety but belong to Codex Alimentarius. The Cartagena Protocol came into force in late 2003 for the international transfer of living modified organisms. Its ratification has been slow because of a lack of understanding of the risks involved and time required for governments to formulate domestic policies. UNEP has developed elaborated guidelines in biotechnology. At the FAO, there is a draft Code of Conduct on

Biotechnology moving through the procedural channels calling for, among other things, determination of risks and benefits of each individual GMO and a case-by-case approach.

The mandate of Codex Alimentarius is to "protect the health of the consumers and ensuring fair practices in the food trade". In 2003, it produced a first set of international guidelines for assessing and managing health risks posed by GM foods. The "precautionary principle" of the Cartagena Protocol allows a country to ban imports of living modified organisms in case of scientific uncertainty. This is in conflict with the WTO agreement whereby GMO-exporting countries had proposed a plan that would restrict the right of a country to refuse shipments of living modified organisms, easing the restrictions of the Cartagena Protocol. At the third meeting of the Parties to the Cartagena Protocol in early 2006, there was ultimately a consensus decision on documentation requirements for shipments of living modified organisms for food, feed or processing. The significant exporters warned about too many details affecting the global commodities trade. The "precautionary principle" conflicts with the independent WTO requirement that health and safety restrictions should be strictly based on science. Although a separate agreement, it depends on Codex Alimentarius for baseline safety standards. Such standards are promoted by the EU but opposed by the United States. The problem is, however, that crop commodities intended as food or feed may end up as seed for growing crops. In fact, the WTO ought to follow the norms of the Codex, not vice versa. The documentation policy for living modified organisms may not be permanent and its implementation is to be evaluated in 2010 in preparation for a final decision in 2012. This streamlining process may take time. In general, the overall framework of regulations and national requirements in various countries makes it very difficult for target countries in the South to be well informed.

Need for National Policy Coordination
National policy coordination is a major issue to be urgently considered with reference to the Cartegena Protocol. All GM foods need to be assessed for safety prior to marketing approval and analysed for possible long-term effects. This calls for a national policy, a recognition gradually emerging. In the Philippines, the national agricultural biotechnology strategy provides legal grounds for poverty alleviation. In mid-2004, the Government of India initiated work on a national biotechnology policy and a framework for research and business institutions. It may include a body to resolve conflicts over IPRs. It also signed cooperative research agreements on biotech with other countries, for instance the Netherlands, China and the United States. Brazil is planning to set up a national mechanism for the approval of GM crops. In Ethiopia, a World Bank loan is including funds

for a biotechnology institute on crops and animals to become operational in 2006. Malaysia has announced a national biotechnology policy and created the Malaysian Biotechnology Corporation for oversight, although inadequate IPRs remain an obstacle. The Chinese Government has announced similar plans, including a high-level policy committee for national coordination since resources and responsibilities were split among various ministries. Long ago, a Swedish Commission stressed exactly this need and recommended, among other things, a central policy committee to be established within the Swedish Government and led by the Ministry for Foreign Affairs in recognition of the highly political nature of biotechnology (Ds, 1996:73). This was based on the fact that seven ministries, 20 government agencies, the private sector and more than 30 university departments were involved in various aspects of biotechnology research. The principle idea was to strengthen coordination within the government bureaucracy and not to create another government agency. Sweden has national rules for the use of GMOs in laboratories, greenhouses and industrial processes. Others are to be expected on food, feed, seed and forest cultivation. In early 2004, even the Swedish biotech industry found such a national policy committee a necessity to be decided upon by the government. In a recent study by the Swedish Agency for Innovation Systems (Vinnova), there are proposals of a national strategy on biotechnology and the creation of a central council to be composed of all stakeholders but external to the various ministries.

There is a more complicated framework for national regulations in the United States. The USDA regulates field trials through its Animal and Plant Health Inspection Service. The EPA controls pesticides and all GM crops producing toxins but no other effects. No external authority is controlling the GM field trials, a task handled by the commercial companies themselves. They carry out the tests so specific crop data may not always be revealed because of competition. In 2004, however, a US federal court ruled that the USDA must reveal the locations of all GM pharmaceutical crop field trials in the State of Hawaii. Locations of "pharma" crop field trials are not confidential business information. This rule may be expanded to other states. Recently, the USDA has also tightened isolation rules for trials of pharma crops and increased the frequency of field inspections. To prevent accidental contamination, several US organizations recommend the planting of pharma crops in a completely contained facility removed from the food system. The Food and Drug Administration (FDA) regulates food and pharmaceuticals but not meat. So far, the FDA is not making safety assessments of GM foods in the United States, still claiming they are safe for human consumption. With a recent change of policy, the process for approval of new drugs will be shortened. The FDA will work more closely with the pharmaceutical companies. This has raised the issue of its independence, particularly since it is also controlling food security. The US system, relying heavily on the biotech developer, has been questioned by critics who argue that GM crops are not thoroughly tested, regulated and proven safe.

Labelling

Free market choices would require full knowledge of all substances, including those of a biotech nature. This requires transparent rules and a regulatory system for controlling both exports and imports of food products, live animals, feed and animal products. These aspects are political and administrative obstacles but relevant to the overall research process. Complications may arise, as exemplified by the NAFTA agreement. It led Mexico to import from the United States one third of the maize it consumes without any restrictions on GM imports. Since about 40 per cent of US maize is genetically engineered, it has been suggested that maize should be ground when exported to Mexico. This would protect local crop varieties.

Labelling is one method for controlling food products, not only GM foods. Currently, some 40 countries have planned for or have mandatory GM labelling, accounting for one third of the world population. Probably this requirement must remain for quite some time, although disputed by governments with a strong biotech industry as working against free trade. However, labelling would allow for long-term epidemiological studies. This is relevant for the whole food industry, in particular if chemical additives are continuously added. Then, research on long-term, low-level consequences is essential for both human health and the legitimacy of the food processing industry. Both drugs and industrial chemicals threaten future food. According to the Union of Concerned Scientists, contamination of the US food system by pharma crops may have taken place and would become likely in the future. The EU has agreed on requirements for labelling of GM food and grains, except milk, egg and meat from animals being fed GM feed. Labelling is mandatory. Now that the EU has lowered the limit of involuntary contamination to 0.9 per cent, more products must be labelled. In Sweden, the limit will be 0.5 per cent unless the EU has approved the product. Russia has approved some GM food products to be labelled at 5 per cent.

Some people may argue that labels may not be read over time. Labelling may imply claims that there is something to hide. Recently, a US dairy in Maine decided to label its products as free from rBGH. But Monsanto took the owner of the Oakhurst Dairy to court, arguing that labelling would deceive consumers into thinking that one kind of milk is safer than another. Another feature of labelling relates to costs. In the Philippines, a recent study concluded that mandatory GM labelling would result in additional food manufacturing costs of some 10-12 per cent, which will be passed on to consumers. Still, a standard on labelling is another issue to be resolved not only as a component of national biotech policy but also for other chemicals, additives or nanotechnology products under Codex Alimentarius.

Can Transgenic Crops Help Resource-poor Farmers?

With a growing population, many observers have argued that GM foods can help meet the needs of the poor. One example would be "Golden Rice", since 400 million people in developing countries have Vitamin A deficiencies and 3 billion women have iron deficiencies. Three genes from a bean are doubling the iron content of rice. The effect of beta-carotene was achieved through the introduction of four other genes. The work began in 1992 and the final research product includes some 70 patents, given away freely in a humanitarian spirit. Once finally developed, this rice will be made freely available to resource-poor farmers by 2007. According to UPOV-91, this Golden Rice is a plant variety carrying multi-dimensional components. But sceptics argue that a person must consume 3.5 kg of dry matter rice to get the daily Vitamin A content, equivalent to 9 kg boiled rice. Wild and cultivated leafy greens growing along paddy fields and rich in vitamins also serve as a good and cheaper source. Two new GM potato varieties were developed in Scotland in 2004 with beta-carotene content at six times the normal level, higher than in Golden Rice. Moreover, coloured rice may not be accepted in areas where white rice is common. Recently, the Syngenta Company reported it had developed a new strain of Golden Rice, producing more than 23 times as much beta-carotene as previous varieties.

Almost 100 countries are food insecure. It is illusory to believe that future food production can be based on GM-free agriculture in the long-term perspective. Still, no visited farmer in Trinidad, Ethiopia and Sweden had used any GM crop variety up to 2005. More time is necessary for adequate risk assessment, the resolution of scientific uncertainties and the creation of effective national and international regulatory systems. The argument that biotechnology alone can solve world hunger is oversimplified. Improved seed is only one factor of agricultural production. Biotech research on crops grown by the majority of the resource-poor producers has only begun. Three hypothetical scenarios (Rosy Future, Continental Islands and Biotechnology Goes Niche) outlined in a study by the USDA in 2005 conclude all that the poorest developing countries will find themselves marginalized. Much of the benefits and profits of biotechnology over the next decade will go to those having done the research and provided the sophisticated services. A technology-driven approach by the private sector may not suit poor farmers. Small-scale farmers in Argentina could not afford the machine used for direct drilling techniques for soybean. Direct drilling is also said to have led to soybean rust, which appeared in 2001 for the first time. This calls for agricultural research to be close to policy development.

Although very important, biotechnology research is hardly the prime solution in places lacking roads, markets, and agricultural inputs and characterized by depleted soils and millions of resource-poor farmers. Farmers would like improved seed and would be much interested in buying it initially but then would make use of their own for consecutive plantings over some years. With higher incomes, they will gradually begin to purchase seed every year. This will not happen generally in the next one or two decades for most resource-poor farmers. In Africa, a non-existent market is a drawback, as is overall government policy for food production. Other issues are IPRs, a regulatory system for GM food, trade liberalization and institutional safety nets. A concentration on GMOs neglects other relevant research and political avenues that may provide more rapid developments. Resource-poor farmers are quickly in need of less complicated inputs and incentives. Experiences of the SG 2000 in partnerships with 14 sub-Saharan countries show that technology is available to double and triple food crop yields. Nonetheless, incentives are lacking in real life for making use of available technology, even if it is relevant. The efforts during the Asian Green Revolution were successful because they were fully supported by the political leaders. Their governments invested funds in agriculture and agricultural research and so did the donor community. This is in sharp contrast to the current situation in too many developing nations.

Transfer of Power to the Private Sector

Over the last two decades large financial investments have been made in biotechnology research, in particular by the private sector in the United States. Five TNCs are the main financial investors in research on GM crops but governments in larger developing countries also make such investments. In India, public annual spending on GM research is some US$ 15 million and private investment is US$ 10 million. According to the International Service for Agric-Biotech Applications, China invested about US$ 110 million in 1999, a figure expected to have increased fourfold in 2005. In Argentina, five companies (Cargill, Archer Daniels Midland, Bunge, Toepfer and Dreyfus) handle almost 80 per cent of all exports of wheat, maize and soy flour, 95 per cent of the soybean oil and 99 per cent of the sunflower seed oil (Penhue, 2004). With a sustained rate of growth, the world market of GM products may reach a value of some US$ 210 billion by 2014, almost a fivefold increase from US$ 44 billion in 2004. More than in the past, actions by the private sector will determine future technical change in agriculture (Falck Zepeda and Traxler, 1998).

The corporate sector is gaining more power over genes. In early 2001, Syngenta reported the complete genome mapping of rice. Praising this effort, the publicly funded IRRI could only hope that the results would be made available at no cost to subsistence farmers. One year later, rice scientists published their findings on the rice genome in *Science*. But those who wish to have access to the gene sequences must sign an agreement with Syngenta. Furthermore, the completed rice map provides a template for other crops. If Syngenta's pending patent application, in about 100 countries, for ownership of specific rice genes is approved, it may acquire monopoly of a range of plant species. In 2005, Monsanto filed two applications with WIPO for patents on the use of biotechnology to breed herds of pigs and the offspring that result, a broad claim on various genes, traits and methods used to create pigs. Dow AgroScience very recently announced it had won a legal victory against Monsanto over the patent for broad and enabling technology for GM crops. That is the end of a decade-long dispute over a method of transferring Bt genes to plants, a method that has been used for several years. This decision may imply more than just pursuing royalties from competitors using Bt technology.

Some corporations have wished to donate some of their research findings, for example the US Orion Genetics. It donated to public researchers all of its proprietary gene-enriched DNA sequence data of sorghum. Syngenta has donated a substantial part of its *Arabidopsis thaliana* functional genomic seed collection to Ohio State University. Likewise, Monsanto has donated the rights on the Bt potato, resistant to the Colorado beetle, to Russia. Possible profits were to be reinvested in Russian industrial biotechnology. The same potato has been offered to various countries, since it was not becoming a commercial product in the United States.

Patenting remains an area of concern, although the GM industry probably considers the regulations the major obstacle. To fight patent monopoly, India's National Botanical Research Institute developed and licensed two Bt cotton genes to a consortium of seven Indian seed companies. The intention was to break Monsanto's monopoly on GM cotton in India. The legal status of the relevant gene has become questionable. It may fall under the category of "essentially derived varieties", making them ineligible for registration under India's 2001 Plant Variety Protection Act. Monsanto had already co-licensed that gene to nine Indian companies whose products are at different stages of development. Another feature is that Monsanto has been taking part in the drafting of new legislation to regulate the sales of GM seeds, for instance in Argentina.

Another aspect of the power shift relates to less transparency in public debate on GMOs. Monsanto protested vigorously when Arpad Pusztai, an experienced researcher at the Rowett Research Institute in Aberdeen, Scotland, claimed on television in 1998 that rats fed with GM potatoes had reduced growth and were harmed in their immune defence system. He was suspended after more than 35 years of service. Six months later, other studies showed that the immune defence system was affected. Seven years later, Michael Meacher, former Minister of Environment, argued that the UK Government had been very slow in conducting tests on the effects of human health of both GM products and a number of other chemicals people are inhaling or consuming. In the US state of Florida, two reporters on the Fox TV channel were fired when they refused to follow orders by their management to accept demands by Monsanto. They revealed synthetic hormones had been used to improve milk production. Monsanto also threatens US farmers with legal action if they sign pledges but do not use rBGH (Smith, 2003). Since 1997, Monsanto has filed more than 100 lawsuits against US farmers who have grown patented GM seed without permission. Individual scientists have also been threatened, harassed and denied promotions in retaliation for work indicating possible drawbacks from their biotech research.

Plant Protection

Crop Losses

Annual global crop losses have been estimated from time to time (FAO, 1975; Pimentel, 1992; Oerke et al., 1995). Average estimations indicate, quite consistently, losses of about 13 per cent from insect pests, 12 per cent from diseases and 13 per cent from weeds. The IRRI has concluded that weeds reduce rice harvests by an average of 10 to 15 per cent, despite manual weeding by farmers. In addition, there are post-harvest losses. They can be substantial; horticultural commodities, for example, involve losses ranging from 20 to 40 per cent of the produce. Continuous cropping increases pest populations. Ten years of double-cropped rice at IRRI showed losses of one third compared to 13 per cent under a wet season crop (Conway, 1997, p. 209).

It was not until World War II that pesticides came into common agricultural use in the United States. As an innovation, the first synthetic pesticide (dinitro-o-creosol) was marketed as early as in 1892. Today, most pesticides are applied to fruits and vegetables, rice, maize and cotton and soybeans. In developing countries, pesticides are mainly sprayed on high value crops such as rice and potatoes but less on staple food crops. Insecticides are common, cotton being a large consumer. About one third of global pesticide applications are made in Western Europe. Developed countries account for 90 per cent of the use of herbicides with North

America as the largest consumer. Maize and soybean account for almost two thirds. Without chemical weed control in wheat fields in the United States, yields may be reduced by 5 per cent (Yudelman et al., 1998). During the 1990s, the annual increase of sales of herbicides was around 20 per cent, a good market for the private sector. For instance, in the United Kingdom the use of pesticides has increased by 30 per cent in 10 years in spite of declining arable land.

Efficiency and Long-term Effects of Pesticides

Both toxicity and the amount of insecticide use have increased tenfold. New formulations of pesticides have become more acutely toxic at smaller doses. Application rates of the active ingredient per hectare have fallen since 1965 for all agrochemicals. Low persistence leads to less soil contamination but it is not clear whether less persistent pesticides are more harmful to natural enemies than older ones. Application techniques are important, since less than half of a pesticide reaches the pest under ideal conditions and seldom under aerial sprayings. In general, most application techniques are not very effective.

The fundamental question is whether pesticides provide a long-term solution for farmers. Long-term data from US agriculture indicate serious doubts. Crop losses due to insect damage doubled from 7 per cent in the 1940s to 13 per cent at the end of the 1980s (Steingraber, 1997). But in 1950, less than 10 per cent of maize fields were sprayed with pesticides compared to 99 per cent in the early 1990s. Still, losses from insects increased from 6 to 13 per cent and losses to plant pathogens from 10 to 12 per cent, whereas the losses from weeds decreased from 14 to 12 per cent (Pimentel, 1995). Wheat, potato and barley were most severely affected. Similar trends can be discerned at a global level, except for coffee. Increased resistance is part of the answer to why this has happened.

A side effect of the use of agrochemicals is the appearance of secondary pests. They develop as a result of the poisoning of insect predators and parasites. This is highly relevant since insects are also valued as human food in some societies (locusts, grasshoppers, caterpillars, ants, termites). Besides, there are now many reports on polluted water and incidence of cancer due to pesticide residues. Food is contaminated and the pesticides cause human deaths, leading to more than 3 million cases of severe poisoning annually. All these concerns justify much more research on alternatives to pesticides. An international, volunteer Code of Conduct for public and private institutions has gradually regulated trading, testing, registration, viability, packaging, labelling and distribution. It has, however, been less effective in halting the trade of banned pesticides. Products banned in the North are exported to and used in the South. Streamlining of rules leads to inconsistency, for instance on paraquat. Banned in Sweden in 1983, its use was again approved by the EU 20 years later, a decision now to be followed by the Swedish Government. Up to now, the EU has approved 13 pesticides previously not allowed for use in Sweden.

Screening of persistence requires experiments on the long-term consequences of pesticides. Such comprehensive tests of industrial chemicals have been less frequent for a variety of reasons. They have to be simple and preferably founded on biological grounds, as demonstrated by an innovative experiment on picloram residues (Ebbersten, 1983). A winter wheat crop in Sweden was sprayed with Lontrel Combi. The wheat kernels were ground to flour and mixed into soil in pot experiments with pea plants. After growth, the pea plants showed symptoms of herbicide damage. In a parallel test, rape plants did not show symptoms at any doses since rape is not sensitive to dichlorpicolin acid but is sensitive to both MCPA and Mecoprop. These findings in the 1980s led to serious concerns in the Swedish agrochemical industry and to the Federation of Swedish Farmers, which marketed these agrochemicals. After the wheat flour was stored for more than 30 years, the same experiment was repeated, with exactly the same results as in 1980 (Ebbersten, 2003). Screening of other agrochemicals with long persistency using biological methods to lower costs would be highly relevant for agricultural researchers, in the public and private sector.

Resistance: the Other Side of the Coin

Biological organisms normally develop resistance over time and an increased use of pesticides speeds up such a process. In 1938, there were seven insect and mite species known to be resistant to pesticides. Since then, resistance has been developing at a high rate (Table 5). Most insects were of agricultural importance and only 3 per cent were beneficial. Prior to 1970, no weeds were known to be resistant to herbicides. They create secondary effects out of weed species. Shade-tolerant grasses, formerly weeds of minor nature, have emerged as major problems because of the absence of competition. In 2002, Australian farmers observed ryegrass resistant to glyphosate on land previously planted with cereals. In Argentina, weed communities have shown increasing tolerance to glyphosate. In the United States, horseweed is becoming resistant to Roundup. It was first noted in the state of Delaware in 2000 but is now common in 10 other states, including California. The active ingredient, glyphosate, has been used for decades in the United States. Its annual use has increased with the spread of GM crops from 5,000 tons in 1995 to some 45,000 tons in 2005. To date, about 15 weed species (including types of ryegrass, bindweed and goose grass) have been identified as resistant to glyphosate. The weeds have either developed resistance or have some strains with natural resistance.

In view of past trends one can predict that for every decade to come, the number of resistant species will increase with a continued use of pesticides. The reluctance of revisited farmers in Trinidad and Ethiopia to use insecticides seems both rational and sensible. From an environmental point of view, the transgenic crops may be a good development in the short term. It is argued that US farmers using transgenic crops have saved about US$ 100 million in pesticide use between 1995 and 2000. However, Bt cotton will not eliminate pesticide use but reduce it. A widespread use of GM plants containing the Bt gene may increase the general level of exposure to the toxin among pest populations. This will hasten the evolution of resistance also in transgenic crops over time. Resistance to Bt in the diamondback moth has already been reported from Hawaii and Asia. In the United States, at least eight insect species were reported to be resistant to Bt toxins a decade ago (Tabashnik, 1994). Disease resistance may break down if based on one gene only. In fact, Bt rice varieties should have two Bt toxin genes rather than one (Cohen et al., 2000). Bt proteins may also have negative effects on non-target organisms, albeit much less so than broad-spectrum insecticides.

Inserting herbicide-resistant genes in a crop will not per se lead to more sustainable agriculture. A number of other aspects must also be taken care of. Besides, intensified use of herbicide-resistant crops may also promote the appearance of resistant strains of weeds. Every cell of each plant is producing a toxin. If the effect is based on a single gene, as in insects, it may fade out. Resistance in insects is quick to develop, but weed resistance may take longer.

Table 5. Number of species with resistance to agrochemicals between 1938 and 2000.

Period	Insects	Weeds	Fungi and bacteria
1938	7	-	-
1950	< 20	-	-
1960	137	0	-
1977	364	?	?
Mid-1980s	n.a.	50	150
1990	500	?	?
1997	n.a.	273	?
1998	n.a.	About 900 species of insects, weeds and pathogens	
2000	537	> 250	?

Source: Carson, 1962; Georghiou, 1985; Schulten, 1990; Farah, 1994; Steingraber, 1997; Yudelman et al., 1998; UN Millennium Project, 2005c.

More research ought to be given to approaches on how to preserve and make better use of arthropods rather than eliminate them. There is also a need to learn more on the biology of the most harmful insects. This includes their behaviour and response to insecticide proteins, temporal and spatial expression of insect proteins in the plant, strategy for resistance management, impact of insecticide proteins on natural enemies and an effective mechanism to deliver the new technology to resource-poor farmers (Sharma and Ortiz, 2000). Preferably, this work could be conducted in collaboration between the private sector and public research institutions. In short, more advanced models of biological control should be given more research attention with resource-poor farmers as the major target group. This does not imply that all pesticides should be banned but they should not be used as a routine. They are urgently required for emergencies, for instance on locust outbreaks and other heavy insect infestations.

Cultural practices provide biological alternatives, often practised by resource-poor farmers. Improved weed control can be achieved by growing crops in combinations such as bean or sweet potato together with maize. Cover crops and/or agroforestry in new designs are other options. On smaller plots, hand weeding will remain a major approach for millions of resource-poor farmers for many years. Through plant breeding, crop plants may be designed with broader leaves to increase light absorption and shading, thereby inhibiting weed growth.

Another area for more research is on biological measurements of environmental pollution by agrochemicals. Honeybees already perform a vital service of pollination. In addition, they could be used for environmental monitoring. If pollutants are sprayed over an area of some 3-4 km from the apiary, the bees will collect the pollutants also and transport them to the apiary, where they can later be analysed. Such pollutants include chlorinated hydrocarbons, organic compounds such as DDT, PCB and other biocides, some heavy metals, increased carbon dioxide and radioactive isotopes (K Ebbersten, 2004, personal communication). Airborne dust may also be detected by analysis of pollen. Another option would be to use specific plants to trace certain chemicals. In Germany, ryegrass is used to trace sulphur, flour and heavy metals in the air. Other plants include nettles and tobacco, which are being tested in the Brazilian forests.

Integrated Pest Management (IPM)

The original concept of IPM was designed in the 1950s but has been further refined. It looks at each crop and pest situation as a whole. Based upon that analysis, a programme is developed integrating various control methods under the local ecology. The system is complex for farmers and calls for effective methodologies. Besides, an IPM programme requires strong political and financial support from the government. On the whole, there have been few activities to promote IPM programmes on a larger scale. Pesticides have been the simplest solution. In the 1970s, IPM became a powerful tool to control the brown plant hopper on rice in Indonesia. The subsidized pesticide sprayings were stopped since they also killed the natural enemies that usually controlled the brown plant hoppers. The Indonesian Government banned 57 of the 66 pesticides used on rice and reduced the subsidies. Pesticide use was reduced by half between 1987 and 1990 and the savings invested in an IPM programme. The programme in turn led to a reduction in the number of sprayings by 60 per cent and yield increases of 15 per cent. When it became a national programme in the early 1990s, too few farmers could be trained and the quality of the farmer-to-farmer training gradually deteriorated. The FAO has promoted IPM in nine Asian countries, leading to a reduction of insecticide applications in some 8,000 villages. Since its start in 1989, thousands of farmers have been trained on IPM at more than 100,000 Farmers' Field Schools, 90 per cent of them in Asia. In devising research approaches, the International Centre of Insect Physiology and Ecology (ICIPE) has allocated about one third of its budget on IPM strategies for resource-poor farmers. In contrast to insect control, there has been no strategy on integrated weed management until quite recently (Holmes, 1994). It must be both simple and ecologically sound.

Bio-pesticides or "Green Chemicals"

Bio-pesticides do not leave harmful residues, are target specific and promote the growth of natural enemies of pests. The market for bio-pesticides is less than 1 per cent of the total pesticide market. Certain plant products can be useful, for instance, extracts from the neem tree (*Azadirachta indica*), which has been used for centuries in India. Neem extracts are considered the only plant-derived pesticide whose active compounds meet criteria elaborated for promising bio-pesticides (van Lautam and Gerrits, 1991, p. 6). Neem has been effective on some 150 insect species and some species of mites and nematodes. Azadirachtin is the most important chemical component in the neem seed. Acting as an anti-feed, it makes crops unpalatable to pests. Neem degrades fairly rapidly in sunlight. So far, resistance to neem has not been reported. Applications must be repeated frequently in tropical areas. The neem tree is highly suitable for agroforestry. The number and complexity of the compounds in neem extracts may preclude the economic synthesis of the full mixture. Other natural plant compounds are extracted from custard apple, croton oil tree, turmeric, castor oil, Simson weed and chilli pepper (Pretty, 1996). However, more research is required on their massive use and their possible effects on both humans and the environment. Recent findings seem to indicate that neem products are not as safe to natural enemies as once believed. Neem applications for the control of chickpea pests reduced about one third of the natural enemy populations (ICRISAT, 2006). The future search for industrial products as "green chemicals", meaning fully degradable chemicals, would be highly relevant in R&D.

In Arkansas in the United States, rice strains have recently been found to exude a chemical that keeps weeds at bay around their roots. Such weed research indicating allelopathy has also been conducted in Brazil. Considered a controversial subject, it has been given little research attention, although some herbicidal effects have been investigated. Many allele-chemicals affect various enzymes and metabolic processes in higher plants. They also adversely affect the level of growth hormones in plants. Allelopathic interactions in tree-crop association may have more bearing on crop production under agroforestry than on agriculture alone (Rizvi and Rizvi, 1992). This area may benefit from more agricultural research.

Biological Control

The ideal approach is to encourage natural enemies to pests. In ancient China, special persons seem to have been responsible for insect control. The consumption of frogs was not allowed, as they eat insects. In traditional farming there is often a mixture of crops. Diverse cropping systems result in fewer serious pests. In home gardens, the diversity of insects is encouraged. So far, intensification efforts have led to monocropping in opposition to a biological concept.

Biological control of pests means that the problem is solved once and for all. The first case of modern biological control dates from 1888: the introduction of a predatory beetle from Australia for the control of cottony cushion scale infesting citrus crops in California. A recent successful example is the introduction of a parasitic wasp from India to Togo to control the mango mealy bug in West Africa. It took research and implementation over 15 years on biological control of the mealy bug on cassava in Africa. Up to the mid-1990s about 5,000 different introductions of biological control agents into a new ecology were made, according to the Commonwealth Agricultural Bureaux International. They are from almost 100 different countries, half of which are developing countries. However, there are difficulties with the introductions of new species and they require very strict testing. Pheromones constitute another area for more research of great potential. The Future Harvest Centers of the CGIAR have given little attention to pest control, which accounts for less than 10 per cent of their total budget mainly on the use of pesticides. In contrast, ICIPE has had a clear biological focus since inception. Since the private sector has concentrated on pesticides, it would have been more appropriate if much more public funding had been directed to biological approaches, for instance to ICIPE. It would have been consistent with the political rhetoric on sustainability.

Agriculture and Medicine

Even today, two thirds of the world's population rely on the healing power of plants. In Uganda, some 80 per cent of the population use herbal medicine. Such drugs can be quite potent, as illustrated by artemisinins against resistant malaria in China. Based on sweet wormwood, it quickly cures fever and lowers blood-parasite levels. It grows wild in the United States but is cultivated in China, Vietnam and India. As a plant it cannot be patented. Now, chloroquinine is virtually useless against parasites but is still cheap so people continue to buy it. The "njavara" cultivar of rice is traditionally used in ayurveda, an ancient health care system of Kerala, India.

During the last decade, the sales of natural drugs, alternative medicine and vitamins have increased significantly in industrialized countries. Ten years ago, some 2.5 per cent of Americans purchased herbal remedies, a figure that now has increased to more than 15 per cent, equivalent to more than US$ 20 billion in 2003. In Sweden, sales have increased tenfold since 1990. Also the use of non-prescription botanical drugs is rapidly rising. As of 1990, quality standards were introduced in the EU. Simultaneously, there has been declining confidence in the pharmaceutical industry due to increased resistance to drugs for several diseases, among other reasons. Globally, 2 billion people are infected by tuberculosis, most of them in Asia. The disease has become resistant to modern antibiotics. Conventional drug development is expensive and public funding has declined.

Over the years, some 10,000-20,000 plant species have been used in medicine but past approaches of screening techniques have been replaced by more "rational drug design." The pharmaceutical sector finds limited incentive to base drug development on plants, which are considered unreliable and difficult. Breeding is complicated since the pharmacology of most medicinal plants is not fully known. Generally, it takes about 3-5 years to develop a functioning crop management procedure. Pure chemical substances from medicinal herbs have come from a rather small sample of plants. In general, it takes some 10 years to develop a pharmaceutical drug at a cost of more than US$ 1 billion, so the drug companies make their investments in the most profitable areas.

With increasing demand of herbal remedies, declining confidence and increased resistance to drugs, there might be an increasing space for agricultural production of herbal plants and land use. Medicinal plants constitute a new area for future agricultural research. One third of the revisited Ethiopian farmers identified spices and/or medicinal crops as supplementary to food crops. But in times of food shortage vegetables (onions, carrots, potatoes, etc.) were a preferred alternative. Seven farmers in southern Sweden grew 7 ha of chamomile and 4 ha of red sunhat in the early 2000s. Another farmer shifted from conventional crop farming on 100 ha to the production of spices on 10 ha. Tests have also begun with blueberries as a new crop in northern Sweden as a result of increased demand from the pharmaceutical industry.

Another example is senna (*Cassia angustifolia*), a source of laxative. It had good potential in both international and national markets. Most of the ayurvedic and allopathic pharmaceutical companies require senna leaves in bulk. This plant is drought-hardy and withstands harsh climates, such as the arid conditions of India. South African scientists believe that three species from the eastern Cape have potential as a low-cost treatment for diabetes. An interesting discovery was made in the mid-1990s when a mutant natural poppy was found to contain no codeine and morphine but still gave pain relief. According to the Australian Commonwealth Scientific and Research Organization (CSIRO), it is now grown on a large scale for the pharmaceutical industry. It may be a practical alternative to the current poppy production in Afghanistan, estimated at 100,000 ha in 2004 according to the US Central Intelligence Agency. About 7 per cent of the Afghani population rely on poppy for all or part of their income.

Another aspect of medicine of relevance to agriculture is bio-pharming. As of 2003, open-air trials of pharmaceutical crops had taken place in 14 US states. This included more than 300 trials of GM crops to produce fruit-based hepatitis vaccine in tobacco leaves for AIDS drugs. More than two thirds of the plant-based medicines are tested on maize. This is of concern to the food industry, since it raises the question whether the final product is food or a medical drug and highlights the issue of ultimate responsibility within the US regulatory system. So far, it seems as if safety concerns have been stronger in agriculture than in pharmaceutical biotechnology. In 2003, environmental groups argued for a ban on pharmaceutical uses of food crops and demanded that the plants be contained in greenhouses. According to Monsanto, such a ban may set back industry 12-20 years. On the other hand, the production of pharmaceuticals in GM crops is not as commercially successful as previously thought. In late 2005, the Large Scale Biotechnology Corporation in California, using tobacco plants, ceased to operate after almost 20 years. Major reasons were the general controversy over agricultural biotechnology and difficulties in getting approvals for plant-produced drugs from the FDA.

Still another area for future agricultural research is the use of animals to produce pharmaceuticals on special farms. At Genzyme Transgenic, transgenic mice produce Beta interferon for insulin or to be used against multiple sclerosis. The US-based Alexion Pharmaceuticals is one of a few leading companies working with xeno-transplantation, the replacement of organs in humans with organs from another species, such as pigs. To give new organs to all Americans who need them would require an additional

100,000 pigs. In 2004, the FDA announced there would be no requirements to label cloned animal products. This was good news for Smithfield Foods, which sold cows and pigs for xeno-transplantations. Farm animals are not excluded from patenting and technologies can be patented. This market may become quite profitable and may over time become important in other countries.

Fisheries and Aquaculture

Fish is a major source of animal protein for about 1 billion people. It represents about one third of the total protein intake for people in Asia and more than 25 per cent in Sub-Saharan Africa and 10 per cent in Latin America. According to the World Fish Center, Africa is the only region of the world where per capita fish supplies are falling. Its wild fish stocks are on the verge of extinction. In the mid-1960s, annual global catch of fish was 52 million tons. Then, most authorities spoke about a maximum of 80 million tons, though there were also several more optimistic estimates. In 2002, the FAO reported a total capture production from marine areas reaching 93 million tons with 70 per cent from developing countries. This has led to over-fishing. Besides, some 20 per cent of total world fish production consists of small pelagic species used for making fish meal for pigs and poultry and in shrimp and salmon aquaculture. There is a growing demand worldwide for this product, relevant to agricultural production but seldom considered in a systems perspective. In 2000, the worldwide production of fish, crustaceans and molluscs reached around 130 million tons, one fourth coming from aquaculture. Fish consumption is expected to increase by some 57 per cent from 1997 to 2020. In developed countries, there is an expected increased demand of only 4 per cent.

Globally, the percentage of global fisheries over-fished or fished at their biological limit is estimated at 75 per cent. Over-fishing is an old concept coined by the Danish biologist C G L Petersen in 1890. The depletion of trawling grounds was a theme discussed long ago in a report by a Royal Commission in Great Britain in 1900. According to the Chr. Michelsen Institute in Bergen, Norway, annual subsidies to fishery fleets around the North Atlantic Ocean are about US$ 2.5 billion. They are one reason for over-fishing. Fish protein is imported from West Africa or South-East Asia. Although it provides income to some, it is an export of protein from regions deficit in protein. On a more optimistic note, there may also be new species for future catches. In 2004, more than 100 new marine fish species were discovered according to the Ocean Biographic Information System. These findings indicate future potentials but illustrate also a need for further research.

From the beginning, fishing was completely free and large-scale operations began only after World War II. In 1974, the rights to the sea were discussed. Thereafter, several states declared they should apply a zone for their fishery of 200 nautical miles and certain rules replaced free fishing. Subsequently, various international attempts to regulate fishing have been implemented. Fisheries have been exposed to the dynamics of the tragedy of the commons. Private ownership and auctioning of rights can be questioned since groups can control such assets in many developing countries (Shiva, 2002). Thus, there might be an optimal level for catches provided there would be proprietary rights to fish, a problem area hitherto given little attention.

Since the mid-1980s, the annual output of aquaculture has increased by 10 per cent. This is in contrast to capture fishery, which increases by 1.6 per cent per year. African aquaculture accounted for only 1 per cent of the total aquaculture production in 2000. Predictions for 2020 indicate that aquaculture may constitute almost 50 per cent of total fish production, most markets being close to larger cities. It may supplement agricultural production as a means of diversification. Atlantic salmon farming is a special case, currently a US$ 2 billion industry (Montaigne, 2003). According to the Worldwide Fund for Nature, the Atlantic salmon are endangered or threatened in 60 per cent of their range. A genetically homogeneous salmon from aquaculture could be ill suited to life in many rivers because of both old and new diseases.

Dissemination of Agricultural Technology

Over the last four decades, agricultural extension has been a key element in providing new technology to farmers. A great variety of approaches have been tried. Nonetheless, there are good experiences from Vietnam, China, and to some extent also India and Indonesia. China has been successful in farmer empowerment but ought to expect a challenge. The UN Population Bureau predicts that between 2000 and 2030, the rural Chinese population will decline by 300 million people. In general, past agricultural extension approaches have been ineffective and the extension agents stretched too thin. Usually, the best students do not study agronomy and the best ones may not end up in government service for agricultural extension. They are usually paid low salaries and are not always provided with transport, vehicles or even fuel. In Sub-Saharan Africa, the extension services face a low level of staff training compared with their research counterparts.

Other extension approaches have been tested, including the training-and-visit system. The World Bank promoted it heavily over more than 15 years. Although it had a strong field orientation, it became costly and limited in coverage and had excessively supply-driven messages. Another example is the Farmers' Field Schools disseminating knowledge by involving farmers in the decision-making process, in particular for IPM. Under certain conditions, this schooling system has been quite successful. Both the public and private sector must intensify the use and improve the quality of their agricultural radio and television programmes with information about innovations and provide practical instructions for improved practices. Mobile phones could be used for quick messages, for instance on availability and prices of agricultural supplies.

New approaches of disseminating agricultural technology and education are required. They must be decentralized, flexible, suited to local requirements and based on the use of demonstration plots, such as the one spreading to 14 African countries under the joint SG 2000 and the Sasakawa Africa Association. Existing local field organizations can also serve as another option. In particular, voluntary organizations seem to have important experience in working at a grassroots level with limited funds. Often, they manage to reach vulnerable groups, although they sometimes lack necessary technical skills. On the whole, this experience has not been much reviewed. These organizations might also be given some technical training by well-informed research establishments and possibly the agricultural universities. A promising initiative is a mid-career extension programme. Originally designed as a scholarship programme at the University of Cape Coast in 1992, it has changed towards institution building in ten African countries. In total, 921 students have graduated (841 in mid-career BSc and Diploma courses and 80 for higher degrees) (Sasakawa Africa Association, 2006). As of October 2005, almost 600 students were under training.

There is a need to involve religious communities in campaigns on anti-hunger and for the dissemination of agricultural technology. In certain cases, more use can be made of NGOs at the local and national levels for the dissemination of new technology. The NGOs emerged as a possible salvation in the 1980s and their number is now impressive, almost 40,000 internationally. In principle, the involvement of NGOs is good since they have been highlighting critical issues. Too many NGOs lack technical knowledge and have had limited capacity to draw upon. In Ethiopia, there are more than 280 registered NGOs, more than one tenth of them in agriculture. The Ethiopian Hunde Foundation is nowadays a project for credit and studying farmers' needs from their perspectives, a long process. Recent reports indicate that this approach is preferred by the people but disliked by government staff, being too competitive. But hundreds of NGOs at the international level are non-governmental Individuals with a mandate of their own. Maybe their exclusiveness and high profile have come to an end. Some have recently been accused of being complacent and self-interested as well as ineffectual.

For quite some time, contract farming has been applied in livestock and for cash crops involving both the public and private sectors. This pattern will continue with increasing globalization. An efficient agricultural extension system requires competent and well-educated staff with motivation and enthusiasm. Work must be directed to substance rather than planning and reporting. It is critical to learn from farmers about patterns of existing production systems, how they change and their major obstacles to food security. Agricultural extension agents should not work with technology alone but must also have time for dialogue and reflection, particularly in view of the MDGs. Still another issue is to make better use of the local traders as partners to assist the farmers in purchases.

Risk Assessments, Values and Ethics

In modern agriculture, a key issue is cheap food to consumers. Economic factors have predominated, most other things being left out and eloquently classified as externalities. The incidence of BSE highlighted the need to reflect on ethics in agriculture and health risks. Countries have different approaches to risk assessment. In the United States, products are considered safe until proven risky. In France, products are considered risky until proven safe. In other countries, products might be sold as safe even when proven risky. The precautionary principle assumes wrongly that there is no risk in making a decision. In fact, most decisions have often complicated consequences. Policy-makers turn to risk assessment for securing public trust in safety, thereby avoiding actions. However, risk assessment is not a science. It works with incomplete data and has no replications.

In many developing countries, the lack of adequate legislation, weak institutions for correct practices and the implementation of laws about food safety are important bottlenecks in food production. In many instances, traditional values remain as guiding principles for ethics in food and animal production. In new intensive production systems, old standards of right and wrong, as measured before by culture or by law, may no longer be recognized or adhered to. Nevertheless, people may not accept a lack of cultural and ethical values just for the sake of economic profitability. In the old production systems they used resources over which they had full control. Thus, they trust introduced products and feed as safe as their own used to be. This has implications for R&D in terms of globalization and large-scale efforts in reaching the MDGs.

Today, one can argue there is a growing value crisis in so-called developed societies. Criteria such as right or wrong, skills, impartiality, knowledge and experience were all guaranteed by the state in the past. They were also within the principles of the Humboldt university system. The nation-state financed critical thinkers. Their financial dependence on the state was firm, providing intellectual independence. The global economy is changing these values since the agricultural universities are also adjusting to the market. Even if agricultural researchers can think freely today, they still have to get some funding from some external source. They are then hardly neutral or strictly objective in describing reality. They make observations within the traditional philosophy of science and assumptions about reality and use current theories of science.

Long ago, Max Weber argued that science must be separate from values. Results of research can never be predicted so it is impossible to decide whether a scientific project is ethically acceptable with regard to practical applications. The need for ethical concerns has been somewhat highlighted by, among others, the Union of Concerned Scientists. Ecosystems cannot just be repaired; future life-styles have to be compatible with their long-term survival. Some ethical principles have been suggested at the systems level in biodiversity, building adaptability into agro-ecosystems (McNeely, 1999). This calls for an understanding of the politics of community resource management (Colchester, 1994). However, little attention has been given to political and land management systems of local people.

There is a need for better awareness of ecosystem services, a conclusion also reached in the report by the Millennium Ecosystem Assessment in 2005. Such awareness is a long-time practice of both Buddhism and Zen. Ethical values differ at various levels of agricultural production. Farmers may hold other values than scientists and corporations with reference to their ecological setting and history. Then, the threats come from the breakdown of community political systems, allocation of rights and systems of land tenure. In Sub-Saharan Africa, 2 per cent of the land is in the public domain and less than 5 per cent of the land is accurately registered. But investments generally do not require ownership. Other factors are more important, such as small number of private enterprises and weaknesses in the legal systems. Therefore, one should not change customary laws but adjust to them in business operations, as exemplified by the Malian Textile Development Company.

The GDP is not value-free, since it is ruling out a vast amount of unpaid human activity. Many ecological problems are mainly technical, although there are ethical aspects in relation to humankind and nature. Ethical problems are about means and goals, leading to questions about what means are acceptable and which goals are legitimate. Such issues are not value-neutral and require some common ground for all involved partners.

Merely by stating that science should be value-free, a scientist expresses a value judgement. Values inescapably enter into the choices of technology and they cannot be value-neutral, in particular with the MDGs. Rather than getting rid of values in science one has to clarify what values science has or should have.

Ethical concerns are a set of standards by which a special group decides to regulate its behaviour, differentiating what is legitimate or acceptable in pursuit of their aims from what is not. There is no international standard but there are business ethics, medical ethics, and so on. The concept of ethics is loose and variable. Ethics are based on a value system embedded in a culture. They are part of everyday life, starting with the family and ending with TNCs. According to the Oxford English Dictionary, one definition of ethics is "the rules of conduct recognized in certain limited departments of human life". Economics, political science and sociology began as subsets of ethics but left, searching for a "value-free objectivity". Up to now, the specialists have not wanted to be confined in their research by ethical considerations. Although people raise mostly ethical questions when little is known about a subject, they then usually turn more narrowly intellectual. After an improved understanding, new questions again may appear as ethical. Consumer choices have already led to the introduction of terms such as "fair trade" and "eco-labelling".

Ethical aspects are also embedded in agricultural problems for research. Bio-ethics means the consideration of ethical issues raised by questions involving life with reference to developments in medicine and biological sciences. In general, Muslims, Buddhists and Hindus see genetic modification as a "meddling with God's creation". Christians may have fewer problems with biotechnology in case safety standards are upheld and human, plant and animal welfare is safeguarded. Sikhs and Jews hold similar views, though Jews may see genetic modification as necessary in case of live-saving products. But most consumers have not demanded GM food products, even though the producers argue that if consumers do not want them, the products will disappear. The reservations by many consumers may reflect a concealed criticism of the way that food is produced. Many conflicts of ethics stem from a lack of understanding and clarity about the impacts of research, for instance on genetic resources for communities with values different from those of agricultural researchers. This problem may accumulate because agricultural researchers argue they are always objective and do not attach values to their research. Without doubt, humility remains a necessary characteristic for scientists, particularly for those trying to tackle various aspects of food security and food safety.

Few new scientific or technological developments can claim immunity from ethical scrutiny (Reiss and Straughan, 1996). In 1998, the CGIAR adopted ethical principles relating to genetic resources that were revised in 2005. It seems urgent to introduce agricultural ethics in a general sense but also because of the commercialization of science. There is an urgent need for an ethical code for agricultural researchers stating that products of research should not be harmful to farmers, farm animals, consumers or the environment. This need can be illustrated by an example from Toronto University in Canada. When negative consequences of a new product emerged, a scientist was forbidden to reveal them to patients, the public or even the scientific establishment. Her publication was stopped and she was later fired in an attempt to conciliate the firm, thereby rescuing a donation from the company to that university. Since future agriculture will come closer to industry and food processing and be associated with medicine this will require universities to balance ethical principles and commercial interests. In 2003, half of Swedish companies had a policy on ethics. It may now be appropriate to establish a code of ethics for both the global and national food industry and its R&D. Even governments may take initiatives to ensure that bio-ethics is integrated in discussions in national biotechnology policy committees to lead the discussions rather than becoming defensive about new prospects and developments of food production.

One must differentiate between moral and ethical concerns. Moral concerns pertain to whether a course of action is right or wrong and how we feel what is right or wrong. Moral concerns may not, however, have ethical significance. Ethics and morals have been much influenced by religion. In Western Europe, these concepts were rejected over time with the appearance of Niccolo Machiavelli, who coined the concept of a strong nation-state. Since then, social engineering has tried to create a society so superior that there is no need for a person to be good. Many observers will argue that industrialization has contributed to prosperity and modern development with a belief that ethical problems can be solved without ethics. With the declining power of the nation-state, people may wish to turn to humanities and infuse poetry, ethics and contemplation into science and philosophy rather than merely materialism or more consumerism. The next phases of development may include more religious features as was illustrated in discussions with revisited farmers, both Muslims (Ethiopia and Trinidad) travelling to Mecca and Orthodox Christians (Ethiopia). About 1.3 billion people belong to Islam, 2 billion to Christianity, 900 million to Hinduism, and 360 million to Buddhism. To a large majority of people, religion is and will remain a player for a long time in the planetary crisis.

International Agricultural Research through the Consultative Group on International Agricultural Research (CGIAR) and the International Centre of Insect Physiology and Ecology (ICIPE)

A System of International Agricultural Research Centres

Apart from the private sector there are various actors in international agricultural research. The specialized Future Harvest Centers of the CGIAR play a major role. Established in 1971, the CGIAR is one significant actor in publicly funded international agricultural research. It is an alliance of 64 countries, international and regional organizations and private foundations. As a federation of the Future Harvest Centers, 15 CGIAR research institutes focus on global research problems with the primary objective of benefiting developing countries. They employ some 8,500 scientists and support staff working in more than 100 countries with an annual budget of approximately US$ 450 million, an increase from US$ 331 million in 2000. About half of the budget is allocated to activities to benefit Sub-Saharan Africa. Even though the CGIAR is a significant global actor, its budget is only slightly larger than that of a single private agricultural corporation. Above all, the CGIAR budget is small compared to investments in agricultural research conducted by developing countries themselves. In this context, a major issue facing the CGIAR is what exact role it should play in the future.

Prior to the formation of the CGIAR, the Ford and Rockefeller Foundations had established four international agricultural research institutes. Other donors quickly became convinced about a need for a set of international research centres. Between 1960 and 1974, 10 CGIAR institutes were created. In addition, donors agreed to fund three non-CGIAR international research institutes. The trend continued during 1975-1990, when donors created three new CGIAR centres as well as seven research institutes outside the CGIAR. Later, some non-CGIAR institutes were accepted by the CGIAR (ICRAF, ICLARM, IIMI and INIBAP).

In the early 1990s, the CGIAR decided to establish one international forestry research institute (CIFOR). In 1995, ILCA and ILRAD were merged to ILRI with headquarters in Kenya. The International Network for the Improvement of Banana and Plantain (INIBAP) was later incorporated into IPGRI. In 2001, the International Board on Soil Research and Management (IBSRAM) was integrated with IWMI. Two years later, CGIAR decided to dissolve the International Service for National Agricultural Research (ISNAR), transferring some of its activities to IFPRI. Some non-CGIAR institutes were also created such as the International Centre for Rain Forest Conservation and Development in Guyana and the International Center for Biosaline Agriculture (ICBA) in the United Arab Emirates. It screens for salt-tolerant species and techniques for using saline irrigation water in agriculture. Its gene bank contains some 6,000 accessions of salt-tolerant plant species.

Exceptional Growth

The CGIAR began with a modest budget of US$ 20 million in 1972 but reached almost US$ 140 million 10 years later. Great confidence in the CGIAR system and its approach led to an exceptional growth. Priorities were clear with a focus on increased food production in developing countries. Funds were effectively managed and without much bureaucracy. Although the number of donors had doubled, four of them contributed almost half of the total financial resources. Contributions from the CGIAR research institutes prior to and during the Green Revolution led to a higher political recognition of investments in agricultural research in developing countries. When the CGIAR came under fire as a major advocate for the Green Revolution technologies several small donors emphasized the need to focus on resource-poor farmers. That view was not very popular at the rather short and informal annual meetings. A period of consolidation followed but continued growth was anticipated. At a Bellagio meeting in 1986, agroforestry research was considered to be included and new CGIAR entities were to be developed in three major regions of Africa. The ICRISAT Sahelian Center was envisaged to become an independent CGIAR entity, quite a difference from its present status as a mothproof bag. Collaboration with national programmes was to improve. But no directors general attended that meeting and only a few donors were there. Above all, funding had started to decline and the majority of small donors argued for consolidation. Up to this time, the CGIAR was probably considered one of the most effective and productive international activities of development aid.

As a financial crisis loomed in the early 1990s, the Technical Advisory Committee of the CGIAR carried out an extensive study on both an expansion into new research areas and preliminary ideas for restructuring. Another idea was a shift towards a global academy for research on natural resource management. An *ad hoc* CGIAR consultative meeting in 1992 on strategic programme matters avoided discussions on restructuring the system. The committee's work led to the inclusion of research on agroforestry, forestry and fisheries in the early 1990s. At that time, the committee concluded it was hardly justified in funding WARDA at the level required to make a viable rice research institute but rejected an idea of a joint board of trustees of IITA and WARDA.

Table 6. Year of establishment and budgets of major international agricultural research institutes.

Institute	Year of establishment	Budget in millions of US dollars				
		1976	1983	1992	2000	2005
IRRI	1960	9.7	20.2	30.6	32.6	33.4
CIMMYT	1966	8.7	17.5	27.1	39.0	38.8
IITA	1967	9.4	19.9	25.0	30.1	40.2
CIAT	1967	6.3	21.7	26.5	29.5	42.4
ICIPE	1970	1.1	7.2[a]	11.0	10.6	10.1[c]
WARDA/Africa Rice Center	1970	0.8	2.8	6.4	9.4	10.9
AVDRC	1971	2.0	3.6	7.6	11.4	10.0[c]
CIP	1971	4.1	10.1	15.2	20.2	22.0
ICRISAT	1972	6.8	21.0	27.7	23.3	28.4
IBPGR/IPGRI	1974	0.9	3.6	7.1	21.5	34.6
ILRAD/ILCA/ILRI	1973/1974	8.9	19.8	33.7	26.5	32.2
IFPRI	1974		7.8[b]	13.0	11.3	39.7
IFDC	1975	0.8	3.8	12.5	21.2	20.5[c]
ICARDA	1975	1.5	19.7	21.4	23.4	29.1
ICLARM/World Fish Center	1977		1.7	4.0	10.4	15.2
ICRAF/World Agroforestry Center	1977		5.0	15.0	20.7	30.0
ISNAR	1979		3.0	7.7	8.2	-
ICIMOD	1983		1.0[b]	2.7	n.a.	n.a.
IBSRAM	1983		1.7	5.0	-	-
IIMI/IWMI	1984			10.0	8.9	23.1
INIBAP	1984			3.3	-	-
CIFOR	1993				12.6	17.5
ICBA	1999					2.4[c]

Source: Annual reports; CGIAR, 2006. [a]1984. [b]1985. [c]2003.

Attempts at Increased Democracy in a Decade of Stagnating Funds

Because of internal pressure from CGIAR centres and certain donors the necessary process of restructuring was interrupted. Instead, a renewal of the system was launched at a ministerial meeting in Lucerne in 1995. It led to a new vision and strategy of work, stating that the centres should be more focused on poverty eradication, natural resource management and an improved procedure for priority setting. Biotechnology was added to the research agenda. In conjunction with the 25th anniversary of the CGIAR, a global forum for all stakeholders was institutionalized for discussions of global agricultural research. CGIAR membership was extended to more partners from developing countries and even from the private sector. In addition, it was decided that a system review should be conducted. Again, several donors and members of the CGIAR cautioned about its increasing size, since the donors had to ensure matching budget approvals and fund availability by avoiding further special project funding.

The CGIAR Chair also launched a range of new committees in addition to the CGIAR Secretariat. This was a democratic move to empower the members. Comprising 10 different committees, it led to too many unstructured discussions. Many critical issues did not find their way to ultimate decisions at the business meeting. By 1995, the Oversight Committee noted concerns among donors about the number of new committees. Gradually, the growth of the system led to further complexity, more new members, less oversight, stagnating funding, longer annual meetings and a strong desire of all involved to feel full responsibility for all activities. During this period, the Finance Committee was probably providing the most valuable work to the system.

In 1998, the CGIAR again adopted a new mission statement proposed by the CGIAR System Review Panel. It focused on food security and poverty eradication through research, promoting sustainable agricultural development. A recommendation of a central governing board was found too radical but a temporary Consultative Council was set up. Tough decisions about structure and the need for major changes were avoided. Donors did not take full responsibility and the mode of decision-making by consensus did not work very well with dwindling funds. Budget proposals by CGIAR centres in 1999 fell short of US$ 50 million. Furthermore, unrestricted funding decreased from 51 per cent in 1999 to 37 per cent in 2002 but such a trend had slowly begun in the mid-1980s. In its recent Meta-Evaluation, the World Bank has concluded, not surprisingly, that the CGIAR needs unrestricted funds to be capable of operating as a system rather than as a group of independent institutes.

Towards Improved Decision-making

In 2000, a new vision was again adopted: a food-secure world for all. The new mission is to achieve sustainable food security and reduce poverty in developing countries through scientific research and research-related activities in the fields of agriculture, forestry, fisheries, policy and environment. The CGIAR should continue to serve as a catalyst, organizer, coordinator and integrator of global research efforts. The idea of a federation of the CGIAR institutes into Future Harvest Centers was accepted. The ICIPE was to be considered a virtual partner of this new federation.

Several CGIAR members expressed concerns about the implications of the vision strategy for CGIAR governance and structure. The Forum for Agricultural Research in Africa noted that the CGIAR spent 40 per cent of its resources in Africa but with limited success. Ideally, there should be two regional CGIAR institutes in Africa, assuming a time perspective of 40 years, at least. In addition, there should be a limited number of global centres for germplasm and biodiversity, advanced crop and forestry research and policy research. In the Durban Statement of 2001, there was a new political recognition of agriculture. It was seen as the engine for improved African livelihoods and economic development with a target growth rate of 6 per cent per annum. The need for increased funds for the CGIAR was reaffirmed, partly since the World Bank was considering an "exit policy" by reducing its general support to the CGIAR and instead investing in research activities of its own interest. IFAD became a new co-sponsor. The adoption of a more programmatic approach through Challenge Programmes was also seen as one way of getting funding from new sources. The Challenge Programmes were an attempt to attach the work of the CGIAR to global issues such as climate warming, HIV/AIDS, desertification and water scarcity. Other recent changes included the creation of an Executive Council, the transformation of the Technical Advisory Committee into the Science Council and the establishment of one System Office for coherent communication strategy and fund raising.

In 2003, the Syngenta Foundation for Sustainable Agriculture became a new member of the CGIAR. In addition, Israel, Morocco and Malaya joined the CGIAR. The NGO Committee announced a freeze of its membership due to its unease with the CGIAR position on GMOs that biotechnology can solve world hunger. Since then, it has remained dormant. In 2004, the G8 Action Plan encouraged CGIAR to increase its research efforts in Africa and its funding for the Challenge Programmes. It also called the CGIAR to develop more projects with the African Agricultural Technology Foundation. In mid-2005, the Sub-Saharan Africa Task Force presented its report on proposed programmatic and structural alignment in the CGIAR system. Some of its recommendations included, for instance, corporate services, holding board meetings at the same time, providing common library services and aiming for one West and Central Africa Global Entity. The research agenda for 2006 is expanding, requiring investments of almost US$ 490 million, including all the Challenge Programmes.

Even though important steps have been taken for improvements, issues still remain, implying the CGIAR has lost some of its focus and possibly its efficacy. Bureaucracy has grown and so has the competition between centres. Other alarming features are a declining dominance in the work on genetic resources, slow responses to change (environmental issues, biotechnology, IPRs) and high transaction costs. Major attention has been given to more immediate concerns, often through added bureaucratic layers rather than visionary thinking and leadership. The current mix of activities may not reflect CGIAR´s comparative advantage on the international scene towards 2015, and in accordance with the MDGs. Certain developments on the international political agenda have also made the technical work of an association perceived to be non-political much more difficult since UNCED. The move towards a federation has hardly resolved this issue. Likewise, the recent formation of an Alliance Board of the CGIAR-supported centres is probably only a new internal administrative layer.

Challenges and Unresolved Issues for the CGIAR and Its Future Harvest Centers

Contributions to Overall Impact

When the first international agricultural research institutes were once established, the basic idea was that they should produce research results for some 20 years and then be transferred to the national governments. Significant contributions made by IRRI and CIMMYT during the Green Revolution led many donors to view the centres not only as research institutions but even as development agencies. The centres were gradually pushed into consultancies to assist donors in their bilaterally supported research projects of three to five years' duration.

Since its inception, CGIAR funding to the overall research agenda amounts to almost US$ 6.9 billion between 1972 and 2004. Nearly one third of this has come from the World Bank, one third from the United States and 37 per cent from the European members. Developing countries and members with transition economies have provided about 2 per cent. Like other research entities, the research results from the CGIAR are presented in scientific journals and various reports. Since public funds are used for development assistance, there is another legitimate requirement that the research should contribute to problem solving in developing countries. Starting in the late 1970s, questions have been raised about the CGIAR and its impact. This led to an impact study proposed by Sweden in 1982. Impact considerations were not unknown in past tropical agricultural research. In 1911, half the British West Indian sugar acreage was planted with improved varieties, giving 10-25 per cent higher yields. The profit from using the new cane seedlings on one estate in British Guyana more than covered the entire cost of the breeding work in Barbados over 26 years (Knowles, 1928). The introduction of a new sugarcane variety to Barbados in 1948 was calculated to have increased the annual income of the island by US\$ 3.5 million (Masefield, 1972).

In 1985, the Impact Study concluded that most tangible economic benefits of the CGIAR were traced back to improved rice and wheat varieties. Annually, those varieties yielded more than 50 million tons more than the old varieties, sufficient to provide food grain for about 500 million people in the mid-1980s. New maize varieties planted on 6 million ha were beginning to have an impact on food production. Increased productivity accounted for about two thirds and expansion of cropped area for one third. Both urban and rural food consumers had benefited. The CGIAR centres had been instrumental in training thousands of research personnel in developing countries. Another outcome of the study was that annual funding to the CGIAR increased by some 10 per cent over a number of subsequent years.

The work on genetic resources has been the strength of the CGIAR. Therefore, the spread of high-yielding varieties based on germplasm derived from the CGIAR institutes is one useful approach in showing impact. It should, however, be recognized that this spread of new germplasm is also a result of interventions by national government institutions and their staff. Some 2,600 improved varieties of wheat and rice have been released with CGIAR-derived material. Many countries have had close research cooperation with CGIAR institutes, for instance China with more than 250 crop varieties derived from CGIAR genetic resources. In pilot cultivation of the "super hybrid rice" the crop produced 10 tons/ha. This is interesting in a global perspective since three quarters of rice production is still coming from irrigated rice.

High-yielding varieties are now used in Asia on 84 and 74 per cent of the wheat and rice areas, respectively (Borlaug, 2004). These changes are dramatic compared with those of the mid-1970s. Then, only 44 per cent of the total area under wheat and almost 28 per cent under rice in developing countries were planted with high-yielding cultivars. This overall development has helped to preserve more than 300 million ha of forest and grasslands. Incomes of farmers have increased. Had there been no new germplasm from IRRI, rice prices for consumers could have been up to 41 per cent higher and rice-producing nations would have had to be importing some 8 per cent more food (IRRI, 2000). Although world rice production increased to 599 million tons in 2000, an increase of 150 per cent since 1966, the population growth requires a global rice production of 800 million tons by 2025 (Virk et al., 2004). In that scenario, one may question that the CGIAR spending on improving crop productivity declined annually by some 6 per cent in real terms during the 1980s and that of training declined by 1 per cent. In 2004, the CGIAR expenditures on germplasm improvement (16%) and germplasm collection (13%) were low compared to sustainable production (33%), enhancing national agricultural research organizations (21%) and policy research (17%).

Impressive figures of the spread of higher-yielding varieties are mainly confined to IRRI and CIMMYT, although many CGIAR crop institutes have been in operation for three decades. Besides, Africa has largely been excluded, a trend that FAO predicted in its Indicative World Plan in the late 1960s. One exception for Africa is about 100,000 ha of improved NERICA as a result of breeding work by WARDA. Assessment studies by CIAT indicate that both the release and adoption of CIAT-related varieties in Latin America have gone up over the years. Some 100 million ha of improved forage grasses have also been noted in Latin American countries. (Further details of accomplishments by individual centres are given in Annex V.)

On policy research, IFPRI's 2020 Vision has greatly influenced the debate on international food policy and food security. Policy research by CIFOR has contributed to the setting of a global agenda on forestry since the World Bank, FAO, and International Tropical Timber Organization have made use of its data. CIFOR research also influenced the Indonesian Government to become the first country in the world to explicitly include forestry crimes as a predicate offence in its new money laundering law. CIFOR has also been somewhat involved in the discussions on Brazil's forest concession law. So far, neither CIFOR nor ICLARM have attained a status at the annual CGIAR meeting to be asked to present their views on global forestry or fisheries in the way that IFPRI regularly presents its views on agriculture and food security.

IPGRI led the campaign for a mechanism for implementing the goals of the Global Plan of Action and the ITPGR. It played an important role in clarifying technical issues in the FAO Treaty negotiations, although its briefing was mainly to other centres and seldom vice versa. As a research institute, it seems to have been less proactive in providing policy options to maintain and strengthen the integrity of the CGIAR on genetic resources rather than promoting the placing of its *ex situ* collections completely under a political bureaucracy. In view of the MDGs, it would have been useful if IPGRI had been more proactive in searching for and setting priorities on minor crops of special relevance to resource-poor farmers. A reorientation towards a mission to serve the poor was recently elaborated by CIMMYT on wheat and maize (CIMMYT, 2004). The IRRI is also reformulating its strategy to include the rain-fed environments and to focus on the reduction of poverty in Africa and Asia.

224

Although the importance of impact assessment has grown over time, it was given no attention by the CGIAR System Review in 1998. This may explain why that review did not result in any increased funding of the CGIAR. The Impact Assessment and Evaluation Group, created in the mid-1990s, generally failed to add new data on actual accomplishments by the system as a whole. Nor are examples of impact a regular feature in the External Programme and Management Reviews. Annual reports of the CGIAR institutions seldom present an overall picture of accomplishments or impact at the field level, with the recent exception of the 2004 Annual Report of IPGRI.

In 2002, the Standing Panel on Impact Assessment and CIMMYT Economics Program stressed that impact assessment has become more complicated with many more players involved. Still, external partners would greatly benefit if each centre were to devote one page of the annual report to a discussion of impact of its research and specific accomplishments in collaboration with other partners. Recent attempts by the Science Council to introduce a pilot system of performance indicators at the centre level may serve this purpose. At the system level, there ought to be tools for monitoring selected parameters of impact, since it will take time to wait for all centres to present a comprehensive picture, as now proposed by the Science Council. Today, the CGIAR web page exhibits some impact.

The time perspective is important when assessing both impact of research and overall development at the country level. The spread of improved varieties takes time. In the early 1990s, about 90 per cent of the area under wheat in India was planted with improved varieties. That figure was only 26 per cent in 1937/38. Except for sugarcane, all crops presented in Table 7 are on the CGIAR research agenda. The area under high-yielding pearl millet increased sixfold in India in 30 years, now accounting for nearly 6 million ha. The yields of millet more than doubled from 300 to over 700 kg/ha. During this period, the area sown by sorghum decreased from over 18 million ha to about 11 million ha with yield increases by 500 kg/ha, now reaching about 800 kg/ha. About 110 chickpea varieties were released up to 1999 but only three between 1948 and 1970 (Shiyani et al., 2001).

Table 7. Percentage of acreage under improved varieties in India in 1937/38 and the early 1990s.

Crop	1937/38	Early 1990s
Wheat	26%	90%
Rice	5%	68%
Sorghum and millet	1%	53%
Maize	n.a.	42%
Sugarcane	76%	n.a.

Source: Imperial Council of Agricultural Research, 1938; Evenson et al., 1999.

One aspect of impact by the CGIAR relates to the benefits to developed countries. Parent material of durum wheat from CIMMYT was introduced to Italy in the late 1970s. Twenty years later, CIMMYT-derived varieties were grown on about 60 per cent of the Italian wheat acreage. The higher production was estimated at about US$ 300 million per year (Bagnara, 1992). CIMMYT germplasm has also been very important to the United States. In the early 1990s, about 20 per cent of all US wheat land was sown with varieties derived wholly or in part from germplasm developed at CIMMYT (Pardey et al., 1996). It was almost 100 per cent for spring wheat in California. The additional wheat produced in the United States and derived from CIMMYT was estimated at US$ 3.6 billion for the period from 1970 to 1993 (Alston et al., 2001). In recent collaboration with China on canola (rapeseed), Australia has benefited by more than four times the financial investment made by the Australian Centre for International Agricultural Research in canola research.

Globally, chickpea and pigeon pea are the third and fifth most important pulses, respectively. They are mainly grown in developing countries. Chickpea accounts for 10 per cent of the total pulse trade (Joshi et al., 2001). India has been a major importer of these two crops. In recent years, Canada has replaced Australia as the leading chickpea exporter to India. In 1995, Canadian farmers tried chickpea as a new crop on 80 ha, replacing less profitable wheat and canola. Two CGIAR centres provided chickpea germplasm and five years later Canada was the largest exporter of chickpea in the world. In the early 2000s, Canadian farmers planted half a million ha of chickpea in drought areas of eastern Canada, using ICRISAT-derived germplasm.

At the national level, new crop varieties are under steady development. The process of the release of new crop varieties can be rapid. In Tanzania, 16 research agricultural research institutes released 30 new crop varieties between 1999 and 2003 (Table 8). It should, however, be remembered that this is not a sign of impact. In fact, improved crop varieties can be officially released but only spread if farmers wish to adopt them. That requires incentives to farmers as part of government policy to support agriculture.

A Rigorous Review System with Diminishing Influence

The CGIAR has an elaborate system of review mechanisms both for the system itself and individual centres. Long ago, the Technical Advisory Committee introduced a procedure of regular reviews of individual centres, now the External Programme and Management Reviews (EPMRs). They focus on centre performance on research results, quality and relevance of research, vision and strategy and management efficiency. Sometimes, they have been split into one review on programme matters and another on

Table 8. Crop varieties released by Tanzanian agricultural research institutes in 1999-2003.

Crop	Year of release					
	1999	2000	2001	2002	2003	Total
Maize			7			7
Sorghum	1			1		2
Rice		2	1			3
Wheat				1		1
Barley			1			1
Cassava	1			2	1	4
Sweet potato		5				5
Bean				1	1	2
Soybean				1		1
Cowpea					1*	1
Pigeon pea	1			1	1	3

Source: Kullaya, 2005.

management. The reviews have been supplemented with *ad hoc* external reviews commissioned by the centres themselves and with great variations in both quality and utility. On the whole, they have given little attention to board governance, performance and efficiency. This may be due to omission of these aspects in the terms of reference or lack of criteria on which to adequately assess the performance of boards.

There has seldom been lack of scientific understanding on the part of a review panel, though there are variations in the way the individual boards react and operate regarding the recommendations of the reviews. According to the CGIAR Secretariat, reviewing 12 reviews between 1992 and 1998, only six centres had failed to implement, at least partly, one or more recommendations on science and strategy-related recommendations of previous reviews. Eleven recommendations had not been implemented at all, the reasons being budgetary constraints, difference of opinion with the panel, slow action or on-going process.

In general, EPMRs have been well received by the boards of trustees, although there have been exceptions such as the 1997 IPGRI review. It underlined the need for IPGRI to initiate a consultation process with other CGIAR centres to reassess the CGIAR genetic resources conservation responsibilities. Its genetic resource commitment lacked coherence and direction. The review did not approve IPGRI's plan to add a genetic resource policy research unit. This reaction is interesting in view of the CGIAR decision to establish a Genetic Resources Policy Committee (GRPC). The committee infringed upon the responsibilities of the IPGRI board.

The first CIFOR EPMR in 1998 contained recommendations on the need for a more transparent and systematic priority setting process, features actually under debate within the board since the centre began formulating its strategy. Recently, the second CIFOR EPMR, noting the high-quality research, recommended a revision of CIFOR's strategy, vision and goals.

The 1997 ICRISAT EPMR recommended that the strategic and global germplasm should be concentrated in the ICRISAT Asian centre. It should focus on networks and concentrate its research on integrated natural resource management research in Africa, a recommendation rejected by the board. About the same proposal reappeared in the 2003 ICRISAT EPMR, stressing the institute had to be reconstituted with its headquarters in Africa and finding its new vision lacking in specificity and priorities. Again, the ICRISAT board took no strategic decision. Agreeing to all the recommendations, the management concluded that "we are not really listening to any of them", according to an ICRISAT press release of 14 May 2004. Donors to ICRISAT hardly reacted but continued funding research projects of their own choice and under their own control. During a period of seven years, ICRISAT had four directors general and four deputy directors general.

End-of-meeting reports from the CGIAR annual meetings have seldom highlighted critical remarks on the EPMRs expressed by the CGIAR. This is a result of past procedure in which the detailed discussions of EPMRs were delegated to working groups because of shortage of time. Their reports to the business session of the CGIAR were generally approved without discussions. The 2002 ISNAR EPMR and the 2003 ICRISAT EPMR constitute exceptions since external task forces were set up, a sign that the board had not fulfilled its obligations. Ever since its establishment, ISNAR has had great difficulty in defining both its role and ultimate mission so it might have been rational to dissolve it after the 1996 EPMR. Another illustration of slow action by the CGIAR governance system relates to ICRISAT and illustrates the dilemma for the CGIAR system with a system of independent board of trustees. By 1997, the ICRISAT board had mounted its own task force on the rationalization of its locations. An implementation plan for activities in Africa got board approval the following year and ICRISAT was to close down the Niamey West Sahelian Centre. That decision was not implemented. Two years later, the board again requested the management to review the Niamey location. Its future role was of great relevance to ICRISAT's global strategy. Although an exit strategy was presented to the board, it failed to take a decision, despite a declining budget. As of early 2006, neither the CGIAR nor the ICRISAT board had yet resolved the issue.

In the early 2000s, turmoil has continued to surface at the boards. The recent International Potato Center (CIP) EPMR concluded there was need for vision, strategic planning and priority setting, major requirements for any organization. Although the IRRI board was in agreement with all the 10 recommendations by the recent IRRI EPMR, it refrained from drawing conclusions about rice research in Africa and IRRI's future role there. The fifth CIMMYT EPMR reiterated early recommendations on programmatic issues already identified by the fourth EPMR and identified major issues on governance. This illustrates a general need for improved monitoring of the implementation of EPMR recommendations. Secondly, there are major problems with board oversight, slow action within the system and a need for improvements on financial management. Members of the CGIAR can only defer the resolution of issues identified by an EPMR to the respective board, anticipating it will ultimately resolve the issue.

The current procedures can be seriously questioned. The dilemma is accentuated by the fact that donors thoroughly review their restricted project contributions to individual institutes. In addition, the Challenge Programmes will require certain monitoring which may or may not be a task for the CGIAR alone or primarily the sponsors of a Challenge Programme, just like the former "special projects" at CGIAR centres. The EPMRs are demanding of time and funds. Most likely, there are too many reviews and they fail to provide the system with up-to-date information on emerging issues. The independent boards are hardly the most effective mode of future governance in the CGIAR.

Since its inception there have been three reviews of the CGIAR system. The first review in 1976 recommended, among other things, a standing committee to solve problems between annual meetings. The majority of donors rejected this recommendation. The second review in 1981 concentrated on strategy, policies and procedures and had little influence. To many observers, the third system review in 1998 was a great opportunity to resolve major issues that the CGIAR had failed to cope with over quite some time and address new challenges. However, the review panel failed to provide guidance in spite of high costs and much effort. The NGO Committee bluntly stated that there were no recommendations for radical structural changes of the CGIAR. It was a strategic mistake to leave it to the review panel itself to decide the precise terms of reference. Although the panel held a range of meetings with various stakeholders, conclusions from these deliberations were seldom drawn within a contextual framework. Although there was no progress report, after a year of work, all members of the CGIAR remained silent at the Mid-Term Meeting in Brasilia in 1998. The Oversight Committee "did not wish to intervene" on substance. Much later, the CGIAR has taken a few steps to reform governance, for instance the creation of the Executive Council in 2001, similar to the recommendation of a Standing Committee three decades earlier. The council is composed of some 20 members, almost the number of the CGIAR members in the 1970s. Finally, and this is an in-built complication, the representatives of donors did not wish to make real changes, thereby diminishing their own governing power and influence. Most of them had had a long association with the CGIAR, they liked the atmosphere and the dialogues with Centre staff and the influential role they could play by selecting and funding projects. Still, the responsibility for reform lies ultimately with the members of the CGIAR. Even in 2005, the CGIAR lacks legal personality, a persistent weakness for effective governance in the 21st century.

Governance

Leadership is an issue at both the level of the CGIAR system and at individual centres. The Chairman of the CGIAR has always been a Vice President of the World Bank. He has served as a spokesman for the system and chairs the business meetings. In the past, rumours usually carried the message of leadership hiccups at the level of a centre, leading to lack of confidence for that centre but gradually affecting the system. Silence has prevailed with reference to boards of trustees, since problems were not debated to learn lessons. Being governed by independent, autonomous, self-perpetuating boards of trustees, the centres are the only legal entities of the CGIAR. They have overall responsibility, including the hiring and firing of the director general of a centre. They may or may not react to decisions reached by consensus at the annual business meetings.

The creation of the Oversight Committee in 1993 led to great expectations and that committee rapidly got a full agenda with governance issues at ICLARM and IFPRI, followed by ICRISAT, ISNAR, ILRI and CIP in the mid-1990s. The 1996 ISNAR EPMR concluded that the institute should "define and interpret its niche both to guide its own future planning and to improve the understanding of its role by its clients and the donor community". This was a real wake-up call for an institution that had been operational since 1980. However, the Oversight Committee could only defer the problem to the ISNAR board and did the same with ICRISAT, noting that its board had shown "variable performance" in areas of setting vision, strategy and policy and providing financial and management oversight. Its donors were recommended to express their concerns directly to the board and attend board meetings. At ILRI, senior staff resignations led the Oversight Committee to request adequate centre mechanisms for settling of grievances by staff. The first ILRI EPMR in 1999 recommended that the board should focus on policy and oversight and ILRI should revisit "its vision, strategy and priorities and redesign its planning processes to position the Institute at the core of the international animal agricultural research agenda". After five years of operation, this was a less optimistic tune than at its creation, when success with marker-assisted selection in animal genetics was expected within five years. There is another case of slow action by boards. For years, CIMMYT and ICARDA have been recommended by several EPMRs to divide up the responsibilities on wheat research and streamline activities in wheat improvement in Central and West Asia and North Africa. The issue seems to have been finally resolved in 2005.

In general, the boards have failed to anticipate the funding crisis and almost all CGIAR centres had funding difficulties during the 1990s. Even today, certain centres experience difficulties in funding. Restricted funding has increased so the inevitable question is how much this depends on unproductive performance of centres, the directors general, the boards or non-adopted technologies. Recently, a board showed lack of interest or ability to shift a director general after a long tenure. Another centre had a core budget of some 10 per cent of its total budget in 2004. It still remains to be seen whether the new Executive Council can exert more power than the Oversight Committee over the independent boards. This applies also to the follow-ups to EPMRs, an issue raised for discussion by the Executive Council in 2004. Then, the members were asked to submit suggestions to the CGIAR for actions that ought to be taken in response to non-compliance.

The Challenge Programmes are a new feature of the CGIAR. The IWMI was the leader among a total of 18 partners in the development of a five-year Challenge Programme on Water and Food. It started in 2003 with a budget of US$ 100. Harvest Plus is another Challenge Programme on bio-fortification coordinated by IFPRI and CIAT. It has a budget of about US$ 80 million and is financially supported by the US Ministry of Health and a donation by the Bill and Melinda Gates Foundation. Considered a 10-year project, it is seeking partnerships with private and seed and biotech companies as well as national research institutions. Its focus is on remedying micronutrient malnutrition by adding iron, zinc and vitamin A to a variety of crops. In late 2004, the Generation Challenge Programme was launched. It is to propel the use of plant genetic diversity and genomic research to create staple crops to meet the needs of small farmers in developing countries. Emphasis on genomic research may imply more involvement of the private sector as with Harvest Plus and the recent development of GM *Arabidopsis* plants with heightened levels of folate at the US Donald Danforth Plant Science Center. The most recent Challenge Programme is Securing the Future for Africa´s Children, developed by the Forum for Agricultural Research in Africa.

The Challenge Programmes are justified in themselves. As specialized programmes, their governance is most likely facilitated by their having their own directors, unlike the previous CGIAR system-wide programmes. Besides, they are time-bound. They will involve less donor influence and there will be complications between the governance of individual Challenge Programmes and the overall CGIAR system. A new, diffuse layer of decision-making has been added to an already complicated structure. What is the exact governance role of the CGIAR and what role may existing boards play in case their area of expertise is involved? Ultimately, the CGIAR may simply serve as meeting ground for the Challenge Programme sponsors. In fact, the programmes, once established, could easily operate outside the CGIAR system, a reminder of how the non-CGIAR research institutes once came into being. Then, some donors had special interests, and funds.

One aspect of board performance is the annual assessment of the director general and a board evaluation process. In the early 1990s, the Committee of Board Chairs affirmed to the CGIAR that all centres made an annual evaluation of the director general and most boards had an annual process for the evaluation of the chairperson and some process of board self-evaluation. Several EPMRs highlighted, however, that such changes had not been introduced or followed. In 1996, the CGIAR endorsed specific reference guides for the centres and their boards. Nonetheless, some boards have not adopted or implemented these guidelines, IFPRI being a recent example according to the EPMR in 2005. The new approach by the Science Council to strengthen the responsibility of boards in monitoring and evaluation by more self-assessment of programme matters is ideally a good move. Again, implementation depends on the individual boards, requiring decisive action by chairpersons.

More transparency in board operation is necessary. In the early 2000s, a consultant was banned from a meeting organized by the FAO at one CGIAR centre. Some boards have made formal decisions not to allow donors to attend open board meetings to avoid exposure to the difficulties of a centre, a highly counter-productive action. Members of the CGIAR have generally given attention to board composition and the gender issue. The procedure by which the system could approve three members of each board has generally been less effective and is still undergoing revision. Today, the number of female members of the board has increased and there is also a collective balance in the number of members from developing and developed countries. A matter of principle relates to composition, since the CGIAR is to benefit developing countries. Developing countries should then be in a majority. If independent boards of each centre are to remain in the future, they ought to have a three-quarter majority of individuals from developing countries. The expenditure for the operation of the boards seems high, costs reaching 2-3 per cent of the total budget. In contrast, the costs of the board of AstraZeneca, a large pharmaceutical company, amounted to 1 per cent of its turnover in 2004.

In the past, the views of directors general, who are ex-officio members of a board, have largely been ignored at the system level. Not surprisingly, they have rejected a central board that would have reduced both centre and donor autonomy. If the CGIAR is to operate as a system, their individual views and experiences must be accorded more weight. Their new coalition, the Alliance of Future Harvest Centers of the CGIAR, was formed in 2005 and is perceived as "the third pillar of the CGIAR System". It is now leading to the addition of a new administrative layer and an Alliance Executive to head the Future Harvest Alliance Office. In reality, this may be a conservative, internal reaction to necessary changes, giving little guidance on the future direction of the system.

Responses to Environmental Issues, Poverty Reduction and New Challenges

Over the years, most budget allocations in the CGIAR have been given to research on rice, wheat and maize. These research priorities may have been right during the first two decades. But wheat and even maize are also important to developed countries. This ought to influence the future research agenda. As early as in the late 1980s, the Technical Advisory Committee made an effort to include the concept of sustainability in the CGIAR goal statement, a paper that remained a draft. It is unclear whether any individual centres made changes at that time. The present issue is what kind of knowledge is needed and for whom. The Science Panel of the System Review had great difficulty in concluding whether CGIAR science should be poverty-oriented or not. Recently, the CGIAR has decided on an orientation in line with the MDGs. This will require significant changes, as was illustrated in the recent IWMI EPMR. The review noted a shift from primarily knowledge generation through high-quality scientific research to knowledge generation and its application on the ground. The CGIAR must prove that its research can reach the poor: a huge challenge. Agriculture provides employment for about 60 per cent of the working population in Asia. In Sub-Saharan Africa, agricultural production declined by 5 per cent between 1980 and 2001. At the same time, the number of hungry increased by 50 per cent. Some 500 women farmers demonstrated recently outside ICRISAT, demanding that the institution work for them, using their germplasm as a heritage, rather than assist the TNCs (www.ddsindia.com/observation_gwa.htm). These facts must prominently influence the future work of the CGIAR.

So far, the CGIAR has been lagging behind private seed companies and universities in exploiting genetic resources for food and fibre production. While private companies invested some billions of US dollars in agricultural biotechnology in the 1990s, the CGIAR spent some millions. The Private Sector Committee questioned quite early on whether the CGIAR had a credible critical mass of biotech scientists. In one way, the CGIAR's wealth of germplasm might be a treasure for licensing arrangements. It has even been advocated that the CGIAR should register or patent its intellectual property in order to ensure access to what is called public goods. To the private sector, the major concerns were bio-safety, a need for information exchange and dialogue on market segmentation. To some donors, a strong CGIAR involvement in biotechnology has been questionable and the NGO Committee expressed serious reservations about collaboration with the private sector. It was not until 1999 that the CGIAR decided that the Technical Advisory Committee should conduct a system-wide review of plant breeding across the centres regarding both biotechnology and conventional breeding. These discussions led to few actions and vague policies (on IPRs, material transfer agreements, benefit-sharing, cooperation with the private sector). While discussions have continued, progress has been slow. In contrast, several national agricultural research systems in developing countries have moved quickly, for instance India, Brazil and China.

The new Science Council has started an impressive and transparent process in trying to work out the new strategic research priorities between 2005 and 2015. Its model of connecting the research priorities to eight performance indicators at the centre level can be useful. Outputs, outcomes and impact are among the indicators. In reality, this approach may take too long a time to satisfy members of the CGIAR with data of impact in the field and results of their financial investments, particularly if they are to refrain from special project funding. The five priority areas recommended for research are quite on target but those areas are appropriate to any proactive agricultural research organization with an outlook towards the future. They serve better as a preamble and require specifications on areas where the CGIAR can make its unique contributions compared to other actors. A succinct message of future research priorities is important to convey to policy-makers, convincing them about the advantage of the CGIAR. Then, 20 research priorities may be too many (Science Council, 2005). The priority areas are less specific about what the CGIAR is not to do. The new idea of allocating 20 per cent of the financial resources for exploratory, innovative research is attractive. But it is less clear why not all budgetary resources should be given such an orientation, since the Science Council concludes that research by the CGIAR is for development in the long term, not for development per se. Development work is also necessary and probably in great demand by all member countries. A focus on the MDGs and resource-poor farmers means also quick results, that is, by 2015. This may require more research, not less, on aspects such as cultural practices, biological control, IPM and bio-pesticides, features that may become highly relevant in view of an emerging oil crisis. It may also imply less research on wheat but more efforts on the genomics of under-exploited crops, where there is no private sector. A future focus on GM crops will require a different set of institutional arrangements from the current system, more division of labour between the centres and the national agricultural research systems (NARS), recognizing their varying degrees of capabilities in different continents. Finally, there is need to define the division of labour between the work on germplasm by the CGIAR and the agricultural private sector.

Collaboration with National Agricultural Research Systems

The CGIAR centres were originally seen as a focal point for technology transfer to developing countries but a discussion started early on about the nature and content of collaboration with NARS, an issue that has remained active over the years. Some centres initiated research collaboration quite early. For instance, the year 2000 marked the 40th anniversary of collaboration between Thailand and IRRI. Long-term collaborative research efforts were suggested by Sweden by 1979 and based on a SAREC consultation with agricultural scientists from developing countries. Another consultation was arranged as part of the second review of the CGIAR, followed by a round-table dialogue between African leaders of NARS and the CGIAR in 1992. For the first time ever, the Technical Advisory Committee devoted a full session to discussions with representatives of NARS at its meeting in 1994. Then, African agricultural research leaders called for more involvement with the CGIAR institutes. The principal issue concerns both the content of research collaboration and the need to give equal rights to all involved partners to direct the research.

Although discussed at several CGIAR meetings, specific conclusions for the CGIAR and its collaboration have rarely been drawn. One reason is administrative, since members and donors have two budget lines, one for multilateral research and another for bilateral research activities. But the individual boards decide independently for their centres and the recipient NARS have usually to simply accept proposals for cooperation and financial contributions, in separate bilateral negotiations. A change requires hard decisions on the design, implementation, guidelines, time perspective and other aspects. Furthermore, what countries should be targeted, small or large ones, that is, those with weak or strong national research capacity? This in turn requires a transfer of power in decision-making to the level of NARS rather than by the donors and the CGIAR centres. For the future, each of the CGIAR centres should select some 30-40 of the NARS in key countries in the South with reasonable research capacity for long-term collaborative research arrangements with mission-oriented research in a perspective of 15-20 years, thus allowing effective cooperation to solve jointly identified agricultural research problems. The CGIAR centre should withdraw after the set time period and turn to other countries by rotation. Bilateral funding to the country level may supplement this work by the CGIAR. By this step-by-step approach, institutions might genuinely be strengthened over the next three decades. It will require priority setting also on collaborative research arrangements in Africa as compared to the stronger NARS in Asia and even Latin America. A few centres have taken some steps in this direction. For the majority of developing countries, networking, exchange of genetic material and information and other arrangements should be maintained as they are.

The Private Sector and Intellectual Property Rights

Several CGIAR centres have turned proactive to collaborate with the private sector on genetic resources, CIMMYT and IRRI being among the first ones. In India, 13 seed companies are involved in the ICRISAT hybrid sorghum programme and 16 seed producers work with its pearl millet programme. As members of a Hybrid Seed Consortium, the companies pay an annual fee per crop to participate in the breeding programme. The membership provides companies with five years' exclusive access to ICRISAT hybrid parental lines, some 80 ha of land and the gene bank at ICRISAT. The seed companies have pledged financial contributions for breeding disease-resistant, pure pigeon pea lines and hybrids. Cooperation has also been initiated with Bioseed on cotton, a non-mandated crop for ICRISAT.

The debate on IPRs in the CGIAR is of a rather recent origin. In response to a Swedish request for a policy on plant genetic resources in 1980, the Technical Advisory Committee two years later referred to Plant Breeders' Rights as the avenue for a CGIAR policy. At that time, few developing countries were members of UPOV. Six years later, the CGIAR adopted a policy statement that "collections assembled as a result of international collaboration should not become the property of a single nation, but should be held in trust for the use of present and future generations ... and be available to bona fide users." Although IPRs on genetic resources were to be debated at UNCED in 1992, a first working document on IPR policy of the CGIAR was finalized only after intensive discussions during the Mid-Term Meeting in Turkey a few weeks prior to the start of UNCED.

The new scenario set by the CBD led the Technical Advisory Committee to discuss changing responsibilities and roles for plant genetic resources within the CGIAR. In the early 1990s, there was a brief discussion in the Committee of Board Chairs but a position was avoided on the implications of the CBD for the CGIAR centres. The involvement of the private sector in CGIAR affairs elevated the issue of IPRs and patenting. Today, stronger contract enforcement and stronger patent rights lead to larger private investments to R&D, other things being equal (Alfranca and Huffman, 2001). The stripe study commissioned by the Technical Advisory Committee recommended a system-wide programme on genetic resources and recommended that IPGRI become an International Agricultural Genetic

Resource Institute. Although such a proposal violated the autonomy of centres there was a need for unified policy, accountability, visibility and the accommodation of new genetic resources activities such as animal and aquatic organisms. The *ex situ* collections should be held in trust in accordance with CBD but policy responsibilities should rest with an inter-governmental authority. The committee rejected the idea of the International Agricultural Genetic Resource Institute and the directors general suggested a lead centre for a system-wide programme. Then, the committee proposed IPGRI.

A proposal by GRPC on guiding principles on intellectual property and genetic resources was adopted by the CGIAR in 1996. Again, it was an "interim working document". The CGIAR centres should promote access to germplasm and the products of research, recognize sovereign rights of states over their genetic resources and farmers' rights and not claim legal ownership or apply intellectual property to the germplasm they hold in trust and require the recipient to do the same. Intellectual property rights were to be sought for centre research products only when in the best interest of developing countries. Bio-safety guidelines of a collaborating country were to be adhered to. There was no discussion on benefit sharing but IPGRI presented a document on access to plant genetic resources and equitable sharing of benefits outside the official agenda at the FAO Commission of Plant Genetic Resources in Rome in 1996. At the Conference of the Parties of the CBD, it was agreed that a framework for access and benefit sharing, including a "certificate of origin" should be developed by 2006. Another set of IPR guiding principles was proposed by the GRPC in 2000 and for adoption on the last day of the CGIAR meeting. After considerable debate, the principles were rejected and a more general statement was adopted. It was in support of a multilateral approach to obtain germplasm, and all crops within the CGIAR should be part of that system.

During this long process of various negotiations both internationally and within the CGIAR, the private sector kept a relatively low profile. In recent years, GRPC turned into a new international mechanism for discussions of general aspects of IPRs and genetic resources rather than a facilitator to support and find optimal solutions to the CGIAR in the ever-growing complex range of international conventions and agreements. In early 2005, GRPC signed an agreement enabling the CGIAR centres to formally join the ITPGR. Neither GPRC nor IPGRI has given attention to alternatives to conventional gene banks in light of the new technologies to match the private sector work on the storage of gene sequences, DNA printings and DNA libraries.

Gradually, CGIAR centres have decided on an IPR policy, for instance IRRI in 2000. Rice nations were urged by IRRI to prepare for laws on plant variety protection and intellectual property. Also CIMMYT and IPGRI adopted an IPR policy and ICRISAT took a decision in 2001, including a Code of Conduct for interaction with the private sector. An example of a recent move forward is ICRISAT's consultation with the Hybrid Parents Research Consortia in 2005. The objectives were not only to identify research priorities but also to strengthen private sector partnerships, a model claimed to be unique within the CGIAR. In this context, it seems very doubtful that the CGIAR can avoid patenting but ensure that such a system be designed to give maximum benefit and profits to developing countries. Such an indirect outcome most likely emerged from the discussions during a workshop on GM contamination organized by the Science Council and the GPRC in 2004. A year later, GRPC agreed on guiding principles for the development of the Future Harvest Centers' policies to address the possibility of unintentional presence of transgenes in the *ex situ* collections of the CGIAR. This move towards work on GMOs may strengthen a perception that CGIAR science for poverty alleviation would imply that transgenic crops ought to be the quick and ultimate solution. In fact, the genetic fix with improved seed alone is not sufficient. Above all, strong government commitment to agriculture is required together with public support for GMOs.

CGIAR and the International Treaty on Plant Genetic Resources for Food and Agriculture

In the aftermath of UNCED, agreements on genetic resources were signed between the FAO and individual CGIAR centres. The vague term of "trusteeship" relied on good intentions. The centres were to hold designated germplasm "in trust" and in the public domain to serve humanity, and maintain them properly and without IPRs on the genetic material as such. The transfers of germplasm were regulated in standard material transfer agreements. As a result of the new Treaty, the CGIAR is to place its *ex situ* collections listed in Annex I according to Article 15 of the Treaty. This means all samples covered by the CGIAR-FAO Agreement, that is, gene bank samples of a large number of species. New material transfer agreements are to be "consistent with the provisions of the revised Undertaking". An interim committee has managed the process until the new governing body took charge in 2006.

Various aspects of the Treaty cause serious complications to the CGIAR. This is somewhat strange, since a review of the Treaty negotiations concluded that IPGRI "was the leading source of the scientific and technical information that provided the foundations of a multilateral system" (IPGRI, 2002, p. 19). Some of the issues and anomalies are as follows:

- The gene banks of the CGIAR have made possible past improvements in crop productivity and increased production. They remain a great potential for future developments in most plant breeding. Acceptance of the new Treaty by the CGIAR and the individual independent boards of the Future Harvest Centers may lead to uncertainty and more bureaucracy for all partners. For a long time, the CGIAR members have refrained from taking a political stand at annual meetings in favour of the CGIAR Centres. The CGIAR had no chance to negotiate the entry of its collections under the Treaty. They are requested to place all their collections under the Treaty in contrast to countries that are signatories to the Treaty. This may take place during late 2006. Individual countries decide themselves what genetic resources they wish to put under the Treaty and can withdraw with two years' notice. The CGIAR collections in the Treaty are in perpetuity. In reality, the Centre boards have been asked to follow the recommendations by the GRPC and seemingly have lost most of their original independence. This is not conducive to reaching the MDGs, since flexibility is required in the exchange of genetic material. Also, there have already been practical consequences, for instance on future collecting missions by individual Centres and their transparent research collaboration with private sector institutions. A continued exclusion of the private sector from active participation in the actual work of the Treaty may not be conducive to necessary collaborative arrangements. The private sector is already involved in work by several CGIAR centres.
- The Treaty does not discuss new potential crops or plants within agroforestry and forestry, which are critical in a context of land use systems. Fish species are not included. Attention is not paid to important microbes, including mycorrhiza. Some of these areas relate to the private sector and they are all relevant to aspects of future food security. For instance, an Indonesian company dealing with horticulture, forestry and plantation products declared a few years ago that it was in full charge of IPRs on mycorrhiza in the country. Notably, the most important genes for herbicide and insect resistance originate from microorganisms and are therefore not covered by the Treaty.
- In mid-2006, the Governing Body of the Treaty reached consensus on a standard material transfer agreement. Under the former FAO/CGIAR Agreement, the CGIAR centres designated gene bank samples and not centre-improved material. For decades, samples have been freely distributed worldwide as a "public good" from the CGIAR. But the Treaty deals with crop species, not samples. During the negotiations it was unclear whether elite or improved varieties were to be included in the multilateral system, since contracting parties may exclude "materials under development" at the discretion of breeders. In 2004,

the GRPC endorsed "the development of system-wide Material Transfer Agreements for Center-improved materials". It will ensure the extension of the Treaty's benefit-sharing provisions on such material, a point of view explicitly highlighted in the fifth EPMR of IPGRI. Since most major crops for food security have improved genetic material from the CGIAR, this has broad implications, in particular since the collections are to go on for perpetuity. All the Green Revolution varieties of rice, maize and wheat have fed billions of people over the past three decades and all varieties derived from them by national programmes may be subjected to restrictions on their future use. Moreover, the expansion of the use of material transfer agreements for exchange and research collaboration may imply that national rights over genetic resources will eventually be signed over.

- The initial FAO-CGIAR agreement allowed CGIAR Centres to return samples to the supplying country without any restrictive material transfer agreement. According to the Treaty, contracting parties will be exempt from agreements on repatriated samples from the CGIAR. But only some 100 countries are parties to the Treaty, in contrast with the 188 signatories of the CBD, an almost complete coverage. Thus, there are complications for a global CGIAR, in particular since even three host countries of CGIAR institutes are not yet parties to the Treaty: Mexico, Colombia and Nigeria. For example, Colombia will have to sign an agreement for repatriated sample of *Phaseolus polyanthus* from CIAT. Non-ratifiers of the Treaty wishing to breed from repatriated germplasm will have to pay into the multilateral system whenever they use samples in conditions where the Treaty mandates payment to the Treaty fund.

- Article 12.3(d) of the Treaty states that "recipients shall not claim any intellectual property or other rights that limit the facilitated access to the plant genetic resources for food and agriculture, or their genetic parts or components in the form received from the Multilateral System". The wording "in the form received" is open to interpretation. Genetic samples cannot be patented "in the form received". They can be eligible for patenting once a modification has been made. Broad patents claim right over a gene or a gene vector, implying the same effect as patenting the whole plant. Although the United States has recently decided to sign the Treaty, this particular aspect, and others, may cause complications as to its final interpretation to be considered by other members of the Treaty.

- Procedures for benefit sharing have been a very difficult and time-consuming process of the negotiations. Various attempts have been made to decide upon mechanisms of benefit sharing and the GPRC was quick to endorse the material transfer agreements. It has been argued that genetic material under the Treaty and protected by UPOV-style Plant Breeders' Rights will not be subject to mandatory monetary benefit sharing. But this may apply only to UPOV 1978. The Treaty requires compensation except "when a product is available without restriction to others for further research and breeding." Since varieties covered by UPOV 1991 are not without restriction, this implies that countries that have signed UPOV 1991 will have to pay to the multilateral fund if using CGIAR genetic material to be put under UPOV regulations on derived varieties. This would apply to countries such as Australia, the United Kingdom, the United States, Sweden and Germany. In contrast, countries that have signed UPOV 1978 are not forced to pay, for instance Canada, Ireland, France, Norway, all being members of the CGIAR. In mid-2006, the Governing Body of the Treaty concluded initial negotiations on two types of benefit sharing and decided on a model agreement with the CGIAR on material transfer agreements. So far, there is little experience as to how such agreements will work in practice and generate revenues through benefit sharing. The time factor is also highly relevant, since revenues may be expected only after 8-15 years.

A specific example is from Ethiopia, where panicles of local wheat and barley were collected in the late 1960s and made available to both Swedish public and private plant breeders. If the Treaty is to be retroactive, this raises questions about the extent to which this material can be tapped for benefit sharing, who should do this, how and on what basis. How is the private sector to participate in benefit sharing, in particular if material has been forwarded also to others through the mergers of seed companies?

- Prior to the adoption of the Treaty, other activities have taken place at bilateral level. Starting in the late 1990s, five-year agreements have been worked out between individual CGIAR centres and, for instance, the USDA. This move can be seen as a practical approach to secure access to plant material without the bureaucracy through an international treaty. The intention was to formalize collaborative partnership in a symbolic way without financial obligation but to allow free access.

- The discussions on setting up a special fund for maintaining the international gene banks within the CGIAR led to the Global Diversity Conservation Trust, now called the Global Crop Diversity Fund, which was launched at the World Summit on Sustainable Development in 2002. The governments of the United States, Switzerland and Egypt and Ted Turners' UN Fund joined the CGIAR in a US$ 260 million endowment campaign to protect the world's major *ex situ* crop germplasm collections. Other countries have joined, including Sweden in 2004. It is unclear how this fund may relieve the CGIAR centres from budget constraints on their active work on genetic resources and serve as a practical improvement in the context of the current CGIAR governance and the overall administrative and political complexities of the Treaty.

Donor Autonomy

Since the CGIAR is not a formal international organization, donors collectively bear much of the responsibility for the prolonged crisis during the last decade. They expanded its goals but did not provide adequate funding or make appropriate changes in structure to accommodate them. Donors wanted the centres to do their pet projects rather than accept the collective wisdom of well-functioning boards of trustees. Donor priorities change quickly with a constant search for new topics. The consensus model of management was not well suited to take hard decisions on structure. In addition, the independent boards could decide to stick to their own agenda, not always consistent with decisions taken by a majority of members of the system. This approach ruined most attempts to operate as a system. For instance, donors saw system-wide programmes as a convenient way of support and more direct control than via the boards. In contrast, the centres considered such programmes an add-on to their normal work. A new Challenge Programme adds on "independent activities" but probably requires more work for new committees and meetings. Transaction costs are high, leading to less centre authority, and donor autonomy is challenged.

For a few centres, funding deficits remained a permanent feature over many years. To them, the World Bank served as a donor of last resort. It protected the system financially but also concealed problems. The sensitivities were often veiled in silence at the system level, at least for donors not well informed. They could merely show discontent by providing restricted funding or drastically reduce their core funding to such centres. Financial reductions were due to various reasons such as domestic budget cuts or a centre might be found less strategic on poverty reduction, unproductive, performing unsatisfactorily or less competitive than other alternatives. Although the Finance Committee attempted to reveal issues, there was not always a fully transparent analysis. Since the donor of last resort protected the boards, they were rather immune to complaints forwarded to them by individual donors. When the system was smaller, some of this time-consuming activity of budget control was done quite well by the Technical Advisory Committee but was given less priority during the expansion period. Today, the Executive or Science Council must carry out this task. Early signals from the donor community imply already that they have now even less time to follow the work of centres to make their own qualitative assessments. Alternatively, the donors may hand over the full responsibility of fund disbursement to another actor, no longer being accountable as to how their development funds are spent.

Another emerging issue is that members of the CGIAR do not seem to recognize the need for accepting long time horizons for most CGIAR research. One example is the research work on high-lysine maize. It potential was discovered in the 1960s and research was conducted at CIMMYT for many years. In the 1990s, it came into farmers' fields through SG 2000 and the Sasakawa Africa Association. This process took more than 30 years. Therefore, it is unrealistic to expect research results and measure impact on the ground from short-term projects of 3-5 years. This misunderstanding must be rectified at the system level where the Challenge Programmes seem to demonstrate the insight of a long-term approach.

Swedish Policy on the CGIAR

Sweden has been a member of the CGIAR since its inception and provided its first grant in 1973. Total Swedish support to the CGIAR amounts to about SEK 1030 million, mainly as core contributions. Over the years, the annual Swedish contribution has been about 3 per cent of the CGIAR budget, declining to 2.7 per cent in the early 2000s but rising again in 2003. Then, the largest donors (the United States and the World Bank) were each providing some 14 to 13 per cent.

When SAREC was established in 1975, it was requested by the Swedish Government to advise on future support to international research, including the CGIAR. Financial support should be given to selected CGIAR institutes on the basis of merit (Bengtsson, 1977). Funds should be given as core contributions. Swedish policy on the CGIAR was much influenced by early dialogues with partners in developing countries and by ideas on poverty-oriented agricultural research elaborated for WCARRD. This policy was later reiterated in two internal reviews, adding that Sweden should stimulate cooperation between the CGIAR centres and NARS. The participation of developing countries in CGIAR affairs should be increased and more attention given to the impact of the research work by centres at the national level. An external study by Swedish reviewers of Sweden's cooperation with the CGIAR was conducted in 1994 but led to no major changes of the overall policy (Lundgren et al., 1994).

8

Table 9. Swedish financial support to the CGIAR and ICIPE 1978-2005.

Centre	1978	1986	2001		2003		2005	
	Core	Core	Core	Projects	Core	Projects	Core	Projects
CIP	3.0	6.5	6.5		6.9		6.6	
CIAT		1.5	3.4		3.7		3.4	
CIFOR			3.0	1.0	3.2	4.0	2.9	3.0b
CIMMYT			2.5		2.8		2.5	
ICRAF			3.4	12.9	3.7	6.0	3.4	8.5b
ICRISAT	3.4	6.0	4.2		4.4		3.8	
ICARDA	1.5	3.5	4.2		4.5		4.2	
ICLARM			2.2	5.2	3.2	2.0	2.4	
IFPRI			3.5		3.7		3.4	
IIMI/IWMI			2.7		3.0		2.7	
IITA		1.3	3.5		3.7		3.4	
ILRI	2.0a	4.9a	7.5		7.8		7.5	
IRRI	0.3	3.2	3.5		4.0		3.7	
ISNAR		0.1	2.5		2.8		1.5	
IBPGR/IPGRI	1.4	2.0	4.2		4.5		4.2	
WARDA	0.9	2.0	3.2		3.6		3.3	
Challenge Programmes					2.4		1.1	
TOTAL	12.5	31.0	60.0	19.1	67.4	12.0	65.0	11.5
ICIPE	0.8	3.9	8.0		8.0		8.0	
Formas							5.0	

a ILCA and ILRAD. b Funds for 2004.

Starting in the late 1990s, Sweden changed its policy of selecting some centres on merit to a general funding of all centres (Table 9). In addition to core contributions, funds were also given to special projects both by the old SIDA to support activities of its own interest and also by the Department for Research/SAREC of the new Sida. At present, Sweden provides almost SEK 100 million a year: two thirds as core contributions and one third to specific projects. Moreover, Sweden has also decided to support Junior Professional Officers and Bilateral Associate Experts at CGIAR centres to maintain a human resource base in Sweden. Through financial support to Formas, Sida is now stimulating the Swedish scientific community to be more international by participating in the CGIAR work. These new features are rather odd, since the objectives of the CGIAR are primarily to strengthen the participation of national scientists and institutions of poor developing countries. They have much fewer opportunities to participate in research activities than better-equipped Swedish research partners. They could easily get access on their own scientific merits rather than by a "donor push", a feature of the old donor community.

Since 1990, Sweden has more than doubled its financial contribution to the CGIAR. Although there is an increase in the volume of assistance, it is less clear why certain centres have been given Swedish financial support and whether this can be based on merit. This will call for detailed monitoring of the accomplishments of centres and oversight, including a watchful eye on the difficulties facing the CGIAR as a whole, based on an overall Swedish strategy. Time constraints and/or lack of competence at Sida may reside behind a floating idea to provide all future funding to the CGIAR through the Science Council. This will further reduce the responsibility of Sida as a sponsor and weaken its own competence. Though it may ease the bureaucratic workload, it raises the question about accountability and how to professionally decide the future size of Swedish financial contributions. To avoid duplication of effort, an even more rational alternative would be to transfer such responsibility to the EU administration for development assistance, including the CGIAR.

Some Conclusions and Suggestions

The original concept of the CGIAR served very well during the first two decades. The centres were in focus, professionally governed by strong and capable boards of trustees. When the conceptual thinking changed to a CGIAR system it led to ambiguities on responsibilities and accountability. After more than three decades, and increasing bureaucracy, it is critical for the CGIAR to turn proactive and meet challenges in line with the MDGs. A focus on poverty eradication cannot exclude other categories of farmers, so priorities must be clear. The role of a growing global private sector in agriculture and forestry calls for, and justifies, modifications in spite of recent changes for improvement. If the CGIAR is to operate as one system, the members must decide to make it a legal international organization with one central board with affiliated centres and/or Challenge Programmes.

The future CGIAR should be responsible for the coordination, collection, conservation, characterization and dissemination of genetic resources and the development of policies and techniques on improved management of natural resources, including training. Future research must produce options to both resource-poor and more advanced farmers, although the products of research should primarily aim for the target group according to the MDGs. This implies a time perspective of two to three decades, at least. Although the focus ought to be on resource-poor farmers, the management of natural resources is a research issue of global relevance. Emphasis in research ought to change from commodities to system issues of water use for plants and animals towards sustainable agriculture with high productivity.

The responsibility for location-specific research should belong to institutions in developing countries, including the development of new crop varieties. In future, CGIAR centres cannot work in all countries requesting collaboration or in response to demands by donors. Instead, long-term, genuine research collaboration should be established between CGIAR centres and selected NARS. In reality, this means a focus on the weakest countries, where national governments have decided on a poverty reduction strategy and where agriculture and land use are major components of the GDP. The time horizon for collaborative research arrangements should be 15-20 years, with an exit strategy. Although the concept of research partnerships has been in fashion over the past few years, there is a need to better define the incentives to be built into them and what specific outcomes are to be expected. For success of such collaborative efforts, national governments must give priority to agricultural R&D.

Germplasm has been, and may continue to remain, a major asset of the CGIAR. The rice genome has already been mapped, a trend that will continue on other crops. Through the new research tools, gene sequences and mapping and the identification of proteins will be increasingly important. DNA fingerprinting of plants is in place. There should be one strong CGIAR institution to balance the private sector research on those crops, animals, fish and trees that are of relevance to resource-poor farmers. At the same time, it is important to realize that biotech research for this category of farmers in most developing countries is not yet in the field. It will take longer and requires political action by national governments on issues of biotechnology to get rapid practical results. It also calls for support by public opinion.

The transformation of the Federation of Future Harvest Centers into an international organization with one central board and an elected chairperson raises the legal issue of independent centres. The same dilemma will also be a constraint in any follow-up of the programmatic and structural alignment efforts as recommended by the Sub-Saharan Africa Task Force. Only the individual boards can make such changes, but CGIAR members can make this transition take place. As long as the World Bank provides general grants, one of its Vice Presidents may chair the central board. The board should be composed of a group of at most 14 well-qualified individuals, a majority with little involvement in other CGIAR affairs. An executive body should, among others, closely monitor the performance of the subsidiary boards of the centres.

One of the existing centres with the most experience and best equipped for biotechnology research should be developed into one specific CGIAR centre on crop genomics, bio-informatics, IPR and training in biotechnology. Such an International Research Centre on Genomics and Bio-informatics could assist other CGIAR and non-CGIAR centres in the acquisition and development of technologies, diffusion of methods, cloning and related activities. This requires a re-examination of the mandate and future role of

IPGRI. Through contractual arrangements, such a new centre may jointly appoint researchers at other research institutions on commission for special tasks, including those in the private sector. This will save costs, increase efficiency and primarily serve the needs of resource-poor farmers in contrast to the private sector's focus on the larger and highly commercial farmers. CIMMYT and IRRI may be good candidates. Eventually, the creation of such a centre may take place even if there is no central board of the CGIAR. An interesting step forward in this direction is the decision taken by the IRRI and CIMMYT boards in 2005 identifying four priority areas for an alliance to meet the MDGs.

A much less attractive option is to maintain the current CGIAR mode of operation with gradual changes. This includes steps such as joint board meetings or providing common services as recommended by the Sub-Saharan Africa Task Force. A system approach may be maintained only if all donors to the system follow a policy of unrestricted funding to meet 75 per cent of any centre budget. Individual centres should compete on merit but be guaranteed a minimum budget based on the average budget for the last five years. If any of the most productive centres are to be maintained as independent bodies their boards must be strengthened and gradually be composed of well-qualified individuals, the majority of whom coming from developing countries. This means that a few centres have to be modified or closed down. Those with budget deficits over the last five years and showing little impact over the last two decades should be reviewed in depth for possible improvements or closed down. Suggestions for some individual centres are given in the following paragraphs.

- ICRISAT has fulfilled its mission in India. By the early 1990s, Indian private research companies were spending almost as much research funds as the Indian Government on pearl millet and sorghum. ICRISAT's Asian programme should be devolved to the Indian NARS, whereas genetic resources of the ICRISAT gene bank may be under the CGIAR. ICRISAT's major research for the semi-arid tropics should be moved to Africa, possibly to the former ICRISAT Sahelian Centre, or administratively integrated as one component of IITA.

- A focus on the MDGs may go counter to the strong CGIAR focus on wheat research. CIMMYT's wheat programme in Latin America can be focused on germplasm only. In 1997, private companies accounted for some 98 per cent of commercial maize seed in Latin America and 75 per cent was seed of proprietary hybrids whose pedigrees contained CIMMYT-derived germplasm. One alternative might be to fund CIMMYT up to 50 per cent using development funds to serve developing countries and 50 per cent using non-development funds for wheat and maize in industrialized countries.

- Overall responsibility for rice research should be given to IRRI. It can then make contractual arrangements with other centres now working on rice research. The steps to create a West and Central Africa Global Entity out of IITA and WARDA neglect the original nature of WARDA. It was established by the African states but later turned into a CGIAR centre. In the long term, it would be more appropriate to allow WARDA's major tasks to be a responsibility of the West African governments rather than let it merge with IITA. Rice research at WARDA could easily be under guidance by IRRI in the short term.

- Over the years, ICARDA has received some 10 per cent of its funding from Middle East countries, IFAD and OPEC, but almost one third from the United States. CGIAR could reduce core contributions, assuming an increased demand from the Middle East region. The CGIAR should confine its core support to match financial contributions from the region.

- Whereas the World Agroforestry Center (ICRAF) deals with genetic resources in agroforestry, this has not been a task for CIFOR. Policy aspects of forest genetic resources ought to be included in the research agenda of CIFOR.

- A focus on the MDGs will require more geographical focus (in Africa, Asia, Latin America, and the Middle East) by some of the existing CGIAR Centres that do not work exclusively on global research agendas. Thus, they may be expected to develop more into R&D institutions, as many donors have perceived them to be for quite some time.

- Fruits and vegetables account for a rapidly expanding sector of production and trade in developing countries, providing cash income to farmers. It is therefore a positive change to note a proposal by the Science Council in early 2005 to include these commodities among the research priorities. This ought to lead to an integration of the Asian Vegetable Research and Development Center into the CGIAR. Other areas for more research are on orphan crops and a search for other potential crops not only to suit demands by resource-poor farmers but also for animal feed and to broaden our base for future food security. Another feature for increased research attention by the CGIAR is on long-term food safety with reference to chemical pollution of water, soils and additives to food products, possible contaminations in GM food products and pandemics.

- If donor's rhetoric is to turn into action for safer plant protection to humans and the environment, more public funds ought to be given to research on alternatives to pesticides. This is not a focus of the private sector and has not been given adequate attention within the CGIAR.

Therefore, the ICIPE should become a full partner of a global research, given the responsibility of coordinating research programmes on IPM, bio-pesticides, pheromones and other subjects. Its stagnant budget of the 1990s ought to be doubled in the next decade, a target that can be met by donors even if the ICIPE is to be kept outside the CGIAR.

Some Issues in University Teaching and Future Role of Agricultural Universities

Growth of Universities and University Teaching in Agriculture

Higher learning is of old date. One example is the library in Alexandria, established more than 2000 years ago. In Europe, the University of Bologna is regarded as the oldest university, although it became a "studium generale" almost one century after its traditional year of birth (1088). The University of Paris was established in 1210, whereas those of Oxford and Cambridge came later. So did the Swedish universities of Uppsala (1477), Copenhagen (1479) and Lund (1666). The overall intellectual objective in university teaching was to gain wisdom and to understand other cultures and their societal institutions. In Europe, examinations and certificates appeared in the late 1700s. Special institutions were established for training professionals. Other sponsors outside the church founded a number of academies.

The creation of university departments gave stability to emergent sciences. Over time, they also became obstacles. New ones had to be formed to facilitate change. Some research areas became redundant but were not abolished. Gradually universities became petrified. In the early 19th century, universities regained their power, in particular in the von Humboldt University in Berlin, where teaching was combined with research. This attracted many US doctoral students. This new philosophy of teaching and research also led to the foundation of two private US universities in 1892. Scientific independence became important. These developments in the United States influenced not only the research work but also non-US universities. The term *public service* was coined, giving rise to both diversity and a pressure for conformity. In turn, this led to the emergence of different schools, independent research institutes or even think tanks. Today, there are public and private universities throughout the world.

Prior to the 1960s, the research capacity of developing countries was chiefly strengthened through financial support to existing tropical research institutes, through scholarships and university contracts, particularly for the agricultural colleges. The contractual arrangements remained for several years, for instance under Title XII of the US International Development and Food Assistance Act of 1975. Over time, these college contracts resulted in a great number of agricultural graduates who could take part in national agricultural development. Gradually, that infrastructure for agricultural education was taken over by the national governments. On the negative side, such contracts became obstacles in delaying the expansion of the national institutional capacity. In the 1980s, institutional linkages and research collaboration were again suggested as an instrument for capacity building of research in developing countries.

In contrast to general academic teaching, agricultural universities have a shorter history. The Swedish University of Agricultural Sciences (SLU) was formally established in 1976 as a merger of three colleges of agriculture, forestry and veterinary medicine. However, higher education in these subject matters had started in 1849, 1828 and 1775, respectively. In the early 2000s, SLU had some 3,300 undergraduates and almost 200 professors. This is a dramatic change since 1937, when the College of Agriculture had 13 professors and 94 undergraduates. Since then, the farming population has declined from one third to a small percentage.

In 1920, a British Tropical Agricultural College Committee recommended the establishment of a college in Trinidad to create a body of British expert agriculturists for the tropics and of scientific advisers in tropical agriculture. It was converted into the Imperial College of Tropical Agriculture in 1923. It was to carry out long-range research applicable throughout the tropical dependencies of the British Empire and focused on banana, cocoa, sugarcane and citrus. In 1960, it merged with the University College of the West Indies, located in Jamaica, forming a Faculty of Agriculture. In 1975, the regional Caribbean Agricultural Research and Development Institute was to work in close collaboration with the Faculty of Agriculture of the UWI, applying a market-driven approach to research. At present, the UWI has three campuses located at Mona in Jamaica, Cave Hill in Barbados and St Augustine in Trinidad and serving 14 West Indian territories. In the early 2000s, the St Augustine campus accommodates some 7,600 students in seven faculties. Its Faculty of Agriculture and Natural Sciences has six departments with 83 academic staff members. Current plans deal with ideas of incorporating a School of Agriculture into the new faculty.

In Ethiopia, Addis Ababa University was established in 1950. Three years later, the Alemaya College of Agriculture started up, based on a contractual agreement between Ethiopia and the United States. BSc classes in agriculture started in 1956. Its agricultural extension activities were transferred to the Ministry of Agriculture in 1963. The college produced only Masters students up to 1978/79. It later became autonomous as the

Alemaya Agricultural University, thereby incorporating the Debre Zeit Agricultural Research Station. In 1988/89, it launched its own MSc and PhD programme in animal breeding. Alemaya University has recently expanded to include both medical and social sciences, comprising some 660 students. In the early 2000s, its Faculty of Agriculture had 231 undergraduates and 47 postgraduates (Alemaya University, 2003, p. 40).

Some Issues for Universities Serving Future Society in Agriculture and Land Use

Focus on Higher Education

Today, most universities have become educational facilities to a large extent. The universities have expanded in number and in size, offering education for more students, sometimes to cope with growing unemployment. Teaching has become a priority particularly in developing countries. A decade ago, the African Regional Post-Graduate Programme in Insect Science concluded deteriorating quality of education at African universities. Overcrowded universities led to poor facilities, impoverished research programmes and overproduction of graduates, including graduates in the "wrong" subjects. In Kenya, 40 per cent of all schoolteachers were untrained and there was a shortage of medical doctors. The Kenyan Minister of Livestock Development predicted that graduates other than those in medicine, veterinary medicine, agronomy and engineering might face unemployment. Also, higher education in forestry is given low political priority, lacks funding, and attracts too few students, according to recent reports by FAO, IUCN and ICRAF. The number of plant physiologists, plant breeders, taxonomist and foresters is declining. The lack of institutional capacities and declining numbers of plant breeders has led to a new, global initiative to be launched in 2006 by the Governing Body of the ITPGR. It aims for a strengthening of the capacity of developing countries to develop better plant breeding and seed delivery systems.

With a population of 120 million, Nigeria has 40 universities and more than 40 polytechnics. Still, its educational system does not provide skills and capacity to solve the problems of development, according to the Nigerian Minister of Science and Technology (Ohadoma, 2002). The number of students has increased tenfold but the standard has fallen to unacceptable levels. Excellence and output have been lost. Curricula must be made more relevant to agriculture in transition from subsistence to commercial production. African universities must also become more flexible in accommodating mid-career professionals from extension and research. It is vital that they move away from foreign, often inappropriate, postgraduate training and build up local capacity. This is long overdue and in many cases the fault of donor agencies. In India, agricultural higher education ought to be examined since more flexibility is needed in the existing course curricula. It must get "out of its mould of formal, organized, rigid framework" and accept a role of continuing education, adjusted to the needs of target farmers (Paroda, 2000). Students should become "locally relevant and globally competitive".

In industrialized countries, the role in serving the agricultural sector has diminished in pace with a declining farming population. Nevertheless, the number of new courses has increased and so has the number of Swedish students at SLU. Between 1993 and 2004, their number increased by 70 per cent. The Swedish postgraduate students and teachers are collaborating in research work for a PhD. The success of a university depends on the quality of the high school students, a prerequisite for the Humboldt model. Although the objectives are formulated according to that model, the actual implementation is difficult. The task of a typical Swedish teacher may too often be to find out what individual students are good at rather than teaching them things they do not know. One fourth of Swedish children spend 15 terms at school and leave without a certificate. To some extent, the current Swedish schools seem to have turned into a system that does not wish any student to be defeated. Implicitly, this means a right of a student to be successful. Still, one fifth of Swedish students do not make progress in their university education. In developing countries, good education is seen as a sure way out of misery and students work very hard to graduate. In recent years, immigrants in Sweden even try to find schools abroad to ensure the best education for their children.

Universities worldwide are producing millions of graduates. During 1989-1998, the number of Swedish students increased by 83 per cent, compared to an increase of 17 per cent for teachers. In contrast, the administrators increased more than 100 per cent. This expansion of the number of students has led to diminishing funds per student, resulting in deteriorating quality (Eklund and Carlsund, 2000). The administrative aspects have gained more importance, a concern of equal weight in both the North and the South. The boards of Swedish universities are currently composed of representatives from diverse sectors of political spectra, staff and students. This dates back to the university reform of 1976, when the Swedish Minister of Education stated that the academic concept was not central, research should not be discussed and the special status of universities was to be eliminated. That reform dealt with administration and direction and less with objectives and content, except that there should be a focus on professional education. It led to a "waiting-room for the best opportunity to get a job" rather than an intellectual meeting place (Husen, 1985).

Today, a university degree per se does not guarantee a well-paid job in many countries. In 2006, there were some 70,000 Swedish university graduates without employment after their graduation. Since similar trends are found in developing countries, it should be investigated why so many graduate students spend time and resources to face unemployment after graduation. In general, an emphasis on that kind of Western-type research and educational system will, at best, produce a limited group of specialists who most likely serve the privileged strata in urban areas. There must be a strategy for a reasonable balance between demand and output, a responsibility for both the agricultural university and the government concerned. This is complicated if universities are getting their financial support on the basis of the number of graduates.

A similar trend refers to the demand to produce PhD students in larger numbers, often resulting in unemployment, particularly in developed countries. In Sweden their rate of unemployment is about 5 per cent. For the last decade, less than 1 per cent of employees in Swedish industry had a PhD, although this figure ranged from 5 to 9 per cent in a few companies. In a number of countries in the South there is only a small private sector. Many PhD students may turn to academic positions but prefer to seek international employment. Often industry wishes to recruit younger persons with a much broader orientation than with the conventional disciplinary basis at universities. Current approach in Western agricultural research training may add to the dilemma, since the research work for a higher degree frequently focuses on aspects of curiosity rather than on topics of social relevance or problem solving. This is a great challenge even in many developing countries. In view of limited resources, PhD work should be of interest to a university department as a whole and not exclusively to the individual if agricultural research at university level is to help solve some of the global problems confronting humankind and provide guidance to farmers and national agricultural development. This is critical in a global context, since there is a failure by schools and colleges to teach rudimentary courses in agriculture (Borlaug, 2004).

Lost Monopoly of Knowledge

Today, universities are losing their monopoly of knowledge. In highly industrialized countries, the private sector has taken over a large portion of industrial, medical and agricultural research. Globalization will force universities to integrate even more with industry and labour markets. In the early 2000s, Swedish industry accounted for 78 per cent of all research compared to 19 per cent by universities and colleges. The largest companies allocated 10-15 per cent of their turnover to R&D but less than one tenth of Swedish companies conduct pure science. The Foundation of Swedish Farmers sponsors agricultural research, mainly applied research. Ten new foundations for public research support became influential in the late

1990s. Their general approach seems to have been "more of the same" rather than a different focus for their research investments.

The trend of commercialization is increasing globally and is influencing the operation of universities and their value system. As of 2004, all Swedish researchers must show their potential commercial interest in their applications for public funding. Recently, legal aspects have kept scientists from "dialogues"; they turn to lawyers on matters of IPRs. Legal issues are becoming more and more critical to universities also. This is a result of the professionalism of science or the bureaucratization of the knowledge industry. Many Swedish universities have established holding companies to assist researchers in patenting, marketing new products and establishing new private enterprises. This will impose restrictions on future collaborative research arrangements between institutions in the North and the South. The growing problem of IPRs will influence research in "the South", particularly in research cooperation with the private sector, for instance on the production of improved seed or fertilizers. Since research results must sometimes be kept secret for commercial reasons, there is a need for transparent decisions on whether and how patenting or proprietary science can be managed in publicly funded research. An obvious implication is declining financial contributions by the governments to agricultural universities, assuming more support from the private sector. In turn, this will affect future research priorities of agricultural universities. Above all, it leads to complications in patenting if research institutions of the North increasingly form collaborative research arrangements with the private sector but still may be asked to collaborate on the same topic with partners in the South. This calls for a reality check by agricultural universities and the development of a strategy to cope with these complications.

In developing countries, a continued demand on higher education may mean even less production of research findings for society. In my discussions over the last two decades with many Ministers of Agriculture in developing countries, they do not generally expect their universities to provide more than higher education. Frequently, they consider them marginal in comparison with more specialized research institutes. In agriculture, there is often no private sector prepared to invest significantly in agricultural R&D. The problems of the universities are sometimes long-standing and persistent, as exemplified from Ethiopia (Egziabher, 1987; Bekele, 2001). Another problem for developing countries is their dependence on project funding by donors, which is confined to three to five years.

Piecemeal Knowledge without Contextual Framework

Over time, academic education has become fragmented and involves little analysis of reality. Such knowledge has broken down into piecemeal elements that can be studied almost anywhere. Specialization is a consequence of the quantitative growth of knowledge. Since it is difficult to gain prominence in a well-established discipline, new scientific areas appear. Greater specialization leads to an even narrower understanding of already complicated biological systems of agriculture and forestry. Certainly, there is a need for different competencies, requiring not only specialists but also generalists. Equal status must be given to both the details and the synthesis of the whole in both research and teaching. In view of the global challenges, this factor must be better reflected in future agricultural university education and scientific discussions.

In fact, students have voiced their concerns. During a period of more than a decade, more than 80 per cent of some 1,600 students at US colleges found activities outside their ordinary course work most rewarding (Light, 2000). Students got most satisfaction from teachers explaining the contextual framework and analysing the training in relation to historical experiences, ethics and values. Those discussions led to a critical mind in students and helped build their ability to make decisions on the basis of values and ethics in their working life. Similar experiences from one or two departments at SLU confirm these findings. These observations by students are of paramount interest if and when future agricultural research is directed towards real-life problems of resource-poor farmers in the context of the MDGs.

One issue is whether students get access to knowledge and critical analytical ability or simply information. Information is available on the Internet. With information technology we have got an information society rather than a knowledge society. Certainly, there is a need for international databases on the Internet that are accessible to all agricultural researchers worldwide. In spite of some international networks for R&D, such as the Third World Academy of Sciences, the International Centre for Theoretical Physics and the International Council for Science, this is not yet possible. Knowledge is developed only when we have studied the information, assessed and digested it and put it into an overall context. This requires hard work and is a life-long process. We need solid knowledge when meeting the unexpected realities of a future agriculture.

There are now many initiatives to make better use of the Internet. In southern Sweden, Lund University has decided to establish a foundation to sell Internet-based education in Asia and Latin America. A special company is composed of 21 universities and the US company Thompson Learning. It was planned to comprise 27,000 students in 2005. Another example is a distance-learning university through a CGIAR initiative. The idea is to foster the development of collaborative research and institutional arrangements. Although this is of general interest, it might have been more constructive to form alliances between a few CGIAR centres and a selected number of agricultural universities in the South. It would have empowered the staff at the centres to publish their scientific findings through the universities, thereby also influencing both their research and teaching. Such an approach might have stimulated discussions at the agricultural universities to focus on real field problems in accordance with the MDGs and the work by the CGIAR centres.

Another aspect is the online term-paper industry, emanating from the United States in the early 1970s. Boston University initiated legal action to crack down on term papers bought by students. Now, many US states have created laws banning term papers. This is now a worldwide problem. So is the alarming development that academic degrees can be purchased on the Internet. Estimates indicate there are some 500 fake universities offering a PhD at an approximate price of US$ 50. This will be a great nuisance in future collaborative research arrangements. It may become a new profitable area requiring careful screening by special consultancy firms such as the US-based Quest Research.

Many students in developed countries may consider the university a service organization. In the past, the major task of the university was to train professionals such as priests, lawyers, medical doctors and agriculturists. Today, most professions require some kind of research training to serve in a rather complex society. By turning towards general sciences, the agricultural universities may lose the linkage between research and qualified professional training. It is crucial to maintain the task of training professionals rather than follow the thoughts of the Bologna process. It may threaten the concept of agricultural research. It calls for a perspective and a training of the students to think independently. Scientific methods can be more important than current truths. This is vital in the training process, since the private sector often defines what students should know rather than how they should know it. The natural sciences are basic at the agricultural universities but characterized by fragmentation. They are very useful in solving many technical problems but alone they may be less adequate for a growing number of complex problems in agriculture and land use. They call for synthesis of researches as well, a task for which an up-to-date agricultural university should have a pivotal role.

The British have their Royal Society and the Americans their National Academy of Sciences to support the political decision-making process. There are 13 national Academies of Science in the 55 African countries. Having been operational for only 10 years, the African Academy of Sciences has probably had a marginal influence on government policy. The situation may be improved through the new African Science Academy Development Initiative supported by the Bill and Melinda Gates Foundation over a 10-year period. It aims at strengthening the ability of the African academies to provide independent and evidence-based advice to the national leaders, for instance on GM crops. In recent times, the academies have generally played a minor role in agricultural research. In the Nordic region, the academies have had a long tradition, also influencing agricultural research and development in the past. They serve primarily as meeting grounds for scientific discussions but too seldom take part in public debate on controversial issues or invest funds in larger research projects in spite of significant funds in contrast to those in developing countries with very scarce resources. Academies and foundations with strong financial resources of their own could be more proactive by sponsoring research on emerging priority areas, leading to more options for policy-makers.

The Illusion of Scientific and Political Freedom

Frequently, politicians request more research for the solution of both political and non-political problems. Since the policy-makers fund the universities, they exert power over the researchers. For political reasons they add items to the research agenda of agricultural universities often without reference to strategic plans. Such plans can be useful for guidance, albeit time-consuming. The overall message to policy-makers is that allocations to research and higher education may assist a government in solving certain social problems. In the early 2000s, strategic plans were worked out for several Scandinavian universities and other Western countries. A major weakness was their lack of operational objectives.

Most scientists usually argue that universities should not be exposed to measurements of impact of research. This would be right, of course, if higher education were the only task. Nonetheless, good education requires significant research activities. In fact, the justification for increased budgets for research is to convince the public about the utility and the necessity of more knowledge of relevance to society. Ultimately, neither the scientists nor the private sector can deny some social responsibility. How much should a government provide in public funds for agricultural research for the sake of curiosity and how far should it restrict funding to problem solving? Agricultural universities in highly industrialized countries ought to focus sharply on problem solving research. As a general rule such research should make up two thirds of the budget, at least. Less than one third could be for curiosity research. In addition, alliances with other universities and the private sector should be encouraged. In developing countries with a large agricultural sector, less than one tenth of the budget should be geared towards curiosity research during the next few decades.

The bureaucracy of the growing EU exerts another, quite gigantic and anonymous coordination probably leading towards more homogeneity within Europe. If current plans for the implementation of the European

Research Council are successful, more funds for basic research will be requested from all member governments. This raises a principal question whether there is actually a need for the existing national research councils in member countries, all of them requesting funding of activities and administrative costs. Probably, this overall research network can be trimmed for improved effectiveness.

In the past, scientists had to "publish or perish". Gradually the emphasis has shifted from the quantity of scientific papers to publication in the "right" journals. Papers are judged not on the merit of what is written but via authorized representatives. Those external to the university will guarantee that the product meets internal professional requirements. Since scientists tend to work for a narrowing circle of colleagues, this leads to less collection of primary data and time for reflection, all time-consuming activities. It acts against originality or new approaches. It causes difficulties for those not ploughing the common furrow. Being too outspoken or having ideas outside the mainstream may involve danger of getting penalized.

In many cases, agricultural universities seem to have given up their own rights to evaluate themselves, their scientists and scientific competence. This means loss of confidence and reliance on external criteria and statements when new professors are employed and staff promoted. Freedom is squeezed. There is only one Swedish agricultural university but there are 40 institutions for university education. For public funding, SLU is under political direction by the Ministry of Agriculture. External scientific experts are no longer members of the SLU faculty board for the selection of new professors. Those who make the decisions are not always qualified to balance the different qualifications in areas outside their own specialization. The tasks of a professor can be questioned and ultimately the administrators may decide his or her future. All this influences how much researchers dare to investigate controversial problem areas, thus adding to "silence". This calls for maximum involvement of international evaluators from the North as well as from the South to push for a more global research agenda.

Declining Trend in Public Funding

During the last few decades, many agricultural universities worldwide seem to have been on the defensive, apparently distrusted by their governments. This has led to confusion about their mission and even legitimacy. In fact, they are victims of their own success through large expansions over the last 35 years. They have had rising costs and less revenue. The benefits of higher education accrue more to the individual than to society and this has led to reforms by subjecting universities to quasi-market disciplines as in Western Europe. For the future, universities may have to adapt to a world where governments in industrialized countries remain reluctant about a continuous increase of funding in agriculture. Probably, scarce resources in financial terms may be a lesser problem than lack of an intellectual atmosphere, a task for the top leadership. There is also a need for strong intellectual leadership by the head of research departments.

Usually, interesting and relevant research ideas attract funding. With improved research quality and availability of part time courses, students may wish to combine their studies with employment. Diversification of funding is still another alternative. Students may have to pay more for education. Other alternatives include reduction of the number of students to improve quality and development of closer collaboration or even mergers with other universities.

In 2004, three quarters of Swedish universities and colleges experienced budget deficits. In this past decade, financial support to university faculties by the Swedish Government declined by 40 per cent. Most of that gap was filled from external sources. External research funding amounted to about 55 per cent, compared to one third in the early 1980s. Likewise, the West Indian governments were providing about 50 per cent of the total revenue for the St Augustine campus of UWI in Trinidad. One third of the budget was external project funding. These declining trends in funding are quite similar to the experience of the CGIAR during the last decade. In Ethiopia, science aid provides most of the resources. University scientists are normally squeezed between declining public funding and the need to prepare research applications to ensure future funding. External financing makes them even more dependent, in particular when only a small minority of applicants are granted funding.

Internationalization

According to the US Institute of International Education, about half a million foreign students studied in the United States in the early 2000s. This was about 4.3 per cent of the total number of students and an increase since 1997. Two out of ten get scholarships from their US university. In contrast, many American nationals leave university and join the work force to make money. In order to stimulate student exchange, the European Erasmus Agreement was established in 1992 so that a Swedish student could read part of his or her university education abroad. Ten years later, only 2,600 Swedish students used this option, while the majority (about 15,000) found their own way for studies abroad. In general, internationalization at Swedish universities has meant scientific exchange mainly with Europe and the United States, amounting to 97 per cent, and more recently with China. International experience has not received much specific recognition in Swedish life.

The UWI attracts students from the entire region, facilitating easy international linkages and visits to other countries. In 2001/2002, it had some 1,500 graduate students, out of whom 400 were post-graduate students from the Caribbean, Africa and Asia. In Ethiopia, PhD scholarships are the normal way to get experience from abroad. At Alemya University, 36 Ethiopian staff members left for training abroad and 20 returned in 2003. Foreign academic staff members are common in both Ethiopia and at UWI, in contrast to SLU, which has few staff members from the South. Staff of the UWI is recruited from all participating territories. In 2003, there were 43 expatriate academic staff members at Alemaya University in Ethiopia.

During the last five years, the number of Swedish scholars at European universities declined by 20 per cent. Major reasons are language difficulties and anxiety that courses studied abroad will not be formally accepted as part of a Swedish degree. At SLU, 4 per cent of the first year students came from abroad in 2002. Swedish agricultural students have a long-standing interest in international affairs. Between 1900 and 1932, 11 per cent of agricultural students were practising 1-2 years in Europe prior to their formal graduation as agriculturists (Table 10). This was much more common for students at the former Alnarp Agricultural Institute in the south of Sweden than at Uppsala. It dropped somewhat during the period 1933 to 1948 (9%), mainly because of World War II. Even after graduation, international work was not uncommon in the early 20th century. Swedish agriculturists worked in Indonesia, Ethiopia, the former British East Africa and other places.

Table 10. Number of Swedish agricultural graduates with at least one international employment, 1900-1988.

Period of admittance to degree training	Number of agricultural students	International employment	Percentage
1900-1909*	150	22	1.5
1910-1919*	344	49	1.4
1920-1932*	580	74	1.3
1932-1939	141	20	14.2
1940-1949	361	43	11.9
1950-1959	477	64	13.4
1960/61-1969/70	591	102	17.3
1970/71-1979/80	957	91	9.5
1980/81-1987/88	996	58	5.8

* Alnarp and Ultuna Agricultural Institutes

Starting in the early 1960s, many Swedish agriculturists with professional experience found international employment in developing countries, mainly on contracts for SIDA and/or the UN specialized agencies, such as

FAO and ILO. Since the demand by SIDA was for agricultural experts with practical field experiences, there were few scientists. Except for those working in the United Nations there was no need for a PhD for employment. The number of SIDA employments started to decline in the late 1970s with a reduction of the number of recruited agriculturists. With the introduction of Swedish science aid through SAREC in the late 1970s, the basic concept was to involve the national scientists of the recipient countries by long-term, collaborative research arrangements. They included short-term exchanges of scientists and PhD "sandwich" programmes. Noragric had demonstrated a similar approach. As a focal point of the Norwegian University of Life Sciences it has been collaborating with selected institutions in developing countries on sustainable agriculture and natural resources management since the mid-1980s.

Future Need and Role of an Agricultural University

In general, revisited farmers in Ethiopia and Trinidad expressed little demand and enthusiasm for contacts with agricultural universities. To them, agricultural research and agricultural education had not been prominent. Revisited Swedish farmers realized the need for more agricultural R&D but they did not find it to be an exclusive task for the national agricultural university. In fact, not only research departments at an agricultural university but also any other university department can tackle researchable problems of social relevance. This questions recent moves taken by agricultural universities to expand into research territories of less relevance than agriculture and land use as illustrated by SLU in Sweden, Ethiopia and Trinidad and Tobago. In consequence, it leads to the principal question whether a special agricultural university is needed in the future and its role.

It is interesting to note an initiative taken by the FAO, WHO, UNEP, UN Development Programme and World Bank. It aims at a long-term look at the needs of developing countries for agricultural science and technology as it relates to hunger, rural livelihoods, sustainability and economic growth. The work began in early 2005 as the International Assessment of Agricultural Science and Technology for Development and is expected to present the first reports in late 2007. It will not, however, focus on developed countries.

In principle, agricultural universities in developing countries will play an important role for quite some time but their orientation must change, as clearly highlighted by the experiences of revisited farmers. They ought to concentrate on problem solving in accordance with political commitments of their governments to reach the MDGs up to 2015, and beyond. This implies quite different approaches from current ones. Their focus must be on agriculture and land use, incorporating forestry, agroforestry and fisheries. They cannot be confined to higher agricultural education but must also conduct research of social relevance. This will make their education more technically relevant and also prepare professionals for development work in line with the MDGs. In fact, the complexity of future, researchable problems will require both so-called basic and applied sciences, requiring both scientific disciplines and interdisciplinary approaches. Then, it would be more appropriate to abandon the artificial demarcation between applied and basic sciences.

Another challenge for agricultural universities in both developing and developed countries relates to the issue of sustainability. This is of particular interest in agriculture, where environmental aspects are closely related to production and food safety. Except for the political rhetoric there have been few significant attempts to deal with this research issue both in the South and in the North. During the 1990s, the Nordic region probably led the world in implementing a local Agenda 21. In 2003, a Swedish Parliamentary Committee concluded that more than 70 per cent of 288 local authorities had adopted their own Agenda 21. A more recent study by Oslo University has, however, concluded that the local Agenda 21 municipal sustainability movement has lost momentum. Progress has turned to retrenchment in Sweden, in particular at the grassroots levels in spite of financial support by government. In this context, the SLU's decision in 2003 that there was no longer a need for a chair in ecological production can be seriously questioned, although officially reported as a scientifically founded rather than ideologically based decision. Instead, all of SLU should work for a sustainable agriculture. Although the current concept of ecological agriculture may not serve as the definitive road map for sustainability, there is no other option, except business as usual. Above all, the concept of sustainability emanated from the Brundtland Commission and then Agenda 21 decades ago, and it was heavily politically based.

Today, sustainable development is widely acknowledged as a scientific field in its own right. In Europe alone, more than US$ 1 billion was spent on such research from 2000 to 2005. Still, the universities lack effective ways to practise and implement principles of sustainability. This applies also to a large number of agricultural universities in the developing world, although much less is known about their actual involvement. One example of a step forward is the creation of special training courses for European and North American universities in 2005 at the European Reference Point for Technology Transfer for Sustainable Development in Hamburg, Germany. Simultaneously, a Sustainable Development Commission recommended that the UK Government drastically reduce consumption patterns. Higher values were to be given to health, social and political aspects and meaningful work rather than consumerism, thereby indicating a new role also for the agricultural universities.

In highly industrialized countries, R&D by the private sector will continue to focus on commercial agriculture. Large-scale farmers obtain new technology from the private sector and can easily search on the Internet for the best agricultural institutions worldwide. The farmers themselves can easily conduct field tests of new technology. Apart from higher education, future target groups for research by an agricultural university would be categories of specialized farmers such as small-scale farmers or those with specific needs in production or marketing, exemplified by ecological producers. Other aspects would include new crops, land use issues, bio-energy and environmental monitoring. Third, people living outside the agricultural sector must be better informed why research in food security and safety is even more relevant than in the past. This is a long-term task and of great importance, since most of the people in the North live far from the countryside. Fourth, collaborative research arrangements with the private sector will expand. Thus, agricultural universities are needed in the industrialized countries as well, but they will most likely be much smaller and sharply focused. This requires a vision of future agriculture and land use, food security and safety and a mission statement with clear objectives, very distinct from those of a general university. A smaller agricultural university will benefit from more collaborative research with other universities or could be part of certain regional or even international networks. It may also focus more on global research issues in conjunction with the MDGs.

Governments can be expected to fund research that otherwise would not be privately funded but will in turn expect results of importance to society. This calls for explicit research priorities for national work. The current tendency to violate scale considerations by further centralization of the administration of agricultural research must be avoided. In industry, the large institutions are being dismantled, since a large organization, tightly controlled, is the death of creative research. This would be in contrast to current strong beliefs in mega-departments still being envisaged to solve outdated problems at some agricultural universities in the North. Another task for governments is to design an appropriate framework for IPRs to allow for joint efforts by publicly and privately funded agricultural research. All governments that signed the WTO treaty pledged to put in place a system of IPRs. It is of worldwide importance, since agricultural universities will probably be collaborating more with the private sector and conduct much less "free" research of their own. One example is research by Umea Plant Science Center in Sweden. It has been of greater interest to American forest companies than to Swedish ones, since GM trees or plantations are not accepted in Sweden.

From College of Agriculture to University of Agricultural Sciences: Perpetuating Soul Searching

The Swedish College of Agriculture began its operation in 1932. Three decades later, it had 20 departments, 300 students and 680 employees. A long-term plan was adopted in the early 1970s, when the Chairman of the Board of the College stressed a need for its proactive role in charting the future direction of the institution. In 1976, the colleges of agriculture, forestry and veterinary medicine merged into SLU. Its primary task is to provide knowledge in agriculture, forestry and veterinary medicine. The Swedish Government gave it strong financial support for new buildings and equipment. A new long-term plan in 1981 emphasized flexibility and increased cooperation both within the university and with external actors. A long period of consolidation followed. In spite of international signals on sustainability (the Report of the Brundtland Commission and UNCED, Agenda 21), the board and management refrained from major changes of the research orientation towards environmental concerns during the 1980s. A professorship in alternative forms of production offered by the Swedish Council of Forestry and Agricultural Research (SJFR) was somewhat reluctantly accepted in 1990. Meanwhile, the Swedish Government signalled a permanent declining trend of public funds, finally leading to "soul searching" by SLU.

In 1995, the board decided on a future strategy on natural resources. In reality, this mainly led to an expansion of the number of students by 50 per cent. The government specified seven priority areas and added environmental assessment. At this time, it was felt that SLU lacked a profile of ecological agriculture. Although there were many future-oriented research projects within this area, the Dean of Faculty found them of a traditional disciplinary nature. A Centre for Ecological Agriculture was established in 1997, followed by an evaluation. It concluded that individual research projects in ecological agriculture were of good scientific standard and so was the Research School on ecological land use. However, the synergy between the research projects was less clear and international collaboration was not prominent. Time had been too short to establish strong interdisciplinary research groups.

The Swedish Interdisciplinary Research Program for Sustainable Production (FOOD21) was originally designed to find approaches towards sustainable agriculture, partly serving as an alternative approach to research on ecological agriculture. After eight years of research, FOOD21 was completed with a funding of SEK 130 million (Nordberg and Nybrant, 2000). It has produced a vast number of scientific articles and research reports. The results are presented in the form of reports by individual researchers and/or research groups but generally without synthesis, although synthesis is explicitly stated as a core target. Thus, FOOD21 has failed to produce a coherent research approach towards sustainability with practical guidelines. Major attention was given to research themes such as sustainable pig production, crop protection and meat protection. One key conclusion of practical relevance is the need for improved crop rotation, a rather well-known lesson of the past. Findings regarding meat and pig production are theoretical, concluding, not surprisingly, that each system can reach some objectives of sustainability but may fail on others. To a certain extent, it seems as if most research projects were defined by the scientific interest of the researchers but without adequate analysis of the fundamental problems for achieving a sustainable Swedish agriculture and the values involved in such a research process.

In the late 1990s, SLU leaders once again advocated time and energy to transfer SLU to the Ministry of Education, hoping to ensure more government funds for admitting even more students. As early as 1955, a government-commissioned study on the future role of Swedish universities and colleges had concluded that their work should relate to social needs and combine research and education. The College of Agriculture should be placed under the Ministry of Education, a proposal that was rejected because of the close connection to the agricultural industry and national experimental activities. In reality, the connections to the food, forestry and pharmaceutical industries have much increased over the years. In 1999, the Faculty of Agriculture attempted to formulate a new vision and the board suddenly decided on a new research strategy. The overall research orientation of SLU turned to more general sciences. In search of its future task and responsibilities as a specialized university, a government-commissioned study refrained from making proposals but recommended further investigations. SLU remained under the Ministry of Agriculture. In 2000, the government requested a strategy for future Swedish agricultural research. The request was submitted to the SJFR, not to SLU. The SJFR identified four priority areas for research. Thereafter, the government closed it down as part of a reorganization of the national institutional framework for research. Its successor, Formas, has not made a difference. In 2004, only one out of five research applications were financially approved. It was claimed that the decisions favoured those agricultural researchers who served as members of its various working committees rather than promote research on new topics and researchers outside the mainstream.

In 2001, the Board of SLU decided on new strategic goals of a rather conventional nature, such as science of international standard in research areas of relevance to SLU combined with good education. A Vice Rector was recruited to develop future research strategy. An external working group made proposals about future activities and organization but had little discussion on the future research orientation. In 2003, the board voted in favour of four instead of three faculties. Another research strategy was again approved and nine problem areas selected for special financial support (SLU, 2003). Plans were devised to merge 72 departments into 20-25 but the major issue remained to be resolved: to define SLU's sector responsibility. Without such a decision, it is quite difficult to justify increased efforts in certain priority areas of a "sector university" for 2010. Once again, new plans are under way to define the sector responsibility, define strategic research areas and consider a name change.

To date, the soul searching continues and so does the research focus on conventional agriculture. More than one third of the buildings and laboratories are not being used, a common picture in developing countries. In 2004 alone, more than 7 per cent of the staff was made redundant because of funding deficits.

In the mid-1990s, SLU decided to recruit academic staff members internationally for research on more global affairs. Based in a specific department, those professors have gradually departed and the department has merged with another one. In a smaller scale, certain activities to stimulate students for international work were already initiated in 1970. Agricultural undergraduate students were given the opportunity to study two to three months in a developing country through Sida funding. Up to 2004, some 260 agricultural students did so with good experience. Gradually, this approach has expanded to other disciplines and other Swedish universities. For the future, it would be equally important to support agricultural students in developing countries to allow them to travel to a country of their interest to gain new experiences. It will counterbalance a Western dominance in agricultural research.

In another effort to turn more international, and acquire extra funding, SLU and Sida agreed in 2000 to collaborate with the CGIAR by offering temporary staff positions at CGIAR centres to young graduates of SLU. Again, this would be much more suited to agricultural staff in developing countries.

Food research in Sweden involves more than 20 disciplines competing for the same funds. A realistic future scenario is for SLU to be the principal institution for teaching Swedish agricultural students and conducting problem-solving research. Effectiveness can increase by closer integration and even mergers of current faculties with those at neighbouring universities in Lund, Uppsala and Umea. If SLU continues with a research emphasis on more general sciences, such steps are urgently required. In a national context this implies a Faculty of Agriculture of Uppsala University. Most likely, government funding of SLU will further decline. If this scenario becomes a reality, the Swedish Government and the private sector should establish a Research Institute of Sustainable Land Use with a clear focus on problem-solving research, including forestry, and financing from public and private funds. This will allow for improved cooperation with the Swedish private sector and globally as well as assist farmers and the food processing industry. It would also facilitate mission-oriented research training for doctors' degrees and better employment prospects after higher degrees.

Another alternative would be to combine certain teaching activities with those in other Nordic countries. One step in this direction has been taken by the formalized approach for one Nordic Veterinary and Agricultural University. Established in 1994, it has a small secretariat but relatively few activities. In 2000 a joint Swedish and Danish programme was initiated in horticulture for the BSc and MSc levels. A formal merger with partners in Denmark and Norway (possibly Finland) may be possible within a decade, using a network approach for selected research and training activities. Such an approach may gradually be used within the EU framework for countries with a surplus agricultural production and a small agricultural sector.

Some Conclusions

- For the foreseeable future, agricultural universities will play a major role in most developing countries. They must make research priorities on what to do and what not to do. Increased public funding should be provided only if they direct more efforts to research problems of relevance to the MDGs. Such an orientation must also influence higher agricultural education, an issue for both the central leadership and heads of individual research departments. With reference to the outlook for 2015, better use must be made of MSc and PhD students by focusing their work on strategic problems of social relevance rather than just subjects of curiosity. This applies to both the North and the South.

- For agricultural universities in developed countries there is need to clearly define their future mission by making a distinction how they can contribute by research to social development and through higher education for current and new professions in agriculture, forestry and veterinary medicine. Instead of general expansion they should concentrate on research for specific target groups in which the private sector is currently less interested. They ought to participate more actively in international research networks and joint educational programmes. This would compensate for declining public budgets. In case agricultural universities continue to concentrate on higher education, governments ought to establish agricultural research institutes for problem solving in the North and be jointly financed by public and private funds. Foundations may be the only context in which the private sector accepts greater social responsibility. This approach will ensure better integration with innovators and entrepreneurs. It will also ease problems with IPRs and related issues in proprietary science. A second objective would be to develop new scientific tools and methodologies for interdisciplinary research to account for real environmental costs over longer time horizons.

- Food safety aspects will become more prominent in the future and there is a need for research institutions working on sustainable development. There is also a need for closer integration between agricultural and medical sciences for research and education on possible plant drug development, safety aspects and pollution of water and soils from drugs.

- Time for reflection has been drastically reduced for staff at most agricultural universities. Reflection on acquired knowledge can provide new outlooks and insights. This requires field experience, theoretical insights, wide reading and an extensive knowledge base. Above all, it requires an atmosphere conducive to open and transparent discussions on policy issues of agricultural research. Research policy analysis might become a subject matter, focusing on synthesis of knowledge and analytical discussions of intricate scientific issues with uncontrolled variables. Agricultural researchers must turn to dialogue with the public about our future food security and food safety.

- Since many high school students have little knowledge of basic biology there is a need for compulsory courses in biology in wealthy nations to improve the understanding of food, food safety and the production of food from natural resources. In developing countries, agricultural education must tackle needs of different categories of farmers, not bypass the resource-poor population. Human sciences must play a greater role and discussions on values must become part of agricultural training and scientific activities at a vibrant university.

- Agricultural graduates should possess a high personal integrity. They should master both academic principles and knowledge in order to solve problems of relevance to farmers and society. They should be able to think as scientists and willing to take a stand of their own. This requires education on both specialized features and an ability to synthesize knowledge.

- With farming shrinking in many industrialized countries, and in light of more knowledge, their agricultural universities might be an asset in focusing on global research and research relevant to people of the South. This will, however, require radical changes from a focus on research for curiosity to a focus on problems faced by millions of poor people. It would also require attitudinal changes in development assistance to universities in the South and in the design of long-term collaborative research in the North. Instead of a perceived superiority of agricultural training and research in the North, there is a need to realize that the work to accomplish the MDGs requires a priority setting and final decisions by the authorities in the South.

Development Assistance and Science Aid to Agricultural R&D

Development Assistance in Perspective: Another Problem Area of Trend Setting

The Marshall Plan after World War II was an early example of development assistance. By providing calories to Europe, it also benefited Canada and the United States. One quarter of the Plan was in the form of food, feed and fertilizer. This led to US Public Law No. 480 in 1954 by which food aid was organized into a permanent feature of both US farm policy and foreign policy. As relief and humanitarian aid it is useful but food aid gives disincentives for domestic food production. It depresses market prices for locally produced staple crops and encourages a change in local food preferences. Still, food aid has remained a minor tool in development cooperation. Over the past two decades, the number of food emergencies has risen, increasing from an annual average of 15 to almost 30 at the end of the millennium (FAO, 2004).

In the late 1950s and early 1960s, industrialized development was a major theme of development assistance. A number of bilateral donors were established. Competition between the two superpowers became central. The emphasis was on transfer of technology. It started with local or regional projects but turned gradually into larger programmes and finally to policy interventions through financial stabilization programmes of the IMF over three to four decades.

In the early 1970s, annual aid for agricultural development was about US$ 1 billion per year. Only a few bilateral donors had invested in agriculture, SIDA being one of the pioneers by promoting the CADU project in Ethiopia in the mid-1960s. After a speech by the President of the World Bank at its annual meeting in Nairobi in 1973, there was a significant policy change by the donor community towards rural development and fighting poverty. In fact, that was hardly surprising since nearly two-thirds of all people on the globe made some kind of living by using land resources. Rural development became a key concept, although there was little theoretical or analytical framework for categorizing rural situations. Some research showed that improved technology was not reaching all the farmers or was not even appropriate to them. Poverty-stricken people had different economic systems, survival strategies and social values. Social scientists became more interested in agricultural research and consultancy companies appeared in development assistance. Some donors established special agencies to support research in developing countries (International Development Research Centre in Canada, SAREC and the Australian Centre for Agricultural Research).

In the early 1980s, the trend towards rural development continued. Some 40 per cent of German technical cooperation was directed to projects in agriculture, health and rural development. Support to agriculture constituted one quarter of SIDA's budget. To cope with issues of inequality, some governments established specialized agencies to identify and subsidize small and marginal farmers, for example in India. Many donors developed strategies on rural development. Based on the experiences in Ethiopia, the strategy of SIDA postulated that only in agriculture can the greatest number of poor be employed in productive work over the next decades (SIDA, 1984, p. 4). This required increased participation of the rural poor. Participation was in reality confined to consultations with poor people rather than assisting them to gain power to govern their own destiny.

Influenced by the report of the Brundtland Commission, the donor community quickly turned to activities labelled sustainable. This focus remained up to UNCED, although many of the development assistance activities did not show much long-term sustainability or institutional capacity building. In the mid- to late 1980s, many donors started to conclude that expensive agricultural schemes and most agricultural projects had not performed well. This resulted in major changes in the size and pattern of agricultural development assistance. It declined from 20 per cent to 14 per cent of the total aid. The United States reduced its share of bilateral aid to agriculture from 30 per cent in the 1980s to less than 15 per cent by 1990. One reason was the push by political pressure groups other than farmers. Policy-makers in recipient countries showed little interest in agriculture. Directed by the IMF and the World Bank, international lending turned towards structural adjustments and policy reforms, anticipated to perform better under macro-economic programmes. In retrospect, these adjustments did not attack the fundamental problems of poverty and a large rural population. Ironically, such a focus was already developed in the Lagos Plan in 1980, although the Western monetary and financial agencies rejected it at that time (Davidson, 1994).

By the 1990s, the donors and most national governments in developing countries had abandoned agriculture. Agricultural researchers were seen as facilitators conducting farmer participatory research. Instead, UNCED made biodiversity a key area for development assistance but donors and governments lost interest after about a decade. Many efforts in biodiversity conservation were often based on the perceptions of the North and donor wishes by making reference to global values. At national levels, other values were probably more important, particularly in a perspective of poverty alleviation. At the end of the 1990s, lending for agriculture amounted to 10 per cent of total development assistance compared to 15 per cent in the early 1990s. By 2000, the World Bank reported its lowest level of support to agriculture in its history. Instead, the donor community had turned their focus on global trade and the WTO. The market was seen as the major helper to developing countries.

Now there is again recognition that agriculture should be high on the development agenda. The UN Millennium Project has probably contributed to this emerging policy shift in development assistance. Lending for African agriculture by the World Bank rose to about US$ 500 million in 2004 compared to US$ 100 million in 2000. The new World Bank President argued in 2005 that agriculture must be seen as a priority again, at least in Africa. In addition to economic growth, key areas are access to health, education, land and political power and freedom. Other donor governments also indicate some policy shifts towards agriculture. For instance, the Canadian Government has released a new agricultural aid strategy with increased funding. In the United Kingdom, the British Department of International Development has produced new guidelines on its work on agriculture.

Another prominent feature of past development assistance is a tendency of donors to stress selected aspects of their own choice in line with views expressed by their domestic political masters or agreed to at international meetings. Ever since the 1970s, these fashions and buzzwords have been common in agricultural project activities including appropriate technology, the role of women, ecology, on-farm research, farmer participation, environmental concerns, energy, genetic resources, gender issues, sustainability, HIV/AIDS, biodiversity and sustainable livelihoods. The MDGs may turn out to be the most recent example. Although all these aspects are relevant per se, they may not always–apart from the MDGs–be in line with the most urgent problems in a variety of recipient countries.

For the last decade, biotechnology research has been in focus. This applies to USAID and the Canadian Government as well as Sida. Funding to developing countries has often been approved in spite of some scientific unknowns or the fact that a national regulatory system is not yet in place for allowing a commercial introduction of GM products. Ten years ago, Swedish science aid was given to build up research capacity in Sri Lanka through a well-equipped laboratory for biotechnology. Much less consideration was given to which development problems were to be resolved by the research. There was less clarity on an existing and functioning regulatory system for potential research products. At the same time, national policy does not allow commercial production of GM crops in Sweden itself. Nonetheless, a "science-driven" research agenda has been common in biotechnology (Brenner, 2004). Much less effort has been given to determining the social needs or demands in an economic sense.

In the past, trend setting by donors has been more common than a genuine problem definition for development assistance. Hunger, malnutrition and improved conditions of life to poor people still constitute a major set of problems recently identified by the Millennium Project. By 1965, the FAO had identified the key problems, such as malnutrition, rapid population growth, the difficulty of applying technology effectively and a slow growth of agricultural export earnings in developing countries. Forty years later, the FAO reiterates the necessity for institutional and policy environments that provide incentives for private investments and that public investments concentrate on activities that cannot attract private sector money. Some lessons for policy-making include extraordinary but uneven production gains in developing countries and projections of further declining food prices. World cereal prices have almost halved between 1950 and the end of the century. Accounting for almost 40 per cent of the GDP, agriculture still remains the most likely source of quick, significant economic growth in Africa and parts of Asia. The majority of African countries depend on agriculture for both food and income. The neglect of agriculture and rural development over the past 25 years has missed opportunities and been a strategic mistake.

International discussions expect development assistance will increase to 0.7 per cent from the current 0.2 per cent of GDP. During the 1990s, development assistance declined, being 20 per cent lower in 2001 than in 1990 (in real dollar value). Public aid constituted about one third of the total capital transfers to developing countries. In the early 2000s, development assistance from the OECD countries was equivalent to US$ 51 billion, whereas annual humanitarian aid had stabilized at about US$ 4-6 billion during the 1990s. To achieve the UN targets in 2015, official development assistance is expected to increase annually by some US$ 50 billion. In contrast, the costs for armaments worldwide amounted to some US$ 900 billion in 2003 compared to US$ 50-60 billion for annual development assistance.

The reports of both the UN Millennium Project and the Commission for Africa focused in 2005 on how the world's commitment to the MDGs should be put into practice. They call for capital investment, increased aid flows, better governance in developing countries and attempts to get rid of corruption. So far, only a few countries have reached the UN target of 0.7 per cent of GDP and the largest donors provide aid in a range from 0.11 to 0.27 per cent of their respective GDP. With such a rate of growth, the target will not be met until 2025 or beyond. Decisions by the G8 countries have already led to debt relief for some 18 developing countries. Still, massive help is required for developing countries in order to fight extreme poverty (Sachs, 2005).

Many aid organizations have made commitments for increased development assistance to achieve the MDGs. They have also recognized a need to make it more effective as formulated in the Paris Declaration of early 2005. The Declaration calls for various efforts by both recipients and donors regarding some 12 indicators including ownership, mutual responsibilities, focus on national priorities, harmonization and improved donor coordination. The idea of an expanded budget support to receiving governments is a prominent feature. It will, however, work only where a recipient government demonstrates commitment and actions towards poverty eradication in line with the MDGs. Moreover, it requires a well-functioning administrative and financial system, acceptable to both the respective government and all donors. This is already a great challenge to governments in many developing countries. If these prerequisites are not met, any substantial budget support will simply be a waste of resources, albeit an easy way for donors to spend increased financial support.

There are some good reasons for increased aid. For instance, the governments of France and the United Kingdom have agreed to double their development assistance in 2005. The US Government has proposed a Millennium Challenge Account and the Swedish Government has decided to increase its aid budget to reach 1 per cent of GDP by 2006. However, a proposed doubling of aid seems doubtful in a historical perspective. In 1963, official development assistance amounted to US$ 5 billion (0.51% of GDP). Ten years later, it had almost doubled in dollars but was only 0.3 per cent of GDP. The financial commitments made by the governments at UNCED did not materialize. In late 2004, some OECD members even suggested the idea that UN fund spending on peace keeping operations or training foreign armies might be allowed to be accounted for as development aid. Rearmament is taking place in certain Asian countries, which constitute the largest importing region of weapons. Major importers of conventional weapons include China, India, South Korea and Pakistan (SIPRI, 2006). In addition, China is the world's eighth largest exporter of weapons, sometimes in exchange for timber as in Liberia. The growing economic and political power of China will be a challenge not only to Japan but also the United States, the EU, India, Pakistan and Russia.

Some Issues for Future Development Assistance

Poverty Reduction

When governments commit themselves to the MDGs it means governance in support of the majority of the poor. The issue is whether and how such a shift of power can take place when globalization also accentuates a reduction of scarcity in a perspective of a "Western-type society". Some countries are no longer developed or developing. Nor are concepts of the South and the North adequate categorizations. They conceal sub-groups of countries, such as the heavily indebted poor countries, emerging markets or transition economies. It is more relevant to speak about low-income countries as the real target for the MDGs (2.5 billion people). For most of them, agriculture is very important. In addition, there are the middle-income countries (2.7 billion people), including China and the high-income countries (almost 1 billion people). To them, agriculture and land use is also important but from quite a different perspective.

As an average, it is believed that some 30 million people may be lifted out of absolute poverty every year. The World Bank has argued that this number may be tripled through good governance and enabling policy environment. To reach the poor by going through the powerful is difficult. Neither lease and rental markets nor the input and credit markets operate in the interest of the poor. If the power structure is not changed, existing patterns of inequality will merely be reinforced. This calls for an explicit policy and government actions in favour of the poor. It requires a commitment by the political and economic elites of recipient countries. Otherwise, aid funds reach the middle and upper classes. This policy shift through demonstrated actions is more important than increasing the volume of aid. Without such an implementation, development assistance can do little in changing power relationships. When reviewing its experience long ago, the World Bank noted that development projects could seldom be sustained without national policies (World Bank, 1988). Although there was merit in the rate-of-return criterion, it may conflict with objectives of building human capacity and institutions in the long-term perspective. Obviously, that lesson was forgotten. In fact, if government action in favouring the poor is not demonstrated, donors should not give financial support. Such a determination would in itself stimulate policy dialogue.

International banks give loans to developing countries with the intention of alleviating poverty. But one must question whether commercials bank are able to empower the poor, the Grameen Bank being a special case. Commercial banks also serve corporate interests. The TNCs benefit from contracts that build infrastructure and lay a foundation for further transnational investments, from policy-based lending and from investment in projects. Some observers have argued for writing off the debts of developing countries. Such actions will favour the nation-state but probably mean little to ordinary and poor people in the short term. Both the World Bank and the IMF have actively attempted to assist heavily indebted poor countries through their joint initiative, which was recently extended for a longer period. The Poverty Reduction Strategy Papers (PRSPs) were initially a tool for debt relief for the poorest countries. They represent a shift from project to programme aid in the form of direct budgetary support and requiring an explicit government policy towards poverty alleviation. So far, there seems to be little action by governments through the work with these papers. They frequently seem to bypass agriculture. In principle, this would mean that aid to governments should be based on similar strategic documents. The issue is hardly more plans but action with an adequate focus on infrastructure, primary health care, primary education, agricultural research and development for institution building in the productive sectors to provide employment and income.

The International Fund for Agricultural Development has a special mandate of combating rural hunger and poverty in the most disadvantaged regions of the world. Since 1978, it has financed some 600 projects in more than 110 countries in both loans and grants. These investments are estimated to have moved about 260 million people out of poverty. According to an external review in 2002, efforts by IFAD to develop local institutions increased during the last decade. The new IFAD Strategic Framework is conceived as part of the MDGs with a new mission of enabling the poor to overcome their poverty. It requires not only strong national government commitment but also a much more proactive role by IFAD on national policy issues rather than at the project level. IFAD has been less successful in setting overall policy frameworks. The rural poor must be given improved access to productive natural resources and financial service and markets. Although agriculture is in focus for IFAD, other activities must be considered in the future such as forestry and land use issues, small-scale industries and job creation. Both urban and rural poor have to be accommodated. For instance, the city of Bangkok, with 12 per cent of the population, produces some 40 per cent of the economic output of Thailand. For the future, IFAD must position itself more strategically, in ways not highlighted by the 2002 Review or in IFAD's Strategic Framework. The shift relates to both its mission and overall management, according to another recent review.

During the last decade, the FAO has changed towards more poverty reduction activities. The Anti-Hunger Programme was an attempt to take the lead in some practical steps to implement the recommendations of both the World Food Summit and the follow-up summit held five years later. Over the last few years, the FAO Special Programme for Food Security (SPFS) has turned into an important practical programme in some countries. It has gained high political visibility in developing countries and promotes national ownership by using participatory processes to empower households. Operational in more than 100 countries, the SPFS still has to show convincingly within the next two years that, in a few countries where political will and action exist, it is possible, applying the twin-track approach of the Anti-Hunger Programme and working with other partners, to make rapid progress. In late 2004, the FAO announced its commitment to work in accordance with the MDGs. In fact, the basic concept of SPFS to involve rural poor people may serve as the approach to work with the MDGs in the field, whereby local communities may develop their own MDGs.

The FAO's proactive role in trying to reach the rural poor contrasts with that of the World Bank, which has discarded agriculture as a theme in its major reports. The strategy of the World Bank focused on the creation of better potentials for the poor through growth and distribution of incomes. Social and educational issues have dominated its work and the poverty reduction theory. As in the past, bilateral donors have followed suit in this general orientation. Both the IMF and the World Bank have dominated the process with prescriptions on rapid changes for open foreign trade, privatization of state companies and utilities, and deregulations. They have seldom allowed recipient countries to influence the strategy by anchoring the responsibility with the government concerned. An old example is from Tanzania. In the 1970s, lengthy discussions on rather unrealistic plans for self-reliance in Tanzania led to no substantive changes until Swedish aid also pushed Tanzania to accept IMF and World Bank policies in the 1980s. To involve civil society in the future, any process of "anchoring" may take longer and may also be done step by step.

International NGOs also operate in the agrarian sector of developing countries. The International Foundation for Science (IFS) is one example. Since 1974, it has financially supported young scientists in their own countries, thus preventing brain drain and lessening dependency. By 2004, the IFS had supported some 3,200 scientists in 100 developing countries. Annually, it approves about 250 grants on scientific merits, about one fifth of research applications received. Although valuable per se, this support, given in isolation to individuals, is small considering the need for research support in three continents. For the future, it is necessary to give attention to institutions with weak research capacity rather than to individual scientists for a few years. Priorities are required both for countries and for research topics. A major issue is how IFS-sponsored scientists can be guaranteed a satisfactory research environment when IFS funding ceases. Recently, social sciences and water resources, in line with an overall trend setting of agricultural research, were added as new areas for support by the IFS. More attention is now given to agriculture, forestry and agroforestry and recognizing research issues on poverty alleviation, for instance through the collaboration with CIFOR on the Poverty Environment Network. Another recent change is the focus on scientists in countries with vulnerable scientific infrastructure, mainly in Sub-Saharan Africa. These changes are reactions to an external evaluation in 2001, concluding that although the IFS mission remained relevant, its policies, methods and strategies had to be changed. An interesting new activity is the joint Sustainable Agriculture Initiative of the IFS and the Council for the Development of Economic and Social Research in Africa, sponsored by the British Department of International Development.

For long, aid officials have argued they know how to "develop" and reach the poor, but the records are not very convincing. Five decades of external advice have shown that the priority to agriculture has shifted from none to the highest at various times. Still, agriculture is, and has been, a major activity in a range of poor developing countries. In Tanzania, agriculture provides 50 per cent of the export earnings and is the livelihood of 80 per cent of the population. About 3.6 million households are engaged in farming. About two thirds of the household budget is spent on food, compared with some 12 per cent in Japan and Sweden. Basic requirements must be met in food, water, education, sanitation and health. The target is not only a matter of food production but also to create new employment opportunities and wealth generation also from non-farm rural activities in order to raise incomes. In a recent evaluation, it was concluded that Sida would need very strong reasons for not engaging more broadly in this area in the future (Booth et al., 2001). Ironically, this seems to be a reminder of the old SIDA strategy of the early 1980s. Another lesson of the Swedish-Tanzanian development cooperation of more than three decades is that it can become more effective, evidence-based and relevant by being more strategic. Without rethinking of aid and the development strategy, there is little chance of reaching the MDGs by 2015 and showing success stories. Real progress requires efforts in long-term institution building with a simultaneous transfer of the major responsibilities to national professionals and policy-makers to ensure sustenance rather than quick and massive spending of increased aid budgets.

Policy Dialogue

For long, the World Bank and the IMF have been proactive on policy matters, in particular in negotiations for loans and structural adjustments. A genuine dialogue on policy may often be tough and may lead to different opinions and even controversy. To the FAO, policy dialogue has been a major task for decades. In reality, the FAO being an intergovernmental agency with no funds of its own, this work has been focused on plans and documentation. Often various FAO units have been operating in the same country. Although it is difficult to trace the impact of this work, most member states have appreciated the FAO's involvement and creditability. In 2004, the FAO re-launched its Policy Task Force in response to an evaluation of its policy assistance to promote a culture of collaboration and communication for streamlined work in the member countries. Experience from Mozambique shows that the Swedish dialogue within the country strategy process was very limited (Wuyts et al., 2001). Nor was it useful in shaping the Swedish development cooperation, since fewer than half of the activities were directly relevant to Sida's overarching objectives of poverty reduction, gender quality, democracy and environmental protection. There is a similar experience from Tanzania, where no coherent programme was developed but the formulated strategy was relevant to Swedish objectives.

Project replicability remains a policy issue of great relevance, recently demonstrated in reviews or evaluations of IFAD and SPFS and by the work of the World Bank. It was also seriously discussed in the UN Task Force on Hunger. Isolated development projects with some autonomy and a relatively small number of farmers are easy to manage. The initial effectiveness of the CADU project was due to its relative autonomy from the Ethiopian Ministry of Agriculture. Progress made by CADU stimulated the government to scale up activities to a national level through the Minimum Package Programme. In 1974, CADU, renamed the Arsi Rural Development Unit, expanded under the Ethiopian Military Government with a focus on peasant organizations, involving some 900,000 farmers. In 1985, the Ministry of Agriculture was reorganized, resulting in additional Swedish funds for a programme for 2.7 million farmers. Then, the power in policy formulation rested with the Workers Party of Ethiopia, ultimately leading to disagreement, so SIDA decided to cease its financial support in mid-1989.

In Mozambique, the Nordic programme (MONAP), involving the FAO as well, was also established as a special unit, separated from the Ministry of Agriculture. Gradually, the number of staff of this unit grew, accounting for one third of the Nordic programme's foreign staff and thus becoming almost an empire. More recent examples include the concerns of scaling up agroforestry activities with improved fallows in Zambia and strips of natural vegetation coupled with the Landcare Movement in the Philippines. The scaling-up requires longer time frames, greater emphasis on fostering policy and the simultaneous strengthening of all relevant institutions and the mechanisms for policy implementation. For the future, one policy aspect relates to the importance given to quick results in the field or an additional focus on institution building and strengthening of the government bureaucracy. This will be highly relevant with 30 or 40 donors operating in the same countries. Scaling up on the national level should imply involvement of 5 million more farmers, each one increasing the yield by 10 to 20 per cent, rather than concentration on the minority of the most progressive ones to double their yields. These two alternatives have very different consequences and will also require very different strategies.

Since it is a political choice, this calls for decisions and action by the government concerned rather than imposition of external ideas of current thinking.

A major weakness on the African agricultural scene has been the neglect by both donors and African policy-makers of the development of domestic institutional capacity. This calls for a national vision or a country-specific strategy in agriculture and what it should look like in the next two decades. In turn, this vision will influence future agricultural research. The vision must be elaborated at the national level with due reference to the overall global context. It cannot be a global approach, dominated by a Western experience of agricultural development, by which some 25 million people produce the food and leave some 2 billion rural people without jobs and livelihood. A vision must be elaborated from within, even if it takes time, and must not be "pushed" by outsiders. It is a political issue and requires a proactive policy dialogue within the country also. Such a policy dialogue must include the work of many national NGOs. Ideally, such a vision should be a prerequisite for donor support and possibly be combined with the development of future PRSPs.

Coordination of External Support at the Country Level

There is little doubt that development assistance can be much more effective. One problem is accentuated by a reluctance of donors and the inability of recipients to coordinate activities at the national level for monitoring and sharing results and experiences. Donors compete for national staff for implementation and frequently wish to hoist their national flags as showcases, mainly for their domestic masters. Donors have to spend the annual funds allocated to a particular country. Government staff in recipient countries is small and may cope with 30 or 40 donors and handle hundreds of development projects, for instance by providing quarterly reports to all its donors, each of them with its own reporting system. The Commitment to Development Index highlighted 1,371 projects in Tanzania during 2000-2002, some donors financing less than five projects, others providing resources to hundreds of projects.

In the past, there have been attempts at donor coordination at the country level, primarily by the World Bank. They served mainly for information exchange. In principle, the recipient country rather than the UN system or the World Bank should lead the process of donor coordination, to ensure ownership. It is more important for donors to accept this responsibility even if it means a slower pace in processing the aid programmes and spending their funds. Donors should be more proactive in the policy dialogue and agree to work collectively. The dialogue between donors and recipients must focus on the priorities of the country and how the group of donors could help tackle the major immediate, short- and long-term problems. Then, the donor community should invest in selected institution-building activities for 10 to 20 years with regular evaluations, also conducted collectively. The PRSPs and other national policy documents on the MDGs may serve as an instrument for donor coordination at the country level.

The current idea of a UN residential coordinator at the national level may facilitate better interaction and coordination among the various UN agencies at country level. Such coherence is necessary for the UN system and would save costs. By promoting joint actions, large "bankable" projects could contribute to a coherent national programme based on the PRSPs. This may lead to certain improvements. However, it is unclear whether UN coordination is to include bilateral donors or how integration will be achieved. It remains to be seen whether the bilateral donors will agree to reduce the burden of the recipient government and even channel more of their bilateral aid through the UN specialized agencies. Still, the recipient government must manage all external support within its own bureaucracy, including the NGOs. Above all, any blueprint can be questioned since there are variations between both countries and donor categories. In developing countries with strong administrative systems there is much less need for a large UN office. All donors should agree to use one domestic focal point of the existing government bureaucracy for coordination. Besides, a power shift from bilateral donors to the national level is a prerequisite for "development".

One issue of coordination refers to the development assistance itself, provided by the EU. Employing some 30,000 persons at the European Commission in Brussels, the EU has a staff of some 1,600 who are involved in development assistance activities. In spite of this, its member countries have their own bilateral and multilateral development assistance. In addition to the UN system, this new bureaucratic layer duplicates efforts and time. The EU aid administrators are doing the same kind of job many staff members do at the national level, for example, the Sida staff of about 850 persons in charge of some 5,000 different activities. As a guess, the national aid agencies in the OECD countries may involve some 20,000 persons. In addition, the UN system has a large staff. For the future it seems doubtful that such large national aid agencies should be maintained in the EU countries. One option would be to give full responsibility to the EU to act on multilateral aid, food aid, relief and emergencies, with a parallel staff reduction at the national aid agencies. This could allow more time for evaluation and impact considerations by the national aid bureaucracies. Alternatively, the EU should not be involved in development assistance at all, and should leave that to its members. In any event, the European donor community can make savings by better division of labour on development assistance. Above all, this would ease problems of coordination at the country level in the South by reducing the number of donors.

Donor Dominance and Low Absorptive Capacity

Major investments in agricultural and rural development come from individual national governments. Over time, the influence of donors has been significant apart from the recent past, when agriculture has not been a priority. Many short-term projects have dealt with contemporary issues rather than long-term commitments to solve the basic problems of a majority of people. Lack of continuity in personnel of both recipients and donors implies that seemingly novel ideas have simply repeated earlier efforts. No doubt, recipient governments refrain from refusing funds, meeting the criteria of the donor even though they may not focus on major problems in the country. Since aid is conceptually political it is less analytical. It may change quickly for various reasons. With an acceptance of the MDGs, developing countries must organize themselves better and shift their own priorities towards the poor groups. The governments of Sierra Leone, China and Brazil are recent examples where serious attempts are made to cope practically with issues of poverty. In the future, donors should maintain policy dialogues only with those recipient governments that have turned to a policy on poverty reduction and demonstrated that policy through political action.

Future aid should not be supply-driven. It is time for reflection, possibly with a one-year moratorium on aid rather than quickly spending the next fiscal budget. When nation-states depend upon development assistance, the donors usually drive the agenda. In Mozambique, about one quarter of treasury funds come from donors and in Uganda it is more than half. A change from project to policy funding and overall budget support is a step forward but offers no final solution. The EU has already observed that, although it is willing to channel half of its aid to Africa, that is not possible since the recipient countries lacked absorptive capacity. It is not a new observation (Lele, 1991). The New Partnership for Africa's Development and its Comprehensive Africa Agricultural Development Programme have a similar problem in lack of domestic capacity. The balance between funds for development and absorptive capacity must be considered in much longer time frames than currently perceived by most observers.

Impact of Aid and Lessons Learnt

The effects of aid have been slow. There have been scattered success stories but seldom major comprehensive analysis of accomplishments by national governments or by donors individually or collectively. In contrast, history is full of examples of well-intended aid efforts that went wrong, for various reasons. Donor-supported agriculture development projects have generally had high incidence of failure, resulting in low institutional impact. The external review of IFAD in 2002 considered several areas of impact but not quality of life for the rural poor. Life improvement ought to include income generation and distribution. The FAO, as an intergovernmental organization, usually refrains from presenting impacts on specific topics and geographical regions. The World Bank has been more transparent about its successes and failures, although there are no exact figures on sustainability of agricultural projects between 1987 and 2001. The Operations Evaluation Department of the World Bank has estimated them at 39 per cent. The outcome of its projects has improved over the past few years but averages 61 per cent during 1982-2001.

In general, the governments of developing countries have been less concerned with impact, assuming a steady flow of new funds. This calls for one system at the country level acceptable to all stakeholders but under the full responsibility of the recipient government, not the UN or donors. There is a need for one central, national institutional memory or knowledge bank that could serve the government, its donors and the UN system. This is an important aspect to be considered in the on-going UN reform process at country level and for the monitoring of achievements on the MDGs.

Overall impact does not seem to be a great concern to bilateral donors. Development assistance has been considered a protected activity with a focus on volume rather than content and long-term effects. The Director General of Sida has stated that the effects of development cooperation would not be presented in its annual reports, mainly because it is a long-term task (Sida, 2000). Nonetheless, that organization has existed four decades and has been arguing that it reached poor people from the start, particularly in the 1980s. As an exception, USAID has recently taken a step in reviewing everything it has funded related to natural forests and communities over the last 25 years (Clausen and Hube, 2004; Clausen et al., 2004). It provides 10 country studies and an overview, showing the changes and drawing conclusions for the future through building up institutional memory.

In general, learning from experience is not a prominent feature in the donor community. This negligence was noted already in the late 1980s in an evaluation commissioned by the Swedish Government of SIDA's ability to learn from past experience. Although donors produce and commission evaluation reports, they may be confined to selected projects or programmes with less consideration of overall lessons learnt. In the early 1970s, the Nordic government jointly launched a large and long-term development programme on agricultural research and training, including a dairy at

Mbeya, Tanzania. In funding the Tanzanian dairy sector at the same time, the Dutch Government launched a similar project. It is unknown what effects these investments have had. In late 1999, an open bidding was advertised for the dairy buildings and equipment and the Dairy Technology Training Institute at Tanga. A large number of small and medium-scale dairy plants faced problems in getting sufficient quantities of milk in the dry season. The Tanzanian dairy industry is seasonal. Instead, imports of dairy products from Kenya and South Africa are growing, whereas most of the large-scale processing capacity lies idle. This single example illustrates a need for better donor competence about the real constraints at the time of planning. A more recent example of a long-term approach in supporting capacity-building relates to research on marine resources at the Institute of Marine Sciences at the University of Dar es Salaam. After 10 years it had become a good research environment with international status. In a perspective of the MDGs, however, it challenges the basic philosophy of support since aid administrators have suggested that further research might be required about "the cultural processes for success" (Ingvarsson, 2003). A more promising approach is shown by SG 2000 in Africa. Based upon experiences of agricultural projects during two decades in 12 countries, that organization will concentrate on four "focus" countries and operate only in them by 2007. Three quarters of staff attention and financial resources will be concentrated on Ethiopia, Uganda, Nigeria and Mali.

For evaluation, many donors rely heavily on external consultants, many of whom are contracted again and again for new evaluations. If they wish to be re-hired, they can hardly be too critical. Above all, this procedure turns technically competent aid staff into bureaucrats with less and less knowledge of substance and overview. It calls for a well-designed internal monitoring and evaluation system for each donor, its size and focus being dependent upon the mechanism in the receiving country. The problem is that most developing countries have neither adequate mechanisms for evaluation nor a central mechanism for lessons learnt. A manual with guidelines for evaluation managers at Sida was published only recently (Sida, 2004).

There is seldom any up-to-date institutional memory bank of technical competence at donor agencies, in particular on agriculture. Such an institutional memory may be confined to individuals. The donors of FAO Trust Funds have stressed the same weakness. Frequent changes of staff of donors are one reason why old ideas are reinvented over time. Existing knowledge is seldom used or modified; instead, work starts "fresh" with a "new" idea from the new desk officer. Use of available knowledge and experiences would require a comprehensive overview of past work, proper analysis and insights about emerging issues. By 1985, the first Swedish grant was allocated to research on HIV/AIDS. When the new Sida decided in 1998 that aspects of HIV/AIDS should be included in all development projects, almost 100 interventions related to HIV/AIDS had already been given Swedish financial support. They were not part of a policy but a result of individual initiatives within the organization (Peck et al., 2001). Being a strategic Sida priority, HIV/AIDS was not defined by concrete goals or indicators against which advances or obstacles could be measured (Vogel et al., 2005).

The internal, technical know-how of agriculture has declined considerably in most donor agencies during the last 15 years. Many forestry advisors of development assistance have left service. From the 1960s to the 1980s, agricultural technical staff gained a great deal of technical and practical experience on long-term contracts with SIDA or FAO. Although attempts were made to collect the views of staff when they left service, it is unclear how the collective wisdom ever became instrumental within SIDA. Even today, Swedish rural development professionals are working abroad, numbering some 180 persons in early 2006 (Currents, 2006). Now, the majority are from the private sector and/or consultancy agencies (62%) or the NGOs (27%), a major difference from the past, when they were almost exclusively recruited from the public sector. Since short-term crises are given priority over long-term development efforts, there is less time and little incentive for reflection. The donor community seldom promotes a critical dialogue on development policy aspects. This problem is accentuated by the brain drain in Africa, in particular. According to the World Bank, about 70,000 African scholars and experts abandon their home countries each year, often attracted by better-paid jobs in the West, sometimes by donor countries.

Corruption

Corruption is a long-standing and widespread problem. At the end of the 19th century, reference was made to the "Gilded Age" in the United States. Capitalists bought political power or support to exploit workers and cheat consumers. Corruption can be due to need or greed. It is not confined to developing countries but includes both the United Nations and the private sector. Corruption has long been associated with development affairs and business transactions. Staff members of the World Bank have been accused of involvement in corruption, for example, with Sida funds in the last few years. Most people with an active and vital morality will accept a mutual responsibility for the common good. But many find duties to be for society alone, not for the individual. If an individual believes others are egoistic, a negative spiral will emerge such that all others will only think of their egos. Such a trend, and greed, may have serious implications for society as a whole. It challenges the current type of welfare states and constitutes a major problem in development assistance. From a societal point of view, it is vital to combine human rights with duties.

The 2003 scandal of the Italian company Parmalat led to a financial loss of some Euro 10 billion. In the same year, corruption was revealed in the Swedish state monopoly of alcohol, an unbelievable discovery to many Swedes. A similar scandal appeared in a large insurance company. According to the 2003 November issue of *China International Business*, some 4,000 officials are supposed to have fled the country up to 2002, taking with them some US$ 5 billion earned from corruption. In 2004, a Chinese document was banned that highlighted how corrupt authorities had cheated farmers.

The World Bank estimates that corruption takes about 5 per cent of the global economy or the equivalent of the total GNP of all poor countries. Other estimates show even higher percentages. In the past, corruption was veiled in silence. In 2003, almost 100 countries signed the first Convention on Corruption, which came into force when ratified by 30 Parliaments. Since 1997, Sida has supported the Palestinian Authority. A classified evaluation report dated in 1997 and made public in 2004 concluded that the Palestinian Authority was corrupt and not very democratic. In late 2003, the Swedish Government approved a new grant of SEK 200 million to Mozambique, although it was delayed due to a bank scandal in Maputo. Funds at the disposal of the Mozambique University, for example, have been used to select unqualified people for training. The internal auditors at Sida recently indicated that corruption had reduced Swedish aid also to Ethiopia and Bolivia because of unsatisfactory control measures. Although the Sida board decided to tighten control, the auditors claimed that one reason for this situation might be that the organization did not want to negatively influence the Swedish aid target. This dilemma was identified as early as the 1980s when SIDA had difficulties in spending all the aid funds provided by the Swedish Parliament. In fact, a number of countries receiving Swedish aid feature quite high on the index presented by Transparency International on corruption in the world.

Future Role of the UN System in Agriculture, Forestry and Fishery

The UN reform process must include substantial changes, including all the areas relevant to agriculture, forestry and the environment. The overall annual UN budget has been stagnant at about US$ 2.6 billion. The total number of staff is high and may not always be adequate and sufficiently competent because of both declining budgets and political appointments. To some observers, the UN agencies, with time-consuming and costly meetings, may be too distant from ordinary citizens. During 2000-2001, the UN agencies held almost 15,500 meetings, generating almost 6,000 published reports in six languages. This makes it difficult for small countries among some 190 members to follow and influence discussions and resolutions. Another dilemma, shared by most government bureaucracy, is fragmented decision-making, confined to sectors with specialized agencies having mandates that were formulated long ago.

A common mistake is to confuse governance with governments. Governance is about decisions and how they are made. With political authority, governments also act internationally through the United Nations. Sometimes, large corporations may even act in place of nation-states. But the United Nations is composed of governments only. Scientists and technical experts have played an important part in UN negotiations, and still do. Negotiations may also include views of the private sector, although indirectly. The final agreements may, however, hide the interests of what the "experts" represent. It would be more constructive to put all issues on one transparent agenda, keeping in mind that different actors take various paths to reach a common goal. A future global governance system is unrealistic. Consensus building among governments and NGOs also has limitations. In the current context of globalization, the United Nations ought to realize that future international commitments must involve not only governments but also the private sector. This requires reform and willingness to enter a practical, pragmatic, although difficult, dialogue. The issue is where such a constructive dialogue can take place.

Long ago, it was argued that the FAO was dominated by the producers and thus in need of reform. Today, trade and globalization greatly influence future food security and safety, and the TNCs play a major role. So far, the FAO has had little dialogue with the TNCs in agriculture and forestry. The global agricultural corporations must be allowed to join in efforts to reach the MDGs. As a first step, senior representatives of the TNCs should be allowed to participate in deliberations of the FAO, WHO, UNEP, UNIDO and other agencies. Another area of mutual benefit would be more exchange of instant information. This is an advantage of the TNCs, although their information bases are not easily available to outsiders, including the UN system. Such access could be useful to nation-states for up-to-date information, for instance on agricultural exports, imports and trade in general. This would be relevant even if competition, and classified information, imply that data sharing might carry some restrictions. Nevertheless, a closer dialogue and information exchange must not imply that the UN agencies shy away from critical research and policy analysis on both the TNCs and individual governments and their environmental, social and developmental impacts in developing countries. Among others, it may include a reactivation of the defunct UN Centre on Trans-National Corporations or a similar body.

For the future, it is urgent to investigate the need for much closer integration of the tasks of some of the UN specialized agencies. Although the FAO is a major technical actor on food and agriculture, the future problem area is more complex. In accordance with the MDGs, there is need for actions on a broader approach, for example, agricultural production combined with safety nets for those with no access to productive resources. It also relates to food aid by the World Food Program. IFAD has a special mandate to reach poor farmers, mainly through development projects. On trade and markets, the work of WTO has implications, in particular on access to genetic resources, IPRs and biotechnology, all areas where the FAO is the major normative actor. UNEP has a mandate for the environment that is a necessary component of the sustainable use of productive resources. Agriculture and land use must be closely concerned with environmental issues to deal with food security. This division has prevented genuine cooperation and led to an isolation of environmental aspects in reality. It would be necessary, and a great advantage, to remodel these organizations. Ideally, there ought to be one powerful UN organization dealing with food, land use and environmental issues in the future. It would save costs, improve efficiency and facilitate coordination at the international and country levels. It should be based on policy (from the FAO), environmental aspects (UNEP), food aid and safety nets (through the World Food Program) and development projects with a focus on the poor (through IFAD).

Another aspect refers to the need to realize that the old category of developing countries is now redundant. Therefore, the UN system and its specialized agencies must operate differently when they work in Asia, the Pacific, the Middle East, Latin and Central America or Asia. There is a great need to differentiate between groups of so-called developing countries. With a focus on the MDGs, the focus will be on Sub-Saharan Africa and parts of Asia, calling for different types of support than in other regions of the world. Also, this concentration of efforts may force the specialized agencies to introduce a mechanism of setting priorities not only on geographical areas but also on what to do and what not to do in light of available human and financial resources.

Poverty Focus in Swedish Development Assistance: Rhetoric and Reality

As early as in the 1960s, Swedish development assistance through SIDA aimed at contributing to raise the living standards of poor people. The agency had six programme countries for its bilateral support. A prominent feature of early Swedish support to research, starting in the late 1970s, was a belief that developing countries have much to learn from each other also, not just from contacts confined to Swedish research institutions. Research collaboration was to be promoted. A SAREC contribution advocated that agricultural research be poverty-oriented at WCARRD in 1979. At that time, there was an intensive public debate, many observers arguing that the Swedish aid should have one major goal, poverty eradication, and not just spend the annual funds. The management of SIDA and several members of Parliament countered that there was no need to change Swedish aid and its orientation. The poor were in focus, albeit with only a few demonstrations of real impact.

In 1988, there were 17 countries receiving long-term financial support from SIDA together with some 40 more countries receiving grants. In addition, financial resources were allocated to multilateral activities and various regional activities, emergency operations, special projects and environmental activities. At about this stage, the Minister of Development Cooperation concluded that Swedish companies should be encouraged to participate more actively in development assistance. They had, as exceptions, been involved for years in the Bai Bang paper pulp project in Vietnam. Other observers argued that the results of Swedish aid had been meagre and the concept of aid was inefficient (Karlstrom, 1991). Now, Swedish bilateral support was directed to 19 developing countries (12 of which were one-party states). In the early 2000s, more than 140 different countries received Swedish development assistance.

Around 1990, SIDA initiated a study on the minimum requirements for rehabilitating some universities in southern Africa. The universities were in serious financial difficulties, having grown rapidly and admitted a large number of students. A few years later, a joint SAREC and SIDA proposal to the Swedish Government offered that SAREC would provide general research support to universities and SIDA should finance basic academic education at selected African universities. The proposal did not discuss how such support could contribute to poverty eradication, prospects of employment after graduation and possible negative consequences of intellectual imperialism, dependence on Western science and academic traditions with external ownership. The basic idea was in contrast to efforts in the Netherlands, where the issue was to find a proper balance between a principle of Southern ownership of research programmes and a principle of mutually beneficial cooperation between Southern and Northern partners (RAWOO, 2001). The issue of ownership has now reappeared prominently in the Paris Declaration. Swedish support to African universities expanded further when the government decided in 1995 to merge all small, independent Swedish aid agencies into a new Sida. The former SAREC became a department for research cooperation. It was perceived that access to more funds expanded the opportunities for development research and for research as part of overall development cooperation.

In the mid-1990s, the Swedish Government also set aside funds for development assistance to the Baltic States. This was a unique move bypassing Sida. Recent evaluations have, however, shown that these funds did not reach the objectives, which were to strengthen Swedish entrepreneurs and create new jobs. Fewer than 250 new jobs were created during a five-year period at a price tag of SEK 4 million for each job. Another questionable approach still practised by Sida and other donors is to employ young Swedes as associate experts in the South. In contrast, immigrants to Sweden with higher academic degrees have great difficulties in finding employment, in particular those born in Africa and Asia. In 2003, only 60 per cent of all the academics born outside Sweden had Swedish employment matching their qualifications.

Sida´s own overall assessment of all its activities indicates that direct effects on poverty eradication constituted 17-21 per cent of the funding during 2001-2003 but less was spent on environment and sustainable development (11%) (Sida, 2003, p. 19). Most financial support was geared towards indirect effects through policy and institutions (44%). Budget allocations to private sector development remained quite small, the projects primarily for support to business at micro level (Sinha et al., 2001). The growing proportion of Swedish development assistance was directed towards peace, democracy and human rights as major goals or sub-goals. The traditional Sida approach towards raising farm productivity has recently changed to larger diversity in programmes, decentralization and market orientation (Havnevik et al., 2003). The withdrawal of the State in Africa led Sida to put more emphasis on the private sector and non-governmental actors.

Budget allocations for research cooperation account for 7-8 per cent of the total Sida budget. In the early 2000s, only 1 per cent of the investments had direct effects on poverty eradication (Sida, 2003, p. 69, 2004, p. 61). The number of projects with direct involvement of poor people was small. Financial support to agriculture and rural development had declined. Most research support was geared towards indirect effects through policy and institutions (93%), such as academic training. In many poor countries, only 1-2 per cent of the university teachers hold a PhD. This calls for efforts that have to be assessed in the perspective of the MDGs rather than just financial support to build up research capacity in a range of countries. Moreover, many new graduates often face unemployment in their own countries. In 2004, the top five countries for research support were Tanzania, Uganda, Nicaragua, Vietnam and Bolivia (Table 11). Another issue of concern is whether the research support by Sida and other donors is contributing to a shift towards the resolution of various limitations identified long ago (Streeten, 1974).

Table 11. Top seven countries receiving Swedish grants for research 1980/81-2004 (SEK in million).

Country	Fiscal year					
	1980/81	1990/91	1998	2000	2002	2004
Uganda	-	-	-	7.1	44.6	20.1
Tanzania	2.0	12.0	28.1	43.2	42.3	37.7
Mozambique	1.6	11.2	31.5	29.0	27.2	-
Ethiopia	3.2	15.6	22.0	13.7	22.9	-
Nicaragua	1.2	11.5	14.5	20.5	21.3	27.5
Vietnam	2.0	6.0	-	-	-	25.2
Bolivia	-	-	-	-	-	19.1

Source: SAREC and Sida Annual Reports.

Those highly committed to development assistance of relevance and good quality had great expectations for reform through the work by a Swedish Parliamentary Commission on Global Policy. It deals with access to global public good, contributing to global development by encompassing all relevant sectors both in Sweden and in developing countries (SOU, 2001). The former goal of development assistance to improve living conditions of poor people was to remain. The number of recipient countries should be reduced to 20. The Commission made no assessment of past achievements during more than 40 years of development assistance. Few thoughts were given to what changes were needed and little analysis on the best approaches to reach the poor. Sida has noted that its new goal means a sharper focus on poverty issues (Sida, 2004, p. 16). The major message was that Swedish development assistance should reach 1 per cent of GNP by 2005. There are many proposals but few priorities. Although the Commission concluded that rural development is one means of eradicating poverty, specific proposals were non-existent. In the early 1990s, only 2-3 per cent of total Swedish aid was directed to food production and rural development. A decade later, Sida's allocations to natural resources and environmental aspects (including agricultural activities) accounted for some 16 per cent. In 2005, the Swedish allocation to development cooperation amounted to about SEK 22 billion. Some 60 per cent of this sum is channelled via Sida. The Ministry for Foreign Affairs and the Export Credits Guarantee Board are other major actors.

The government bill on development assistance in 2003 advocated a broad approach with an integrated policy for global justice and sustainable development. Making reference to the formation of the Swedish welfare state and Western-type modernization, it advocated that labour unions and the private sector were to be involved. This runs counter to the basic notion advocated by the Commission that the situation in the South should be the starting point for a formulation of future Swedish assistance and the problems identified should be the focus. Above all, the bill implies a notion that a Swedish type of development, possibly modified, was to be advocated abroad. This may be questioned since the equity aspects have been given less attention, a trend over several years. In 2005, the average salary increase for Swedish CEOs reached 16 per cent, compared to 3 per cent for the average industrial worker. Moreover, a Swedish type of development may even carry certain dangers of cultural imperialism, such as priority aid to piano playing in India and jazz in China. There is no discussion on conflicting goals in the bill. For instance, a key arbiter on trade success can be the scientific judgement in assessing animal, plant and human health risks. This is a problem considering the great imbalance in scientific resources between developed and developing countries. The costs of sanitary and phytosanitary infrastructure conforming to WTO rules are enormous, estimated at US$ 150 million per country. Finally, the Swedish Government refrains from making priorities and providing guidelines on the involvement of the private sector. In brief, the proposed aid policy means the provision of public "goodies" to a majority of poor people.

Although the OECD recently has praised Swedish aid with its focus on the recipient partners, the persistent focus on volume of aid without analysis of the quality and the effects is somewhat unrealistic in view of the huge target to reach the MDGs. This would call for a long-term strategy on both multilateral and bilateral development assistance and their interaction at national levels. To be effective, future Swedish bilateral aid should not be dispersed to a wide range of countries but concentrated on a maximum of 15 to 20 democratic countries for the period up to 2015. The focus should be to strengthen relevant institutions in low-income countries where governments have shown determined actions towards poverty eradication. It requires intensive professional work by a reformed and slimmer Sida. This would be in contrast to a current involvement in more than 110 countries, although some 45 are classified as target countries. Another important task would be humanitarian and relief aid.

Future Swedish support to international organizations could be handled by the Ministry for Foreign Affairs or delegated to the EU. Future support to research and capacity building may require a specialized agency with a global outlook and two objectives: (1) strengthening research capacity in some 20 developing countries in accordance with the MDGs and (2) supporting research for development, that is, finding solutions to development problems through a network of research institutions. This calls for long-term funding (about 10 to 15 years) to priority areas for development in line with the MDGs. On the basis of 25 years of experience it would be realistic to select 20-30 key research institutions. This will also require a policy on what development problems should be allocated research grants in the context of the MDGs. More research is needed on issues such as land use, natural resource management, energy and water resources, agrobiodiversity, long-term soil fertility, food safety. At the same time, recipient governments should commit themselves to matching funds at an annual increase of 5-10 per cent in order to facilitate the takeover when Swedish research funds are withdrawn. This requires scientific competence, mutual interest in building long-term arrangements and insights about which research support can make a difference to people in the greatest need of help.

Some Conclusions

- Development assistance should be based on true partnership with democratic governments demonstrating political will and action towards the fulfilment of the MDGs. Support to activities should be based on long-term commitments (at least 20 years) for building national institutional capacity. Such commitments must include transparent exit clauses. This requires good preparatory work undertaken jointly by the target country and other donors rather than individual missions of one or two weeks. To nations at war and undemocratic governments, only humanitarian aid should be offered through NGOs.
- Development assistance should primarily be given to low-income countries. If their government policy is failing to reach poor people, aid agencies should stop their otherwise long-term aid programmes. A focus on improved livelihoods means food security and safety, land use issues and rural and urban development as major components of the MDGs.
- To reach the MDGs, it will be necessary to form certain alliances with the private sector wherever possible in agriculture and forestry. Such cooperation must recognize the differences between the public agenda for development and the agenda of the business sector to make profit while contributing to prosperity and development. This calls for clear policy guidelines at the country level. The national policy-makers, not the donors, should decide on the destiny and approaches for the development of their country.
- In addition to poverty reduction programmes according to the MDGs, future development assistance should be confined to long-term research capacity building, the strengthening of national institutions and research on policy issues. These areas are all of a long-term nature. Scientists of recipient countries should design their own research strategies and donors should primarily finance research for the solution

of the most urgent development problems. This may also stimulate the national government to invest more funds in problem-solving research. Many collaborative research arrangements should gradually move towards South-South cooperation rather than tying cooperation to research departments in the donor country.

- Poverty reduction should be regularly monitored and annually reported to the whole donor community by one central agency in the country concerned. Brief reports on accomplishments should be presented every second year in national Parliaments of both recipients and donors.

- One UN agency should be responsible for monitoring and reporting on the MDGs at a global level. This may be a role for a reformed FAO and a significantly strengthened Global Perspective Unit.

- There is a need for national aid agencies to have a strategic think tank for initiating and directing discussions on emerging issues. Such preparedness for the future would keep national policy-makers from determining the future content of their aid *ad hoc* on the basis of current domestic affairs and discussions.

- The on-going UN reform has to review the specialized agencies of the United Nations in order to make them tackle problems with higher complexity. This means that environmental aspects must be integrated with agriculture, health and industry. It would be advantageous to integrate activities of current organizations into one powerful body to obtain efficiency, save costs and facilitate coordination at the international and country levels. It should be based on policy (from the FAO), environmental aspects (from UNEP), food aid and safety net (through the World Food Program) and development projects with a focus on the poor (through IFAD).

- Member countries of the EU ought to thoroughly analyse its role in future development assistance, keeping in mind that they have their own bilateral and multilateral activities in addition to funds they provide to the EU. This system adds to bureaucracy, duplication of work and costs. One option would be that the EU not be involved in development assistance work or manage only multilateral contributions. Alternatively, the number of national aid agencies and/ or their budgets might be reduced, in view of the low absorptive capacity recognized by the EU.

Annex I

Glossary

Agricultural extension service: a non-formal form of education, usually taking place at the farm level, and being a system of transferring innovations to the farmer and research problems at the farm level back to the scientist.

Agricultural innovation: a major mechanical, biological, chemical improvement or a major improvement in agricultural practices and methods that will enhance agriculture.

Agricultural research: a systematic search process for gaining and applying knowledge efficiently to the biological, physical and economical phases of producing, processing and distributing farm and forest products, to consumer health and nutrition and to the social and economic aspects of rural life

Animal breed: usually a group of animals with definable and identifiable characteristics distinguishing it from other groups of the same species.

Bioethics: an ethical principle stating that the biosphere has intrinsic value. This means that it has a right of existence for itself.

Biological diversity or biodiversity: total variability within all the living organisms and the ecological complexes they inhabit.

Biotechnology: any technique that uses a living organism or substances from those organisms to make a modified product, improve a plant or animal or develop microorganisms for special uses.

- genomics: the molecular characterization of all species;
- bio informatics: the assembly of data from genomic analysis into accessible forms;
- transformation: the introduction of one or more genes conferring potentially useful traits into plants, livestock, fish and tree species;
- molecular breeding: the identification and evaluation of desirable traits in breeding programmes by the use of marker-assisted selection, for plants, trees, animals and fish;
- diagnostics: the use of molecular characterization to provide more accurate and rapid identification of pathogens and other organisms;

- vaccine technology: the use of modern immunology to develop recombinant DNA vaccines for improving control against lethal diseases.

Cell or tissue culture: isolation of plant and animal cells to raise them to fully grown organisms capable of independent reproduction.

Crop cultivar: an assemblage of cultivated individuals, designated by morphological, physiological, chemical or other characters significant for the purpose of agriculture, horticulture or forestry, that when reproduced sexually or asexually retains its distinguishing features.

Crop variety: an intra-specific unit, representing a morphological variant of the species, often having its own geographical distribution.

Determinism: a view that every event has a cause. Human action and human destiny are not the outcome of human free will. Outside forces, such as nature or God, economics and history limit what we can and cannot do and achieve. Biological determinism claims that the final cause of human actions lies in the genes. Fatalistic determinism argues that all events could be predicted if we knew fully the laws and processes that cause them. Chance or free will has no influence so whatever happens should be accepted (Pepper, 1996).

Ecosystem: communities of interacting organisms and the physical environment in which they live.

Elitism: the interests of a particular elite that is disproportionately represented and influential. Its decisions become undemocratic.

Emergy analysis: analysis of different forms of energy and materials, human labour and economic services by converting them into equivalents of one form of energy, namely solar energy.

Energy analysis: determination of the energy sequestered by the process of making a good or service within the framework of an agreed set of conventions.

Enlightenment: a European and North American movement flourishing in the 18th century. It started from a basic belief that by applying the laws of nature, the material position of humankind could be indefinitely improved.

Entropy: a measure of the degree of randomness in a closed system; degradation or disorder of the universe.

Ex situ conservation: conservation of a plant outside its original or natural habitat.

Gaia: Greek name for the Earth Goddess. James Lovelock's Gaia theory proposes the earth as a complex system that is "alive". Thus, it can continually reconstitute and repair itself in response to environmental changes.

Gene bank: facility in which crop diversity is stored in the form of seeds, pollen or *in vitro* culture, or in the case of a field gene bank as plants growing in the field.

Gene mapping: identifying the sequence of bases of each strand of DNA.

Gene pool: all the genes and their different alleles present in an interbreeding population.

Genetic engineering: the creation of new plant/animal types through genetic manipulation of the gene pool of an organism by the introduction of non-species-specific genes often from other taxa/phyla.

Genetic erosion: loss of genetic diversity between and within populations of the same species over time or reduction of the genetic base of a species due to human intervention, environmental changes or other causes.

Genetic resources: genetic material of plants, animals and other organisms, which is of value for present and future generations of people.

Germplasm: a set of genotypes that may be conserved or used.

In situ conservation: conservation of plants or animals in the areas where they developed their distinctive properties.

Integrated pest management (IPM): the FAO defines IPM as the careful consideration of all appropriate measures that discourage the development of pest populations and keep pesticides and other interventions to levels that are economically justified and reduce or minimize risks to human health and the environment.

Intrinsic value: a value inherent within something that does not depend on the ideas, preferences or prejudices of an external evaluator.

Laws of thermodynamics:

1st law: in an energy system with a constant mass, energy cannot be created or destroyed though it can change form,

2nd law: energy always moves from hotter to colder parts of the system (the law of increasing entropy or disorder in a closed system),

3rd law: it is impossible to cool an energy system to the absolute zero of temperature.

Livelihoods: the capabilities, assets (material and social resources) and activities required for a means of living.

Materialism: a philosophy that everything that exists is material, occupying some space at some time.

Modernism: "an international tendency" in arts and architecture of the West in the late 19th and 20th centuries (Bullock and Stallybrass, 1988). The term means not only the approaches to knowledge and scientific, technological and industrial developments but also the hopes and aspirations of the period from the 18th to the 20th century.

Nanotechnology: the theories and techniques that permit the production and manipulation of minute objects measuring the size of atoms into new structures, devices and systems of novel properties. A nano is one thousand millionth.

Objectivity: the property of being separate from whatever is being studied. The subject studying the object does not influence in any way the properties of the object. It is also the property of being separate from the interests, perspectives and points of view of any particular social group or from any preconceived notion.

Organic agriculture: an agriculture that promotes and enhances agro-ecosystem health, including biodiversity, biological cycles and soil biological activity. In this regard, the greatest possible use of cultivation methods, whether biological ormechanical, is preferred to the application of synthetic products in order for the overall system to function fully (Codex Alimentarius, 1995).

Paradigm: a set of rules, assumptions or procedures under which people study and investigate the subject matter of a discipline.

Plant husbandry: research and experimental work on the useful plants and weeds together with their abilities under different ecological conditions.

Policy research: scientifically produced knowledge with subsequent recommendations and consequential analysis to facilitate decision-making by policy-makers on how to solve complex, multidisciplinary problems to benefit society in a long-term perspective.

Private sector: a basic organizing principle for economic activity where private ownership is an important factor, markets and competition drive production and private initiative and risk taking set activities in motion.

Private sector development: the process by which the private sector moves along the path to becoming well functioning.

Reductionism: the view that a whole can be understood by reducing it to its elementary and basic constituents.

Research: a systematic process for the development of science and technology. It can be both informal and formal. The traditional technology is based upon empirical knowledge, whereas modern technology is based on modern scientific principles.

Research policy: an attempt to combine the technological improvements, their introduction in society and the changes that may or may not follow for the different categories of farmers and the whole rural setting.

Resource-poor farmers: farmers with little or no access to or control of productive resources, land being one of the most important of those resources.

Rural development: a process of change whereby poverty will be reduced. The poor should have access to the resources of society and the environment. They should be encouraged to achieve control of the resources that are introduced at reasonable costs from outside the rural environment.

Scientific basis: critical review of discovered facts of relevance to society.

Scientific methodology: open, transparent and honest description of relevant facts, facilitating critical review.

Science: any system of knowledge that is concerned with the physical world and its phenomena and that entails unbiased observations and systematic experimentation. In general, a science involves a pursuit of knowledge covering general truths or the operations of fundamental laws.

Subsidiary: the principle that decisions should be taken at the most local level that is appropriate, furthest removed from the centre of power.

Sustainable agriculture: the management and conservation of the natural resource base, and the orientation of technological and institutional change in such a manner as to ensure the attainment and continued satisfaction of human needs for present and future generations.

Technology: a society's pool of knowledge regarding industrial arts and sciences, including the principles of physical and social phenomena and the application of these principles and the day-to-day operation of production.

Urban agriculture: an industry that produces, processes and markets food and fuel, largely in response to the daily demand of consumers within a town, city or metropolis, on land and water dispersed throughout the urban and peri-urban area.

Annex II

Acronyms and Abbreviations

AIDS	Acquired Immune Deficiency Syndrome
AU	African Union
AVDRC	Asian Vegetable Research and Development Center
Birr	Ethiopian Birr; 1.00 Birr is equivalent to US$ 0.48
BSE	Bovine Spongiform Encephalopathy, Mad Cow Disease
Bt	*Bacillus thuringiensis*
CADU	Chilalo Agricultural Development Unit
CARDI	Caribbean Agricultural Research and Development Institute
CBD	Convention on Biological Diversity
CFC	Chlorofluorocarbons
CGIAR	Consultative Group on International Agricultural Research
CIAT	International Center for Tropical Agriculture
CIFOR	Center for International Forestry Research
CIMMYT	International Maize and Wheat Improvement Center
CIP	International Potato Center
CSIRO	Australian Commonwealth Scientific and Research Organization
DES	Estrogen (diethylstilbestrol)
EPA	US Environmental Protection Agency
EPMR	External Programme and Management Review
EU	European Union
FAO	Food and Agriculture Organization of the United Nations
FDA	Food and Drug Administration
FOOD21	Swedish Interdisciplinary Research Program for Sustainable Production
Formas	Swedish Research Council for Environment, Agricultural Sciences and Spatial Planning
GATT	General Agreement on Tariffs and Trade
GM	Genetically modified
GMO	Genetically modified organism
GRPC	CGIAR Genetic Resources Policy Committee
Habitat	United Nations Human Settlements Programme

IBSRAM	International Board on Soil Research and Management, now integrated with IWMI
ICARDA	International Center for Agricultural Research in the Dry Areas
ICBA	International Center for Biosaline Agriculture
ICIMOD	International Centre for Integrated Mountain Development, Kathmandu, Nepal
ICIPE	International Centre of Insect Physiology and Ecology
ICLARM	World Fish Center
ICRAF	World Agroforestry Center
ICRISAT	International Crops Research Institute for the Semi-Arid Tropics
IFAD	International Fund for Agricultural Development
IFDC	International Fertilizer Development Center
IFPRI	International Food Policy Research Institute
IFS	International Foundation for Science
IFSM	Integrated Soil Fertility Management Project
IITA	International Institute of Tropical Agriculture
ILRAD	International Laboratory for Research on Animal Diseases
ILRI	International Livestock Research Institute
IMF	International Monetary Fund
INIBAP	International Network for the Improvement of Banana and Plantain
IPGRI	International Plant Genetic Resources Institute
IPM	Integrated Pest Management
IPR	Intellectual Property Rights
IRRI	International Rice Research Institute
ISNAR	International Service for National Agricultural Research
ITPGR	International Treaty for Plant Genetic Resources for Food and Agriculture
IWMI	International Water Management Institute
LRF	Swedish Federation of Farmers
MDGs	Millennium Development Goals
MPA	Medroxyprogesteron-acetate
NAFTA	North American Free Trade Agreement
NARS	National Agricultural Research Systems
NERICA	The New Rice for Africa
NGOs	Non-governmental organizations
NvCJD	Creutzfeldt-Jakob´s Disease
OECD	Organization of Economic Co-operation and Development
PCB	Polychlorinated biphenyls
PFOS	Perfluoroctansulphonate

PRSP	Poverty Reduction Strategy Paper
rBGH	Recombinant bovine growth hormone
SAREC	Swedish Agency for Research Cooperation with Developing Countries
SARS	Severe Acute Respiratory Syndrome
SEK	Swedish crown; 1.00 SEK is equivalent to US$ 0.14
SG 2000	Sasakawa-Global 2000
SIDA	Swedish International Development Authority (prior to 1995)
Sida	Swedish International Development Cooperation Agency (as of 1995)
SJFR	Swedish Council of Forestry and Agricultural Research
SLU	Swedish University of Agricultural Sciences
SPFS	FAO Special Programme for Food Security
TNC	Transnational corporation
TRIPS	Trade Related Intellectual Property Rights Agreement
TT$	Trinidad and Tobago dollar; TT$ 1.00 is equivalent to US$ 0.16
UNCED	United Nations Conference on Environment and Development
UNDP	United Nations Development Programme
UNEP	United Nations Environment Programme
UNFPA	United Nations Fund for Population Activities
UPOV	International Union for the Protection of New Varieties of Plants
USAID	United States Agency for International Development
USDA	United States Department of Agriculture
UWI	University of the West Indies
Vinnova	Swedish Agency for Innovation Systems
WARDA	West Africa Rice Development Association
WCARRD	World Conference on Agrarian Reform and Rural Development
WFP	World Food Program
WFS	World Food Summit
WHO	World Health Organization
WIPO	World Intellectual Property Organization
WTO	World Trade Organization
WWF	Worldwide Fund for Nature

Annex III

The Millennium Development Goals

GOAL 1: ERADICATE EXTREME POVERTY AND HUNGER

Target 1: Halve, between 1990 and 2015, the proportion of people whose income is less than $1 a day

Target 2: Halve, between 1990 and 2015, the proportion of people who suffer from hunger

GOAL 2: ACHIEVE UNIVERSAL PRIMARY EDUCATION

Target 3: Ensure that, by 2015, children everywhere, boys and girls alike, will be able to complete a full course of primary schooling

GOAL 3: PROMOTE GENDER EQUALITY AND EMPOWER WOMEN

Target 4: Eliminate gender disparity in primary and secondary education, preferably by 2005, and in all levels of education no later than 2015

GOAL 4: REDUCE CHILD MORTALITY

Target 5: Reduce by two thirds, between 1990 and 2015, the under-five mortality rate

GOAL 5: IMPROVE MATERNAL HEALTH

Target 6: Reduce by three quarters, between 1990 and 2015, the maternal mortality ratio

GOAL 6: COMBAT HIV/AIDS, MALARIA AND OTHER DISEASES

Target 7: Have halted by 2015 and begun to reverse the spread of HIV/AIDS

Target 8: Have halted by 2015 and begun to reverse the incidence of malaria and other major diseases

GOAL 7: ENSURE ENVIRONMENTAL SUSTAINABILITY

Target 9: Integrate the principles of sustainable development into country policies and programmes and reverse the loss of environmental resources

Target 10: Halve by 2015 the proportion of people without sustainable access to safe drinking water

Target 11: Have achieved by 2020 a significant improvement in the lives of at least 100 million slum dwellers

GOAL 8: DEVELOP A GLOBAL PARTNERSHIP FOR DEVELOPMENT
Target 12: Develop further an open, rule-based, predictable, non-discriminatory trading and financial system (includes a commitment to good governance, development, and poverty reduction—both nationally and internationally)

Target 13: Address the special needs of the least developed countries (includes tariff- and quota-free access for exports, enhanced programme of debt relief and cancellation of official bilateral debt, and more generous official development assistance for countries committed to poverty reduction)

Target 14: Address the special needs of land-locked countries and small island developing states through the Program of Action for the Sustainable Development of Small Island Developing States and 22nd General Assembly provisions

Target 15: Deal comprehensively with the debt problems of developing countries through national and international measures in order to make debt sustainable in the long term

Target 16: In cooperation with developing countries, develop and implement strategies for decent and productive work for youth

Target 17: In cooperation with pharmaceutical companies, provide access to affordable essential drugs in developing countries

Target 18: In cooperation with the private sector, make available the benefits of new technologies, especially information and communications technologies

Annex IV

Major International Conferences and Conventions of Relevance to Poverty Eradication and Agriculture

Some International Conferences

2006: United Nations Conference on Agrarian Reform and Rural Development in Porto Allegro

2002: World Summit on Sustainable Development in Johannesburg

2001: World Food Summit–Five Years Later in Rome

1996: United Nations 2nd Conference on Habitat (Habitat II) in Istanbul;

1996: World Food Summit in Rome

1995: World Summit for Social Development in Copenhagen

1995: Fourth World Conference on Women in Beijing

1994: United Nations 3rd International Conference on Population in Cairo

1993: United Nations Conference on Human Rights in Vienna

1992: United Nations Conference on Environment and Development, Rio de Janeiro

1992: International Conference on Nutrition in Rome

1990: World Summit for Children in New York

1984: United Nations 2nd International Conference on Population in Mexico City

1979: World Conference on Agrarian Reform and Rural Development in Rome

1978: International Conference on Primary Health Care in Alma Ata

1976. United Nations 1st Conference on Habitat in Vancouver

1974: United Nations Conference on Fisheries in Caracas

1974: United Nations World Food Conference in Rome

1974: United Nations 1st World Population Conference in Bucharest

1970: 2nd World Food Congress in The Hague

1963: 1st World Food Congress in Washington D.C.

Some International Conventions or Protocols of Relevance To Agriculture, Forestry and Fisheries

2005: The Kyoto Protocol to the United Nations Framework Convention on Climate Change

2004: International Treaty on Plant Genetic Resources for Food and Agriculture

2004: Stockholm Convention on Persistent Organic Pollutants

2003: Cartagena Protocol on Biosafety

1998: Convention on Access to Information, Public Participation in Decision-Making and Access to Justice in Environmental Matters (Århus Convention), Århus

1996: Convention on Liability and Compensation for Damage in Connection with the Carriage of Hazardous and Noxious Substances by Sea (HNS), London

1994: United Nations Convention to Combat Desertification (UNCCD)

1993: Convention on Biological Diversity (CBD)

1993: Legally Binding Agreement to Promote Compliance with International Conservation and Management Measures by Fishing Vessels on High Seas

1992: United Nations Framework Convention on Climate Change (UNFCCC)

1991: Convention on Environmental Impact Assessment in a Transboundary Context, Espoo

1989: Basel Convention on the Control of Transboundary Movements of Hazardous Wastes and their Disposal

1989: Convention Concerning Indigenous and Tribal Peoples in Independent Countries (ILO Nr. 169)

1987: Vienna Convention on the Protection of the Ozone Layer/Montreal Protocol on Substances that Deplete the Ozone Layer

1983 and 1989: FAO International Undertaking on Plant Genetic Resources, Rome

1982: United Nations Convention on the Law of the Sea (UNCLOS), Montego Bay

1980: Convention on the Conservation of Antarctic Marine Living Resources (CCAMLR), Canberra

1979: Convention on the Conservation of Migratory Species of Wild Animals (CMS), Bonn

1979: Convention on Long-Range Transboundary Air Pollution (LRTAP), Geneva

1975: Convention on Wetlands of International Importance especially as Waterfowl Habitat (Ramsar Convention)

1973: Convention on International Trade in Endangered Species of Wild Flora and Fauna (CITES)

1972: Convention Concerning the Protection of the World Cultural and Natural Heritage (World Heritage Convention), Paris

1972: Convention on the Prevention of Marine Pollution by Dumping Wastes and Other Matter, London

1966: International Convention for the Conservation of Atlantic Tunas (ICCAT)

1952, 1979, 1997: International Plant Protection Convention (IPPC)

1946: International Convention for the Regulation of Whaling (ICRW)

Some Conventions on Hazardous Substances and Nuclear Safety

1998: Convention on the Prior Informed Consent Procedure for Certain Hazardous Chemicals and Pesticides in International Trade (PIC Convention), Rotterdam

1995: Convention to Ban the Importation into Forum Island Countries of Hazardous and Radioactive Wastes and to Control the Trans-boundary Movement and Management of Hazardous Wastes within the South Pacific Region, Waigani

1994: Convention on Nuclear Safety, Vienna

1991: Convention on the Ban of the Import into Africa and the Control of Transboundary Movements and Management of Hazardous Wastes within Africa, Bamako

1986: Convention on Assistance in the Case of Nuclear Accident or Radiological Emergency, Vienna

1986: Convention on Early Notification of a Nuclear Accident, Vienna

1989: Convention on the Control of Transboundary Movements of Hazardous Wastes and their Disposal, Basel

1983: Geneva Convention on Long-Range Transboundary Air Pollution

1963: Vienna Convention on Civil Liability for Nuclear Damage, Vienna

Some International Agreements

1997: Conclusions and Proposals for Action of the Intergovernmental Panel on Forests (IPF)

1995: Forest Partnership Agreement (FPA)/National Forest Plan (NFP)

1995: Establishment of the World Trade Organization (WTO)

1995: Uruguay Round negotiations under the GATT Agreement include:
- Agreement on Technical Barriers to Trade
- Agreement on Sanitary and Phytosanitary Measures

- Agreement on Subsidies and Countervailing Measures
- Agreement on Trade-Related Aspects of Intellectual Property Rights (TRIPS)

Before the conclusion of the Uruguay Round, most agricultural trade took place outside the normal rules governing other trade. Multilateral Environmental Agreements (MEA) also emerged from Uruguay Round GATT negotiations. Of some 200 MEAs, about 20 contain trade provisions and agriculture was brought into the international trade system.

1995: Code of Conduct for Responsible Fisheries

1994/95: Code of Practice on the Introduction and Transfer of Marine Organisms of the International Council for the exploration of the Sea

1994: International Tropical Timber Agreement (ITTA)

1993: Code of Conduct on Collecting and Transfer of Germplasm

1992: North American Free Trade Agreement Between the Government of the United States, the Government of Canada and the Government of the United Mexican States (NAFTA). In 1993, the North American Agreement on Environmental Co-operation (NAAEC) became a supplement. Both agreements came into force in 1994.

1992: Non-legally Binding Authoritative Statement of Principles for a Global Consensus on the Management, Conservation and Sustainable Development of All Types of Forests–The Forest Principles

1992: Agenda 21

1992: Intergovernmental Forum on Forests (IFF)

1985: FAO International Code of Conduct on the Distribution and Use of Pesticides, Rome

1985: Code of Conduct on the distribution and use of pesticides: Resolution 10/85 of the 23rd FAO Conference, Rome

1979: International Code on Ethics in International Trade in Food

1948: General Agreement on Tariffs and Trade Agreement (GATT)

Annex V

Some Highlights of Impact, Research and Policy Orientation by CGIAR Institutes and ICIPE in the Mid-1970s and the Early 2000s

The following text highlights some prominent contributions by CGIAR institutes and ICIPE not previously mentioned. They were presented in annual reports by the respective institutions in the mid-and late 1970s and the early 2000s (Overview in Bengtsson, 1977; Annual Reports 2000-2003; CGIAR, 1998, 2001 and 2004).

International Rice Research Institute (IRRI)

From its start in 1960, IRRI focused on irrigated rice and in 1976 decided to include rice on non-irrigated land. In 1978, IRRI stressed explicitly the need to strengthen national rice research institutions and started to provide assistance directly, for instance to Cambodia (restoring production interrupted by war) and India (hybrid rice).

The share of nationally released varieties with IRRI ancestors increased significantly between 1965-74 (54%) and 1981-90 (72%). Investigations on rice apomixis began in the early 1990s. IRRI has shown that pesticide use can be reduced by 60 per cent both in experiments and in the field, for instance in Vietnam and China. New training programmes have been established based on findings of pesticide poisoning and have been tested in Thailand and Vietnam. The idea of planting mixed rice varieties to control diseases has been further tested. In addition to a reduced need for agrochemicals, such planting gave higher hectare yields. A major challenge is to increase the productivity of unfavourable rice-producing regions, accounting for 40 per cent of the rice lands in Asia and most of the rice grown in Africa.

Africa Rice Center (formerly the West Africa Rice Development Association, WARDA)

WARDA started in 1971 as a regional cooperative effort in adaptive rice research among 14 West African countries. Initially, the CGIAR only supported the rice trials but WARDA gradually became a centre of the CGIAR. Due to pressure on land in the last 20 years, West African farmers have reduced the fallow periods of forest bush from 12-15 to 3-5 years. About 15 per cent of some 20-30 million ha of the inland valleys of West Africa are under cultivation. Less than 5 per cent is used for rice. In the mid-1990s, the WARDA made a breakthrough in research, developing new genetically stable and fully fertile hybrids between *Oryza glaberrima* and *Oryza sativa* (NERICA). The *O. glaberrima* landraces have developed resistance to several stresses, characteristics that have been transferred into new interspecific hybrids. Their potential is significant since upland rice is grown on 12 million ha in Asia and 4 million ha in Latin America. In 2001, WARDA launched the New Rice for Africa Consortium for Food Security in Sub-Saharan Africa. More than 60,000 West African farmers have begun to plant that new rice.

International Center of Tropical Agriculture (CIAT)

From its inception in 1967, CIAT conducted research on cassava as a major commodity. Other early research areas were low input technology, disease resistance on *Phaseolus* beans and a beef programme. Both CIAT and IITA had initially regional responsibilities for rice research.

Since 1970, national agricultural research systems in Latin America have released 363 bean varieties in 39 countries, based on germplasm provided by CIAT. In the early 2000s, the CIAT gene bank had some 22,000 forage samples. Research on forages in Indonesia and other Asian countries is reported to have contributed to increased cash income from the sale of livestock including labour saving of some 20 per cent. A rice variety fully resistant to Colombian hot spot of rice blast has been grown for a decade without breakdown of resistance. Participatory methods have boosted the adoption of soil conservation practices. Another success story is the research work on developing bean varieties resistant to bean golden mosaic virus. The cumulative global benefits of CIAT-related varieties were recently estimated at about US$ 8.7 billion (in 1990 dollars) during the three decades under study.

International Maize and Wheat Improvement Center (CIMMYT)

In the 1970s, CIMMYT gave major attention to higher yields of wheat, resistance to rust and *Septoria* sp. But yield stability was also coming into focus. Commercial durum wheat varieties were predicted by 1980. Wide adaptation and incorporating day-length insensitivity in short-stature lines were among major developments. The barley programme had run for five years and Triticale was considered a high potential crop. Courses on zero-tillage systems were offered.

In 2001, Triticale was grown on 3 million ha worldwide. After five years of studying 50,000 plants in the mid-1990s, CIMMYT found two plants that closely resembled maize and reproduced asexually (apomixis). This would allow farmers in developing countries to keep their seed pure, unaffected by pollen that often contaminates their seed. New plants were received from crosses between maize and the wild grass, *Tripsacum*, involving companies such as Syngenta and Pioneer Hi-Bred, Group Limagrain. Crosses between bread wheat and goat grass had produced hardy wheat for dry, tougher environments. CIMMYT has also been advocating water-saving zero-tillage within the Rice-Wheat Consortium for the Indo-Gangetic Plains. Wheat was directly seeded into paddy fields just after harvest on some 300,000 ha in the 2001-2002 growing season. No-tillage practices with mulching are reported to have reduced labour and efforts invested by smallholders. They cut production costs and reduced agricultural risks.

Some 90 per cent of the spring bread wheat area in the developing world is planted by CIMMYT-related varieties. In Asia, high-yielding wheat varieties are planted on 84 per cent of the wheat areas. Open-pollinated new quality protein maize varieties are seen as a breakthrough in helping prevent malnutrition. Two new varieties have demonstrated a doubling of lysine and tryptophane amino acids. This research was initiated some 35 years ago. CIMMYT has trained more than 6,000 researchers and technical staff.

International Center for Agricultural Research in the Dry Areas (ICARDA)

Established in 1976, ICARDA was to serve as a regional centre in semi-arid areas of North Africa and the Near East and as an international centre on barley. Other crops were lentils and broad beans.

In all, more than 440 new crop varieties using ICARDA germplasm have been released worldwide. In 2004 alone, collaborative research led to the release of 35 varieties of cereal and food crops and forage legume crops in 13 countries. ICARDA has helped China to improve barley lines, resulting in more than 20 per cent increase in productivity. Its collaborative work with Egypt and the Sudan on faba beans has contributed to a doubling of their production since the 1970s. This was achieved through the adoption of a package of disease-resistant varieties, agronomic practices, sowing date and tillage. ICARDA has reported on major efforts in regional agricultural research priority setting. Lentil lines of ICARDA germplasm were released in Iraq, Turkey and the Ethiopian highlands. A special seed unit of ICARDA has collaborated with national programmes during the last 15 years. ICARDA has also tried innovative irrigation in Kazakhstan, expected to benefit the dry continental climate of Central Asia. ICARDA has trained more than 7,500 persons.

International Crops Research Institute for the Semi-Arid Tropics (ICRISAT)

Founded in 1972, ICRISAT was given global responsibility for the improvement of staple food supplies in the semi-arid tropics. In the mid-1970s it kept some 21,000 accessions of sorghum and 6,600 entries of millets. Work was initiated on resistance to the *Striga* weed. Research on other crops included chickpea, groundnut and pigeon pea.

Resulting from ICRISAT partnerships with national agricultural research systems up to the early 2000s, some 400 improved cultivars have been released in 170 countries (130 sorghum, 71 pearl millet, 115 chickpea, 35 pigeon pea and 54 groundnut). Even today, some 450 million people make a living in the semi-arid tropics. The world's first pigeon pea hybrid from ICRISAT has reached farmers. In the early 2000s, significant contributions have come from public-private research partnerships; Bt pigeon pea is an example. Technology for a model for watershed welfare has been developed. The Institute has conducted successful trials on an integrated approach to soil-water-nutrient interactions. Progress is reported on chickpea research in Nepal, Andhra Pradesh in India and Ethiopia. In late 2003, ICRISAT signed a memorandum of understanding with the Andhra Pradesh Government to establish an Agri-Biotech Park. The Agri-Science Park combines the Agri-Biotech Park, the Agri-Business Incubator and a Hybrid Seed Consortium. The 700 ha Patancheru site was recently planned for eco-tourism.

International Potato Center (CIP)

The CIP was established to develop and disseminate potatoes in the tropics. In 1975, potatoes were grown on 2.6 million ha in developing countries. Sweet potatoes were later added to the research agenda.

Since the early 1960s, the average potato yield in developing countries has doubled. Exports of table and seed potatoes have risen sharply over the last three decades. In 2001, a CIP cross called Cooperation 88 was reported to produce large yield gains over previously favoured potato varieties in China. True potato seed has been seen as a means to tackle late blight of potatoes. In India, farmers were growing more than 10,000 ha of potatoes derived from true potato seed in the mid-1990s. Six hybrids were produced and some were sold to Vietnam, Egypt and the Philippines. In 1997, the CGIAR created the Global Mountain Program led by the CIP. Reports from case studies in the Andes have showed local improvements in the conservation of natural resources. In Asia, farmers produced more potatoes because of the development of short duration cereal varieties.

International Institute of Tropical Agriculture (IITA)

At the inception of IITA in the mid-1960s, cassava was given major research attention. But work was also directed to yams, sweet potatoes and aroids in that order of priority. Worldwide, 562 million tons of root crops were grown on 52 million ha in 1975. Then, IITA reported promising results of maize lines on acid soils. Cowpea was seen as a crop with high potential. IITA cooperated with WARDA on rice and with the International Soybean Program (USA) on soybean.

In the early 2000s, IITA scientists reported new technologies to control the parasitic *Striga* weed infesting some 21 million ha in Africa. By using the strengths of existing farming systems of growing legumes together with cereals the weed could be controlled. *Striga*-resistant maize varieties have been produced. According to IITA, overall maize production in West and Central Africa has increased by 295 per cent since the early 1990s. Work on soybean research has led to new varieties spreading in Nigeria. Scientists from CIAT and IITA have discovered a predatory insect to control the cassava green mites. A bug (*Typhlodromalus aripo*) was found in Brazil and introduced to Africa in the early 1990s. It spread to over 11 countries and can reduce mite infestations by up to 90 per cent. IITA collaborated with ICARDA in Egypt on *Orobanche*, which caused reduced faba bean production.

International Livestock Research Institute (ILRI)

Established in 1974, ILCA was the international livestock centre on production systems in Africa. The initial research was directed to arid and semi-arid rangelands. The Institute and its national partners early on developed a concept of planting forage legumes to help agro-pastoralists in West Africa. Up to 1999, an *ex post* assessment showed there were 27,000 adopters of forage legumes on 19,000 ha in 15 West African countries (Elbasha et al., 1999). The total net benefits to society accrued up to 1997 amounted to US$ 16.5 million, equivalent to an internal rate of return of 38 per cent.

ILRAD became operational in 1974. It was charged to develop effective control methods for trypanosomiasis and theileriosis. Its basic research quickly produced scientific results of high standard. It was more difficult to develop vaccines.

In 1995, ILCA and ILRAD merged into a global livestock centre, ILRI. This institute continued the research on the development of a vaccine against East Coast fever, trying to decode the parasite. Estimates have shown that investments in vaccine research would give a high internal rate of return under certain conditions (Kristjanson et al., 1999). This work is a good illustration of collaborative research between the public (ILRI) and private partners (Institute for Genomic Research, USA). ILRI has made an internal assessment of its research impacts and adoption of 53 individual research projects between 1975 and 1998. However, the assessment is less specific on dissemination of results. Rates of adoption of technologies are not very high. ILRI and its predecessors have trained more than 3,000 students from 80 countries.

International Plant Genetic Resources Institute (IPGRI, formerly the IBPGR)

The IBPGR was established in 1973 in an effort to prevent the threatened loss of diversity of many crops and assure future availability of genetic resources. Its task was to formulate polices and recommend actions.

The IBPGR was later changed from a board into a research institute, IPGRI. It presently highlights the adoption of the International Treaty on Plant Genetic Resources at FAO, in which it played a significant role. It has also been instrumental in a model for international public-private cooperation on cocoa germplasm. Other activities include efforts for the conservation and use of native fruit biodiversity on some priority species. The exchange of *Musa* genetic resources has intensified from being virtually non-existent before the creation of the International Network for the Improvement of Banana and Plantain. IPGRI has collected over 200,000 germplasm samples during more than 500 missions. Over 1,800 national program scientists have been trained by IPGRI, often representing their respective governments at international meetings. For instance, former IPGRI trainees accounted for one third of the developing country delegations to the International Technical Conference in Leipzig in 1996.

Center for International Forestry Research (CIFOR)

The CGIAR appointed the board of a new forestry research entity (CIFOR) in 1992. It became operational in 1993. Unlike in other CGIAR centres, the headquarters buildings were provided fully by the host government. The objective of the Center is to contribute to the sustained well-being of people in developing countries through collaborative strategic and applied research in forest systems and forestry and promote appropriate new technologies. CIFOR has focused its research on policy aspects, concentrating on sustainability indicators, forest fires and participatory approaches on local management of land and forest resources. It has established close collaborative links with some 300 researchers in 30 countries. As one result of its research in Indonesia, where about 55 to 75 per cent of industrial wood is produced illegally, the Asia Pulp and Paper Company signed an agreement with WWF, committing itself to use wood only from legal sources.

International Food Policy Research Institute (IFPRI)

The IFPRI was founded in 1975 as a non-CGIAR centre. It was accepted into the CGIAR in 1980. Its objective is to identify and analyse alternative national and international strategies and policies for meeting food needs of the developing world on a sustainable basis, focusing on low-income countries and on the poorer groups in those countries.

Since the mid-1990s, the IFPRI's 2020 Vision has significantly influenced the debate on international food policy. In recent years, the IFPRI has collaborated with some 50 developing country institutions. Its African focus is today on many fronts, food security being a complex goal, requiring multi-dimensional solutions and strategies for both food and nutrition, including trade policies. As of 2004, IFPRI is decentralizing to one hub in Addis Ababa, Ethiopia and another in New Delhi, India. The component in Ethiopia will study how institutional change can enhance the impact of agricultural research and help strengthen organization and management of agricultural research institutions. It is one activity of the former ISNAR, transferred to IFPRI in 2004.

Founded by CGIAR in 1979, ISNAR became operational in 1980. Its task was to help developing countries bring about sustained improvements in the performance of their national agricultural research systems and organizations. This included master plans, guidelines for strategic planning (Indonesia) or raising the capacity of national research staff (Mali, Senegal, Tanzania and Zimbabwe). Uganda is one example of its long-term involvement at the country level. On behalf of all CGIAR centres, ISNAR hosted the Central Advisory Service on Intellectual Property. Training courses on bio-safety were developed for some 500 participants.

World Fish Center (formerly the International Center for Living Aquatic Resources Management, ICLARM)

ICLARM was founded in 1977 as a non-CGIAR centre. It joined the CGIAR in 1992. Its focus is to improve the production and management of aquatic resources to benefit both low-income producers and consumers in developing countries. ICLARM has reported on new methods to raise productivity, providing 250 to 1,500 kg of fish per ha, maintaining the rice yields. Costs of rice production dropped by 10 per cent. Research on small-scale integrated aquaculture-agriculture is considered a great potential for balanced food security and a healthy environment. ICLARM has produced an improved strain of Tilapia, a freshwater fish from Africa. It grows 60 per cent faster and can yield three fish crops per year.

International Water Management Institute (IWMI, formerly IIMI)

As a non-CGIAR centre, the IIMI was founded in 1984 and accepted into the CGIAR in 1991. Its work is focused on efforts to increase productivity of irrigation by 25 per cent, an area covering some 250 million ha in the late 1990s. But it fosters improvement in the management of water resource systems with irrigated agriculture. It has formed strategic partnerships in a broader research network and increased its presence in Africa. The IWMI leads the CGIAR System-Wide Initiative on Malaria and Agriculture, which was launched in 2002. Two other initiatives are a Dialogue on Water, Food and Environment and the Comprehensive Assessment of Water Management in Agriculture. The IWMI is the leading partner in the new CGIAR Challenge Programme on Water and Food.

World Agroforestry Center (formerly the International Center for Research on Agroforestry, ICRAF)

ICRAF was founded in 1977 as a non-CGIAR centre. Donor pressure for quick results caused early problems since strategic research was given little prominence. Some of the false directions included misapplication of the Diagnosis and Design (Young, 2003).

In 1991, ICRAF joined the CGIAR. Agroforestry is about "working trees for working people". In the late 1990s, it ICRAF noted that agroforestry technology could "bring very significant benefits to farmers" in the Sahel region. That work has been productive since the practice of planting trees on farms has provided some 150,000 African farmers with a way to fertilize their soils naturally without relying on costly chemical fertilizers. ICRAF has also been instrumental in getting seed multiplication started in several countries. It has now adopted an integrated natural resource management approach with the focus on the needs of the poorest farmers in diverse environments and concentrating on the functions of natural capital in agriculture. ICRAF has long attempted to convince policy-makers to introduce higher education in agroforestry.

International Centre for Insect Physiology and Ecology (ICIPE)

ICIPE is a non-CGIAR centre in Nairobi, created in 1970. Its task is to find alternatives to pesticides. In 1979, ICIPE applied for membership of the CGIAR but the Technical Advisory Committee concluded in 1980 that "ICIPE would not be among its first choices within a range of possible new initiatives which the Committee would recommend to the Group to undertake". The Committee also noted that a relatively high portion of the CGIAR resources was allocated to the IARCs already located in Africa. At its first annual CGIAR meeting outside Washington, D.C., the donors for the first time voted on ICIPE´s inclusion. The result gave 13 votes for and 12 against inclusion, but ICIPE was not accepted since the major donors had voted against. Instead, the World Bank volunteered to set up a secretariat for the Sponsoring Group of ICIPE, which is still active but no longer chaired by the World Bank.

In community adoption and adaptive research, the use of ICIPE´s tsetse trap has reduced fly numbers by 99 per cent without chemicals and with the doubling of milk yields. Work is continuing with bio-intensive methods and products from natural sources for control of a range of pests and vectors. ICIPE has trained 167 PhD students in insect science in Nairobi, 122 MSc students at three regional sub-centres in Ghana, Ethiopia and Zimbabwe and more than 13,000 farmers and 1,000 extension workers in integrated pest and vector management. The new strategic plan has a focus on problems of African smallholder farmers and closing the gender gap in both farming and research.

Annex VI

References

Ackerman, J, 2002. Food: How safe? How altered? *National Geographic*, 201(5): 2-50.

Ahmed, G J U et al., 2004. Rice-duck farming reduces weeding and insecticide requirement and increases grain yield and income of farmers. *International Rice Research Notes*, 29.1/2004, IRRI, pp 74-77.

Alam, G, 1994. Biotechnology and sustainable agriculture: Lessons from India. Tech. Paper No. 103, OECD Development Centre, Paris.

Albing, M and Hahn, T, 2001. Grundvattenskydd och jordbruk–motstridig lagstiftning bäddar för konflikter (In Swedish). *FAKTA Jordbruk*, No 8, Sveriges lantbruksuniversitet, Uppsala.

Alemaya University, 2003. *Newsletter*, II(4): 11.

Alfranca, O and Huffman, W E, 2001. Impact of institutions and public research on private agricultural research. *Agricultural Economics*, 25: 191-198.

Alroe, H F, 2000. Science as systems learning: Some reflections on the cognitive and communal aspects of science. *Cybernetics & Human Knowing*, 7(4): 57-78.

Alston, J M and Pardey, P G, 2001. Attribution and other problems in assessing the returns to agricultural R & D. *Agricultural Economics*, 25: 141-152.

Alston, J M and Venner, R J, 1998. *The effects of the US Plant Variety Protection Act on wheat genetic improvement*. Paper presented at the Symposium on Intellectual Property Rights and Agricultural Research Impact, CIMMYT, El Batan, Mexico, March 5-7, 1998.

Ames, B, 1998. Micronutrients prevent cancer and delay aging. *Toxicology Letters*, 102-103: 5-18.

AstraZeneca, 2004. *Annual report* (In Swedish). AstraZeneca, Södertälje & London.

Bacevich, J A, 2002. *American Empire: the Realities and Consequences of US Diplomacy*. Harvard University Press.

Bagnara, D, 1992. Developing countries contribute germplasm to Italian plant breeding programs through the CGIAR Centers. *Diversity*, 8(1): 16-18.

Barzun, J, 2000. *From Dawn to Decadence: 15000 to the Present, 500 Years of Western Cultural Life*. Harper/Collins, New York.

Bekele, E, 2001. Present Status, Future Directions and Tasks Ahead. Research and Publications Office, Addis Ababa University, Addis Ababa.

Bengtsson, B M I, 1966. *A survey of cultural practices and utilization of edible aroids in Trinidad*. DTA Thesis, University of the West Indies, St Augustine.

Bengtsson, B M I, 1968. *Cultivation practices and the weed, pest and disease situation in some parts of the Chilalo Awraja*. CADU Publication No. 1, Chilalo Agricultural Development Unit (CADU), March, 1968.

Bengtsson, B M I, 1977. *Past, present and future Swedish support to international agricultural research*. SAREC Report R2:1977, Stockholm.

Bengtsson, B M I (Ed.), 1979. *Rural development research – the role of power relations*. SAREC Report R4:1979, Stockholm.

Bengtsson, B M I, 1983. Rural development research and agricultural innovations. A comparative study of agricultural changes in a historical perspective and agricultural research policy for rural development. Swedish University of Agricultural Sciences, Department of Plant Husbandry, Report 115, Uppsala.

Bernhard-Reversat, F and Huttel, C, 2001. Soil biological fertility undergoes fundamental changes when fast-growing exotic trees are planted on a poor savanna soil. In Bernhard-Reversat, F (Ed.), *Effect of Exotic Tree Plantations on Plant Diversity and Biological Soil Fertility in the Congo Savanna: With Special Reference to Eucalyptus*. Center for International Forestry Research (CIFOR), pp 57-60.

Binswanger, H P and Ruttan, V W, 1978. *Induced Innovation: Technology, Institutions and Development*. Johns Hopkins Press, Baltimore.

Booth, D et al., 2001. Fighting poverty strategically? Lessons from Swedish Tanzanian Development Co-operation 1997-2000. *Sida Evaluation Report 00/22*, Stockholm.

Borgström, G, 1966. *Mat för miljarder* (In Swedish). 2:a upplagan, Halmstad.

Borlaug, N E, 2002. Technology is available – incentives are lacking. Editorial. Feeding the Future, Newsletter of the Sasakawa Africa Foundation, The Nippon Foundation, 17 April, p 3.

Borlaug, N E, 2004, Feeding a world of 10 billion people: our 21st century challenge. In Miranowski, J A and Scanes, C G (Eds.), *Perspectives in World Food and Agriculture: 2004*. Iowa State Press, Blackwell Publishing Company, Iowa, pp 31-69.

Bread for the World Institute, 2003. *Agriculture in the Global Economy*. Hunger 2003, 13th Annual Report on the State of World Hunger, Washington, D.C.

Brenner, C, 2004. Telling transgenic technology tales: lessons from the Agricultural Biotechnology Support Project (ABSP) Experience. *ISAAA Briefs No 31*, ISAAA: Ithaca, NY.

Brookes, G and Barfoot, P, 2005. GM Crops: The global economic and environmental impact – the first nine years 1996–2004. *AgBioForum* 7(2 & 3): Article 15; http://www.agbioforum.org.

Brown, C, 2000. *The Global Outlook of Future Wood Supply from Forest Plantations*. FAO, Rome.

Brown, M T and Herendeen, R A, 1996. Embodied energy analysis and emergy analysis: a comparative view. *Ecological Economics*, 19: 219-235.

Bullock, A and Stallybrass, O (Eds.), 1988. *The Fontana Dictionary of Modern Thought*. Fontana, London.

Burlington, H and Lindeman, V, 1950. Effect of DDT on testes and secondary sex characters of white leghorn cockerels. *Proceedings of the Society for Experimental Biology and Medicine*, 74: 48-51.

Burroughs, W J, 1997. *Does the Weather Really Matter? The Social Implications of Climate Change*. Cambridge University Press, Cambridge, UK.

Busch, L, 1994. The state of agricultural science and the agricultural science of state. In Bonanno, A, Busch, L, Freidland, W, Gouvcia, L and Mingione, F. (Eds.) 1994, *From Columbus to ConAgra. The Globalization of Agriculture and Food*. University Press of Kansas, pp 69- 84.

Carlgren, K and Mattson, I, 2001. Swedish Soil Fertility Experiments. *Acta Agric. Scand., Sect. B, Soil and Plant Sci.*, 51: 49-78.

Carson, R, 1962. *Silent Spring*. Houghton Mifflin, Boston.

CAST, 1999. *Animal Agriculture and Global Food Supply*. Council for Agricultural Science and Technology, Ames, Iowa.

CGIAR, 1998. Recent Accomplishments of the CGIAR International Research Centers. Consultative Group on International Agricultural Research, CGIAR Secretariat, Washington, D.C.

CGIAR, 2001. *Global Knowledge for Local Impact. Agricultural Science and Technology in Sustainable Development*. Annual Report 2001, Consultative Group on International Agricultural Research (CGIAR), Washington D.C.

320

CGIAR, 2004. CGIAR Annual Report: Impact Everybody's Business. Annual Report 2003, Consultative Group on International Agricultural Research, Washington, D.C.

CGIAR, 2006. CGIAR Annual Report. Science-Based Solutions: the Science behind Growth and Development, Consultative Group on International Agricultural Research, Washington, D.C.

Chevron, 1996. Annual Report, p 11.

Chorley, C P H, 1981. The agricultural revolution in Northern Europe, 1750-1880: nitrogen, legumes, and crop productivity. *The Economic History Review*, 2nd Series, 34: 71-93.

Christensen Wiin, C and Wiin-Nielsen, A, 1996. *Klimaproblemer* (In Danish). Teknisk Förlag, Kobenhavn, Denmark

CIFOR, 2003. *Forests and people. Research that makes a difference.* Center for International Forestry Research (CIFOR), Bogor.

CIMMYT, 2001. *Global Research for Local Livelihoods.* Annual Report 2000-2001. International Maize and Wheat Improvement Center (CIMMYT), Mexico.

CIMMYT, 2004. *Seeds of innovation: CIMMYT´s strategy for helping reduce hunger and poverty by 2020.* CIMMYT, Mexico.

Clausen, R and Hube A, 2004. *USAID´s Enduring Legacy in Natural Forests, Livelihoods, Landscapes, and Governance.* Study Summary, Vol. 1. Chemonics International, Washington D.C,

Clausen, R, Gibson, D, Hammett, T, Nduwumwami D, Rebugio L and Seyler, J, 2004. *USAID´s Enduring Legacy in Natural Forests, Livelihoods, Landscapes, and Governance.* Study Report, Vol. 2, Chemonics International, Washington D.C.

Cockrill, W R and Marsden, A W, 1970. *Editorial.* In Feed and Food, World Review of Animal Production, Official Review of the World Association for Animal Production. FAO Special Issue, No. 26, pp 10-12.

Cohen, M B, Gould, F, and Bentur, J S, 2000. Bt rice: practical steps to sustainable use. *International Rice Research Notes*, 25.2/ 2000, International Rice Research Institute (IRRI), Los Banos, pp 4-10.

Colborn, T, Dumanoski, D and Peterson Mayers, J, 1997. *Our Stolen Future.* Penguin Books USA Inc., New York.

Colchester, M, 1994. *Salvaging Nature: Indigenous Peoples, Protected Areas and Biodiversity Conservation.* Discussion Paper DP 55, United Nations Research Institute for Social Development (UNRISD), Geneva and World Conservation Union (WWF), Gland.

Commoner, B, 1984. Relationship between biological information and the origin of life. In Matsuno, K et al. (Eds.), *Molecular Evolution and Protobiology*, Plenum Press, New York, p 283.

Commoner, B, 2003. Unravelling the DNA myth. *Seedling*, July 2003, pp 6-12.

Conway, G, 1997. *The Doubly Green Revolution.* Penguin Books Ltd, London.

Crommentuijn, T, Kalf, D F, Polder, M, Posthumus, R and van de Plassche, E J, 1997. *Maximum permissible concentrations and negligible concentrations for pesticides.* Report No 601501 002, National Institute of Public Health and the Environment, the Netherlands.

Currents, 2006. Rural development professionals abroad as per 1 March 2006. *Currents* 39, Swedish University of Agricultural Sciences, (SLU), February, Uppsala, pp 46-47.

Dagens Medicin, 1999. *Morgondagens medicin – en blick in i nästa årtusende* (In Swedish). Specialutgåva om 2000-talets medicinska utveckling. Helsingborg.

Danielsson, G, 1981. *Mjölkningsmaskinens införande tog lång tid* (In Swedish). Agrifack No. 9, Stockholm.

Dazie, S Jr, 2001. *Biotechnology in Sub-Saharan Africa.* ACTS Science and Technology Policy Paper, No 1, Nairobi.

Davidson, B, 1992. *The Black Man´s Burden: Africa and the Curse of the Nation-State.* Times Books, London.

Davidson, B, 1994. *The Search for Africa. A History in the Making.* James Curey, London.

Davis, R G, 2004. Agroterrorism: need for awareness. In Miranowski, J A and Scanes, C G (Eds.), *Perspectives in World Food and Agriculture: 2004*, Iowa State Press, Blackwell Publishing Company, Iowa, pp 353-416.

Dawkins, R, 1986. *The Blind Watchmaker*. W.W. Norton, New York.

Dazhong, W, 1995. China. In Foltz, T N C (Ed.), *Choosing our Future. Visions of a Sustainable World*. 2050 Project, World Resources Institute, Washington D.C., pp 102-106.

de Condorcet, J A, 1795. *Esquisse d'un tableau historique des progrés de l'ésprit humain*, Paris. Reprint in 1955 by Noonday Press, New York.

de Grassi, A and Rosset, P, 2003. *A New Green Revolution for Africa? Myths and Realities of Agriculture, Technology and Development*, Food First Books, Institute for Food and Development Policy, Oakland.

de Wit, C T, van Laar, H H, van Keulen, H, 1979. Physiological potential of food production. In Sneep, J and Henricksen J T (Eds.), *Plant Breeding Perspectives*, Centre for Agricultural Publishing and Documentation, Publication No 118, PUDOC, Wageningen, Netherlands, pp 47-82.

Delgado, C, Rosegrant, M, Steinfeld, H, Ehui, S, and Courbois, C, 1999. Livestock to 2020: The Next Food Revolution. 2020 Discussion Paper 28, International Food Policy Research Institute (IFPRI), Washington, D.C.

Denevan, W M, 1992. The pristine myth: the landscape of the Americas in 1492. *Annals of the Association of American Geographers* 82(3): 369-385.

Diamond, J, 2005. *Collapse: How Societies Choose to Fail or Succeed*. Viking Press, New York.

Dibner, M, 1991. Tracking trends in US biotechnology. *Bio/Technology* 9.

Dixon, J and Gulliver, A with Gibbon, D, 2001. *Farming Systems and Poverty. Improving Farmers' Livelihoods in a Changing World*. FAO and World Bank, Rome and Washington, D.C.

Drucker, P, 1994. The age of social transformation, *The Atlantic Monthly*, November 1994.

Ds 1996: 73. Biodiversitet och framtida svensk genpolitik (In Swedish). Utrikesdepartementet, Ds 1996:73, Stockholm.

Dumont, R, 1967. *A False Start in Africa*. Sphere Books Ltd and Andre Deutsch Ltd, London.

Dyson, T, 1996. *Population and Food: Global Trends and Future Prospects*. Routledge, London.

Ebbersten, S, 1983. *Rester av kemiska ogräsbekämpningsmedel med särskild hänsyn till preparatet Lontrel Kombi – några resultat och en principdiskussion* (In Swedish). Institutionen för växtodling, Sveriges lantbruksuniversitet, mimeo, Uppsala.

Ebbersten, S, 2003. *Statement. Hearing on Organic Agriculture, organized by the Committee on Agriculture and Rural Development of the European Parliament*, 12 June 2003, Brussels.

The Economist, 2001. Therapeutic antibodies. Seed v seed. *Science and Technology*, 8 September, p 84.

Edqvist, L E, 2002. Aktörer och sjukdomssituation i EU (In Swedish). *Kungl. Skogs och Lantbruksakademiens Tidskrift*, 141: 12 51-56.

Egziabher, T,G, 1987. *Research Problems and Policy at Higher Learning Institution in Ethiopia*. A draft document presented to Commission of Higher Education, Addis Ababa.

Ehrensvärd, G, 1972. *FÖRE EFTER. En diagnos av Gösta Ehrensvärd* (In Swedish). Aldus/Bonniers, Stockholm.

Eicher, C, K, 1986. *Transforming African Agriculture*. The Hunger Project Papers, No. 4, San Francisco.

Eklund, N and Carlsund, E, 2000. *Resursanvändningen inom högskolans grundutbildning*, 4:e delrapporten om Högskolan (In Swedish). Rapport från Riksdagens revisorer, Stockholm.

Elbasha, E, Thornton, P K and Tarawali, G, 1999. *An Ex Post Economic Assessment of Planted Forages in West Africa*. ILRI Impact Assessment Series 2, International Livestock Research Institute (ILRI), Nairobi, Kenya.

Eriksson, J, 2001. *Halter av 61 spårelement i avloppsslam, stallgödsel, handelsgödsel, nederbörd samt i jord och gröda* (In Swedish). Rapport 5148, Naturvårdsverket, Stockholm.

322

Erwin, D H, 1996. The mother of mass extinction. Disaster struck 250 million years ago with the worst decimation of life in history. *Scientific American*, 275(1): 72-78.

EPI, 1999. *The State of Working America 1996-1997*. Economic Policy Institute (EPI), Washington, D.C.

EU, 2003. *World Energy, Technology and Climate Policy Outlook*. European Union, Brussels.

Evenson, R E, 1998. *The Economics of Intellectual Property Rights for Agricultural Technology Introduction*. Paper presented at the Symposium on Intellectual Property Rights and Agricultural Research Impact, CIMMYT, El Batan, Mexico, 5-7 March 1998.

Evenson, R E, Pray, C E and Rosengrant, M W, 1999. *Agricultural Research and Productivity Growth in India*. Research Report 109, International Food Policy Research Institute (IFPRI), Washington, D.C.

Fagan, B, 2004. *The Long Summer: How Climate Changes Civilizations*. Granta Books, London.

Falck-Zepeda, J B and Traxler, G, 1998. *Rent Creation and Distribution from Transgenic Cotton in the US*. Paper presented at the Symposium on Intellectual Property Rights and Agricultural Research Impact, CIMMYT, El Batan, Mexico, 5-7 March 1998.

FAO, 1970. *Production Year Book*, Vol. 24. FAO, Rome.

FAO, 1975. *Pest Control Problems (Pre-harvest) Causing Major Losses in World Food Supplies*. AGP, Pest/PH75/B31. FAO, Rome.

FAO, 1989. *Production Year Book*, Vol. 43. FAO, Rome.

FAO, 1999. *The Multifunctional Character of Agriculture and Land: the Energy Function*. Background Paper 2: Bioenergy, Cultivating our Futures, FAO/Netherlands Conference on the Multifunctional Character of Agriculture and Land, 12-17 September, 1999, Maastricht, Netherlands, pp 43-78.

FAO, 2000. *The State of World Fisheries and Aquaculture*. FAO, Rome.

FAO, 2002. *The State of Food Insecurity in the World 2002*. FAO, Rome.

FAO, 2002b. *Global Agriculture: Towards 2015/2030*. FAO, Rome.

FAO, 2003. *Report on the Situation of the World's Forests*. FAO, Rome.

FAO, 2004. *The State of Food Insecurity in the World 2004*. FAO, Rome.

FAO, 2005. *The State of Food Insecurity in the World 2005*. FAO, Rome.

FAOSTAT, 2002. *FAOSTAT database*, July. FAO, Rome.

FAO Yearbook,1989. *Fisheries Statistics, Catches and Landings*, Vol. 68. FAO, Rome.

Farah, J, 1994. *Pesticide policies in developing countries: Do they encourage excessive use?* Discussion Paper No. 238, The World Bank, Washington, D.C.

Feeney, G and Feng, W, 1993. Parity progression and birth intervals in China: The influence of policy in hastening fertility decline. *Population and Development Review* 19(1): 95.

Fernandez-Armesto, F, 1995. *Millennium: A History of the Last Thousand Years*. Scribner, New York and Bantam Press, London.

Ferraro, P, 2001. *Global Habitat Protection: Limitations of Development Interventions and a Role for Conservation Performance Payments*. Cornell University, Ithaca, New York.

Feyerabend, P, 1987. *Farewell to Reason*. Verso, London.

Fölster, S, 2003. *Jämlikheten som försvann* (In Swedish). Ekerlids Förlag AB.

Foltz, T N C (Ed.), *Choosing our Future. Visions of a Sustainable World*. 2050 Project, World Resources Institute, Washington, D.C., pp 102-106.

Foster, A M, Kayaayo E and Ssembatya, C, 2002. MAAIF-SG 2000 Technology Transfer Program in Uganda. In Breth, S A (Ed.), *Food Security in a Changing Africa*. Proceedings of the Workshop on Africa Food Security in a Changing Environment: Sharing Good Practices and Experiences, held in Kampala, Uganda, 6-9 June 2001, Centre for Applied Studies in International Negotiations, Geneva, pp 51-60.

Franzese, P P et al., 2006. Energy Analysis and Emergy Synthesis of Selected Human-Dominated Ecosystems: A Comparative View. Paper for the Fourth Biennial Emergy Research Conference, 19-21 January 2006, Centre for Environment Policy, Florida University, Gainesville.

323

Frison, E A, Mitteau, M and Sharrock, S, 2002. Sharing responsibilities for *ex situ* germplasm management. *Plant Genetic Resources Newsletter*, 131: 7-15.

Fukuyama, F, 1992. *The End of History and the Last Man*. Penguin Books, England.

Fukuyama, F, 2002. *Our Posthuman Future: Consequences of the Biotechnology Revolution*. Farrar, Straus and Biroux, New York.

Gaillard, J and Busch, L, 1993. French and American agricultural science for the third world. *Science and Public Policy*, 20(4): 223-234.

Gebrekidan, H, 2000. Alemaya University of Agriculture, Ethiopia. Case Study of Innovative Agricultural Extension Training. In Breth, S (Ed.), *Innovative Extension Education in Africa*. Sasakawa Africa Association, Mexico City, pp 33-38.

Georghiou, G P, 1985. *Pesticide Resistance: Strategies and Tactics for Management*. Committee on Strategies for the Management of Pesticide Resistant Pest Populations, Board of Agriculture, National Research Council, US National Academy Press, Washington, D.C.

Gerholm, T R, 1972. *Futurum exaktum*. Brombergs, WSOY, Finland.

Gerholm, T R, 1999. *Futurum exaktum*. Brombergs, WSOY, Finland.

Gesslein, S, 2001. Cropping system, plant nutrition and soil fertility–18 years of results on a previously exhausted soil. *Kungl. Skogs- och Lantbruksakademiens Tidskrift*, 140(9).

Gill, A S, 2002. Following agroforestry system research helps achieve successful farming. *ICAR News*, 8(4): 5-7.

Goodman, M M, 1998, Research policies thwart potential payoff of exotic germplasm. *Biodiversity*, 14(3&4): 30-35.

Gould, S J, 1989. *Wonderful Life*. W. W. Norton, New York.

Griffin, K, 1974. *The Political Economy of Agrarian Change*. Macmillan, London.

Grove, R, 1990. The origins of environmentalism. *Nature*, 345(3 May): 11-14.

Gura, S and Lpp, 2003. Losing livestock, Losing livelihoods. *Seedling*, January 2003, pp 8-12.

Havnevik, K, Hårsmar, M and Sandström, E, 2003. *Final Report. Rural Development and the Private Sector in Sub-Saharan Africa*. Evaluation of Sida´s experiences and approaches in the 1990s. SLU, Uppsala.

Hawksworth, D L, 1995. The resource base for diversity assessments. In Heywood H V (ed.), 1995, *Global Biodiversity Assessment*. United Nations Environment Programme, Nairobi and Cambridge University Press, Cambridge, pp 549-605.

Hayami Y and Ruttan, V W, 1971. *Agricultural Development: an International Perspective*. Johns Hopkins University Press, Baltimore.

Hayenga, M, 1998. Structural change in the biotech seed and chemical industrial complex. *AgBioForum*, 1(2), Fall.

Hecht, S B et al., 2006. Globalization, forest resurgence, and environmental politics in El Salvador. *World Development*, 34(2), February.

Holmes, B, 1994. A natural way with weeds. *New Scientist*, 7 May 1994, pp 22- 23.

Horgan, J., 1996. *The End of Science. Facing the Limits of Knowledge in the Twilight of the Scientific Age*. Addison-Wesley, Helix Books, New York.

Horton, T, 2005. Saving the Chesapeake. *National Geographic*, 207(6): 22-49.

Hunag, Z, You, J and Cheng, J, 2004. China´s Grain Security and Trade Policies after Entry into the World Trade Organization (WTO): Issues and Options. In Miranowski, J A and Scanes, C G (Eds.), *Perspectives in World Food and Agriculture: 2004*. Iowa State Press, Blackwell Publishing Company, Iowa, pp 323-333.

Huntington, S P, 1996. *The Clash of Civilizations and the Remaking of World Order*. Simon & Schuster, New York.

Husén, T, 1985. Universiteten och forskningen: En studie i forskningens och forskarutbildningens villkor (In Swedish), 2:a upplagan, Stockholm.

ICAR, 1999. Indian Council of Agricultural Research. *ICAR news*, 5(3), July-September, New Delhi.

ICAR, 2000. *ICAR Vision 2020*. Indian Council of Agricultural Research, New Delhi.

324

ICARDA, 2001. *ICARDA Annual Report 2001*. International Center for Agricultural Research in the Dry Areas, Aleppo, Syria.

ICRAF, 2002. *World Agroforestry Centre. Transforming Lives and Landscapes*. Annual Report 2001-2002, International Centre for Research in Agroforestry, Nairobi, Kenya.

ICRISAT, 2001. *ICRISAT Research Impacts*. For informed research targets and technology design, International Crops Research Institute for the Semi-Arid Tropics, Andhra Pradesh, India.

ICRISAT, 2001. *ICRISAT Annual Report 2001*. Andhra Pradesh, India.

ICRISAT, 2006. Neem: the bitter truth. ICRISAT´s monthly e-newsletter. *SATrends*, 67, June. International Crops Research Institute for the Semi-Arid Tropics, Andhra Pradesh, India.

IFA, 1998. *The Fertilizer Industry, World Food Supplies and the Environment*. International Fertilizer Industry Association. Published in association with the United Nations Environment Programme, Paris.

IFPRI, 2005. *New Risks and Opportunities for Food Security: Scenario Analyses for 2015 and 2050*. 2020 Vision Discussion Paper and Policy Brief, International Food Policy Research Institute, Washington, D.C. Imperial Agricultural Bureaux, 1944. *Alternate Husbandry*. Joint Publication No. 6, London.

Imperial Council of Agricultural Research, 1938. *Review of Agricultural Operations*. New Dehli: Government of India.

Ingerslev, M, Mälkönen, E, Nilsen, P, Nohrstedt, H-G, Oskarsson, H and Raulund-Rasmussen, K, 2001. Main Findings and Future Challenges in Forest Nutritional Research and Management in Nordic Countries. *Scandinavian Journal of Forestry Research*, 16: 488-501.

Ingvarsson, A, 2003. Forskning, miljöarbete och fattigdomsbekämpning. *Teknik och Vetenskap*, 3: 36- 39.

IPCC, 2001. *Climate Change, Impacts, Assessment and Vulnerability*. A Summary for Policy-Makers. Intergovernmental Panel on Climate Change (IPCC), www.ipcc.ch.

IPGRI , 1999. *Diversity for Development*. The New Strategy of the International Plant Genetic Resources Institute. IPGRI, Rome, Italy.

IPGRI, 2002. *Annual Report 2001*. International Plant Genetic Resources Institute, Rome.

IRRI, 1994. Rice and climate changes: a glimpse into the crystal ball. *The IRRI Reporter*, 2/94.

IRRI, 1998. *Biodiversity. Maintaining the Balance*. IRRI 1997-1998. International Rice Research Institute, Manila.

IRRI, 2000. *IRRI Hotline*, 10(4): December 2000.

Isaksson, A, 2002. *Den politiska adeln* (In Swedish). Wahlström & Widstrand, Smedjebacken.

IWMI, 2002. *Annual Report 2001-2002*. Partnerships for change, new initiatives, research impacts and outputs and partner perspectives. International Water Management Institute, Colombo.

James, P and Thorpe, N, 1994. *Ancient Innovations*. Ballantine Books, Random House Inc., New York.

Jeffries, C (Ed.), 1964. *A Review of the Colonial Research 1940-1960*. H.M. Stationary Office, London.

Jensen, S, 1966. Report of a new chemical hazard. *New Scientist* 32: 612.

Johnson, A, 1999. *Inte bara valloner - Invandrare i svenskt näringsliv under 1000 år* (In Swedish). SNS Förlag, Stockholm.

Johnson, C S, 2000. Genetic enhancement of crops: the major way remaining to ensure global food security. *Diversity*, 15(4): 22-24.

Jonasson, L, 2001. Tools to achieve a sustainable agriculture. Economical aspects. In Ecologically Improved Agriculture – Strategy for Sustainability. *Kungl. Skogs- och Lantbruksakademiens Tidskrift*, 140(6): 71-76.

Jonsson, E, 1982. Jordbruksnäring med framtidstro. I *Vårt lantbruksuniversitet. En bok till Lennart Hjelm*. Allmänna skrifter 6, Uppsala, s 23-25.

Jonsson, H, 1996. Greening the fields. United Nations Environment Programme (UNEP), *Our Planet*, 8(4): 16-17.

Joshi, P K, Parthasarathy Rao, P, Gowda, C L L, Jones, R B, Silim, S N, Saxena, K B and Kumar, J, 2001. *The World Chickpea and Pigeonpea Economies*. Facts, Trends, and Outlook. International Crops Research Institute for the Semi-Arid Tropics, Andhra Pradesh, India.

Judson, O, 2003. *Dr Tat Sex Advice to All Creation: The Definitive Guide to the Evolutionary Biology of Sex*. Chatto & Windus, London.

Kahn, H and Wiener, AJ, 1967. *The Year 2000 – A Framework for Speculation on the Next Thirty-Five Years*. The Hudson Institute, Inc., Macmillan, New York.

Kaihura, F and Stocking, M (Eds.), 2003. *Agricultural Biodiversity in Smallholder Farms of East Africa*. United Nations University Press, Tokyo.

Karlström, B, 1991. *Det omöjliga biståndet* (In Swedish). Studier och Debatt, SNS Förlag.

Kauffman, S, 1993. *The Origins of Order*. Oxford University Press, New York.

Kerr, R A, 1998. Climate change: greenhouse forecasting still cloudy. *Science*, 276: 5315: 16: 1040-1042.

King, M, 1990. Health is a sustainable state. Viewpoint, *The Lancet*, 336, September 15, 1990, pp 664-667.

Kirchner, J W and Weil, A, 2000. Delayed biological recovery from extinctions through the fossil record. *Nature*, 416: 420-424

Kjaergaard, T, 1995. Agricultural development and nitrogen supply from a historical point of view. In Kristensen, L with Stopes, C, Kölster, P, Granstedt, A and Hodges, D (Eds), *Nitrogen Leaching in Ecological Agriculture*, Proceedings of an International Workshop, Royal Veterinary and Agricultural University, Copenhagen, Denmark, A B Academic Publishers, pp 3-14.

Knowles, L C A, 1928. *The Economic Development of the British Overseas Empire*. 2nd ed. George Routledge and Sons, London.

Koestler, A, 1964. *The Sleepwalkers. A History of Man´s Changing Vision of the Universe.*: Penguin Books, Middlesex, England, p 525.

Kristjanson, P, Rowlands, J, Swallow, B, Kruska, R, de Leeuw, P and Nagda, S, 1999. *Using the Economic Surplus Model to Measure Potential Returns to International Livestock Research. The case of trypanosomiasis vaccine research*. ILRI Impact Assessment Series 4, International Livestock Research Institute, Nairobi, Kenya.

KSLA, 2004. Climate change and forestry in Sweden–a literature review. *Kungl. Skogs- och Lantbruksakademiens Tidskrift*, 143:18.

Kuyek, D, 2002. *The Real Board of Directors: The Construction of Biotechnology Policy in Canada, 1980-2002*. Ram´s Horn, Sorrento.

Lal, R, 1995. Erosion-crop productivity relationships for soils of Africa. *American Journal of Soil Science Society*, 59, 661-667.

Lapierre, D and Moro, J, 2003. *Fem över tolv i Bhopal* (In Swedish). Leopard förlag, Stockholm.

Larsson-Kilian, A, 1916. *Jordbrukarens läsning. (Den mindre jordbrukarens handbok, 1)* (In Swedish), Svenska Andelsförlaget, Stockholm.

Lassen, K and Friis-Christensen, E, 1995. Variability of the solar cycle length during the past five hundred centuries and the apparent association with terrestrial climate. *Journal of Atmospheric and Terrestrial Physics*, 57(8): 835-845.

Leal, W L, 2005 (Ed.). *Handbook of Sustainability Research*. Peter Lang Scientific Publishers, Frankfurt/Main, New York.

Leander, L, 1995. *Tetra Pak. Visionen som blev verklighet*. En annorlunda företagshistoria (In Swedish), Lund.

Lebedys, A, 2004. *Trends and Current Status of the Contribution of the Forestry Sector to National Economies*. FAO, Rome.

326

Lejonhud, K, 2003. *Indian Villages in Transformation. A Longitudinal Study of Three Villages in Uttar Pradesh*. Karlstad University Studies 2003:11.

Lele, U (Ed.), 1991. *Aid to African Agriculture. Lessons from Two Decades of Donor Experience*. The John Hopkins University Press, Baltimore and London.

Liedman, S-E, 1997. *I skuggan av framtiden. Modernitetens idéhistoria* (In Swedish). Albert Bonniers Förlag.

Light, R J, 2000. *Making the Most of College. Students Speak Their Minds*. Harvard University Press, Cambridge.

Lomborg, B, 2001. *Världens verkliga tillstånd* (The Skeptical Environmentalist: Measuring the Real State of the World). Swedish edition, SNS Förlag, Stockholm.

Long, M E, 2002. Half life. The lethal legacy of America´s nuclear waste. *National Geographic*, 202(1): 2 –33.

Lovins, L H, Lovins, A B and von Weizsäcker, E U, 1996. *Faktor vier*. Droemer Knaur.

Lundgren, B et al., 1994. *Swedish Support to the Consultative Group on International Agricultural Research (CGIAR)*. A Quinquennial Review 1987-1992, SAREC Evaluations 1994:1. Stockholm.

Madeley, J, 2002. *Food for All: The Need for a New Agriculture*. Zed Books, London and New York.

Maguire, C., 1998. More people: less earth. The shadow of man-kind. In Maguire, D C and Rasmussen, L, *Ethics for a Small Planet. New Horizons on Population, Consumption and Ecology*. State University of New York Press, Albany, pp 1-64.

Marsden, T, Flynn, A and Ward, N, 1994. Food regulation in Britain: a national system in an international context. In Bonanno, A, Busch, L, Freidland, W, Gouveia, L and Mingione, E, (Eds.), *From Columbus to ConAgra. The Globalisation of Agriculture and Food*. University Press of Kansas, Lawrence, Kansas, pp 105-124.

Marsh, G P, 1864. *Man and Nature*. Cambridge. 1965 Reprint by Harvard University Press.

Martin, H-P and Schumann, H, 1996. *Die Globalisierungsfalle. Der Angriff auf Demokratie und Wohlstand*. Rowolt Verlag, GmbH.

Masefield, G B, 1972. *A History of the Colonial Agricultural Service*. Oxford University Press, Oxford.

Mattson, R, 1978. *Jordbrukets utveckling i Sverige*. Aktuellt från lantbruksuniversitetet, Nr. 258, Allmänt (In Swedish), Uppsala.

McLachan, J, Newbold, R and Bullock, B, 1975. Reproductive tract lesions in male mice exposed prenatally to diethylstilbestrol. *Science*, 190: 991-992.

McNeely, J, 1999. Ethical systems and the Convention on Biological Diversity: setting the stage. *Diversity*, 15(1): 29-30.

Mead, P, Slutzker, L, Dietz, V et al., 1999. Food-related illness and death in the United States. *Emerging Infectious Diseases*, 5: 607-625.

Merchant, C, 1982. *The Death of Nature: Women, Ecology and the Scientific Revolution*. Wildwood House, London.

Miljörådet, 2002. *Miljömålen–når vi fram?* (In Swedish). De Facto, Stockholm.

Montaigne, F, 2003. Atlantic salmon. *National Geographic*, 204(1): 100-123.

Myrdal, J, 2001. *Den nya produktionen–det nya uppdraget. Jordbrukets framtid i ett historiskt perspektiv* (In Swedish), Jordbruksdepartementet, Ds 2001: 68, Stockholm.

Myrdal, A and Myrdal, G, 1934. *Kris i befolkningsfrågan* (In Swedish). Albert Bonniers Förlag, Stockholm.

Murphy, S, 2002. *Market structure and the gains from trade*. Paper for FAO Expert Consultation on "Trade and Food Security: Conceptualizing the Linkages", 11-12 July 2002, Rome.

Naess, A, 1973. The shallow and the deep, long-range ecology movement: a summary. *Inquiry*, 16: 95-100.

National Research Council, 1989. *Alternative Agriculture*. Committee on the Role of Alternative Farming Methods in Modern Production Agriculture, Board on Agriculture, National Research Council, National Academy of Sciences, Washington D.C.

National Research Council, 2000. *Genetically Modified Pest-protected Plants: Science and Regulation*. National Research Council, National Academy Press, Washington, D.C.

Njoroge, G N and Newton, L E, 2002. Ethno-botany and distribution of wild genetic resources of the family Cucurbitaceae in the central highlands of Kenya. *Plant Genetic Resources Newsletter*, 132: 10-16.

Nordberg, M and Nybrant, T (Eds.), 2004. *The FOOD 21 Symposium–Towards Sustainable Production and Consumption*. Extended Abstracts. Swedish University of Agricultural Sciences (SLU), Uppsala (and Report Food 21 1/2005; 5/2005 and 6/2005).

Odum, H T, 1996. Environmental Accounting: Emergy and Environmental Decision-making. John Wiley & Sons, Inc. USA.

Oerke, E C, Dehne, H W, Schoenbeck, F, and Weber, A, 1995. *Crop Production and Crop Protection: Estimated Losses in Major Food and Cash Crops*. Elsevier, Amsterdam.

Ohadoma, C, 2002. *A Science Agenda from an African Perspective*. www.spacedaily.com/news/africa-02b.html.

Opido-Odongo, J, 2000. Roles and challenges of agricultural extension in Africa. In Breth, S (Ed.), *Innovative Extension Education in Africa*. Sasakawa Africa Association, Mexico City, pp 5-15.

Örlander, G, 2000. "Local change" and the Swedish forest. News and Views. *Scandinavian Journal of Forest Research*, 15(5): 2-3.

Padulosi, S, 1999. Criteria for priority-setting in initiatives dealing with underutilized crops in Europe. In Gass, T and Thorman, I, *Implementation of the Global Plan of Action in Europe, Conservation and Sustainable Utilization of Plant Genetic Resources for Food and Agriculture*, Proceedings of the European Symposium on Plant Genetic Resources for Food and Agriculture, Braunschweig, Germany, 30 June - 4 July 1998, International Plant Genetic Research Institute, Rome, pp 236-247.

Pardey, P G, Alston, J M, Christian, J E and Fan, S, 1996. *Hidden Harvest: US Benefits from International Research Aid*. IFPRI Food Policy Report, International Food Policy Research Institute, Washington D.C.

Pardey, P G and Beintema, N M, 2001. *Slow Magic: Agricultural R&D a Century after Mendel*. Food Policy Report, International Food Policy Research Institute, Washington, D.C.

Parkenham, T, 1991. *The Scramble for Africa*. Weidenfeld and Nicolson, London.

Paroda, R S, 1996, The Last Page. *ICAR News*, 2(4).

Paroda, R S, 2000. The Last Page. *ICAR News*, 6(2).

Payne, W J A and Hodges, J, 1997. *Tropical Cattle. Origins, Breeds and Breeding Policies*. Blackwell Science Ltd, Oxford.

Peck, L, Dahlström, K, Hammarskjöld, M and Munck, L, 2001. *HIV/AIDS-related Support through Sida–A Baseline Study*. Sida Evaluation Report 01/14, October, Stockholm.

Penhue, W, 2004. A short history of farming in Latin America. *Seedling*, April 2004, pp 5-11.

Pepper, DM, 1996. *Modern Environmentalism. An Introduction*. Routledge, London.

Pimentel, D et al., 1992. Assessment of environmental and economic impacts of pesticide use. In Pimentel, D and Lehman, H (Eds.), *The Pesticide Question: Environment, Economics and Ethics*. Chapman and Hall, New York.

Pimentel, D et al., 1995. *Pest management, food security and the environment: History and current status*. Paper presented for the IFPRI Workshop on Pest Management, Food Security and the Environment: The Future to 2020, Washington, D.C.

Pimentel, D et al., 1997. Impact of population growth on food supplies and environment. *Population and Environment*, 19: 9-14.

Plucknett, D L, 1993. *Science and Agricultural Transformation*. International Food Policy Research Institute, Lecture Series 1, Washington D.C.

Ponting, C.A., 1991. *A Green History of the World: The Environment and the Collapse of Great Civilizations*. Penguin Books, New York.

Pretty, J, 1996. Sustainability works. United Nations Environment Programme (UNEP). *Our Planet*, 8(4): 19-22.

Prisco, J T, 2000. *The Brazilian Semi-Arid Tropics: New Approaches to Old Problems*. Presentation to the ICRISAT Governing Board Meeting on September 9, 2000 in Patancheru, India.

Purseglove, J W, 1974. *Tropical Crops*. Di-cotyledons. Volumes 1 and 2 combined. The English Language Book Society and Longman, London.

Qaim, M and Zilberman, D, 2003. Yield effects of genetically modified crops in developing countries. *Science*, 299: 900-902.

Qayam A and Sakkhari, K, 2003. The Bt gene fails in India. *Seedling*, July 2003, pp 13-17.

Quist, D and Chapela, I, 2001. Transgenic DNA introgressed into traditional maize land races in Oaxaca, Mexico. *Nature*, 414: 541-543.

RAFI, 1999. *The Gene Giants*. RAFI Communique, Rural Advancement Foundation International, March/April 1999.

RAWOO, 2001. *Balancing ownership and partnership in development research*. Review of 1999 and 2000. Netherlands Development Assistance Research Council (RAWOO), The Hague.

Reid, T R, 2002. The New Europe. *National Geographic*, 201(1): 32-47.

Reilly, J, 1996. Agriculture in a changing climate: impacts and adaptation. In Watson R T, Zinyowera, M C and Moss, R H, *Climate Change 1995. Impacts, Adaptations and Mitigation of Climate Change: Scientific-technical Analyses*, Contribution of Working Group I to the second assessment report of the Intergovernmental Panel on Climate Change. Cambridge University Press, Cambridge.

Reiss, M J and Straughan, R, 1996. *Improving Nature? The Science and Ethics of Genetic Engineering*. Cambridge University Press, Cambridge.

Richards, J F, 1990. Land Transformation. In Turner, B L et al. *The Earth as Transformed by Human Action*. Cambridge University Press, Cambridge, UK, pp 163-180.

Rizvi, S J H and Rizvi, V, 1992. Exploitation of allelo-chemicals in improving crop productivity. In Rizvi, S J H and Rizvi, V (Eds.), *Alleopathy: Basic and Applied Aspects*, Chapman and Hall, London.

Rohrbach, D D and Kiriwaggulu, J A B, 2001. *Commercialisation Prospects for Sorghum and Pearl Millet in Tanzania*. Socioeconomics and Policy Program, Working Paper Series No. 7, International Crops Research Institute for the Semi-Arid Tropics.

Rosenzweig, C and Parry, M L, 1994. Potential impact of climate change on world food supply. *Nature*, 367: 133-138.

Rossiter, M W, 1975. *The Emergence of Agricultural Science: Justus von Liebig and the Americans, 1840-1880*. Yale University Press, New Haven.

The Royal Society, 2003. *Keeping Science Open: the Effects of Intellectual Property Policy on the Conduct of Science*, April, London, www.royalsoc.ac.uk.

Rydberg, T and Jansén J, 2002. Comparison of horse and tractor traction using emergy analysis. *Ecological Engineering*, 19: 13-28.

Sachs, J, 2005. *The End of Poverty. How we can make it happen in our lifetime*. Penguin Press, New York.

Sahn, D E and Steifel, D C, 2002. Progress toward the Millennium Development Goals in Africa. *World Development*, 31: 23-52.

Salam, A, 1989. *Notes on Science, Technology and Science Education in the Development of the South*. Prepared for the 4th and 5th Meetings of the South Commission, 10-12 December 1988, Kuwait and 27-30 May, 1989, Maputo, Mozambique, The Third World Academy of Science, Sixth Imprint, Trieste.

Sanchez, P A, 2004. Reducing hunger by improving soil fertility: an African success story. In Miranowski, J A and Scanes, C G (Eds.), *Perspectives in World Food and Agriculture: 2004.* Iowa State Press, Blackwell Publishing Company, Iowa, pp 75-86.

Sasakawa Africa Association, 2006. SAFE. *Feeding the Future,* Newsletter of the Sasakawa Africa Association, 22: 10.

Sasakawa Africa Association, 2003. *Annual Report for 2002-2003.* The Nippon Foundation.

Sasakawa Africa Association, 2003. Editorial. Time to reconsider fertilizer subsidies. *Feeding the Future,* Newsletter of the Sasakawa Africa Association, Issue 18, The Nippon Foundation.

Sayer, J and Campbell, B, 2004. *The Science of Sustainable Development: Local Livelihoods and the Global Environment.* Cambridge University Press, Cambridge.

Sayer, J A and Campbell, B M, 2001. Research to integrate productivity enhancement, environmental protection and human development. *Conservation Ecology,* 5(2): 12 pp, http://www.consecol.org/vol5/iss2/art32.

Sayer, J, Ishwaran, N, Thorsell, J and Sigaty, S, 2000. Tropical forest biodiversity and the world heritage convention. *Ambio,* 29(6): 302-309.

Schulten, G G M, 1990. Needs and constraints of integrated pest management in developing countries. *Med. Fac. Landbouw. Riksuniv. Gent.,* 55, 2-216.

Science Council, 2005. *System Priorities for CGIAR Research 2005-2015.* Science Council, Consultative Group on International Agricultural Research, October, Rome.

Seckler, D and Amarasinghe, U, 2004. Major problems in the global water-food nexus. In Miranowski, J A and Scanes, C G (Eds.), *Perspectives in World Food and Agriculture: 2004,* Iowa State Press, Blackwell Publishing Company, Iowa, pp 227-2514.

Seckler, D, Amarasinghe, U, Molden, D, de Silva, R, and Barker, R, 1998. *World Water Demand and Supply, 1990 to 2025: Scenarios and Issues.* International Water Management Institute, Research Report 19, Colombo.

Sen, A, 1977. Starvation and exchange entitlements: a general approach and its application to the Great Bengal Famine. *Cambridge Journal of Economics,* 1: 33-60.

Sen, A, 1998. *Development as Freedom.* Knopf, New York.

Seneviratne, S N de S (Ed.), 1993. *Developing the Science Constituency.* National Academy of Sciences of Sri Lanka, Occasional Papers No. 1, September, Colombo.

SG-2000, 1999. *Editorial.* Sasakawa-Global 2000, Newsletter, Issue 14, October.

Shah, T, Alam, M, Dinesh Kumar, M, Nagar, R K and Singh, M, 2000. *Pedalling out of Poverty: Social Impact of a Manual Irrigation Technology in South Asia.* International Water Management Institute, Research Report 45, Colombo.

Sharma, H C and Ortiz, R, 2000. Transgenics, pest management, and the environment. *Current Science,* 79(4): 421-437.

Shiva, V, 2002. *Water Wars: Privatisation, Pollution, and Profit.* Zed Books, London and New York.

Shiyani, R L, Joshi, P K and Bantilan, M C S, 2001. *Impact of Chickpea Research in Gujarat.* International Crops Research Institute for the Semi-Arid Tropics, Impact Series No. 9, Andhra Pradesh, India.

SIDA, 1984. *Lantbruk och livsmedelsförsörjning* (In Swedish). Swedish International Development Authority, Faktablad, September 1984. Stockholm.

Sida, 1999. *A Sea of Opportunities. Research Cooperation 1998.* Department of Research Cooperation, SAREC, Swedish International Development Cooperation Agency, Stockholm.

Sida, 2000. *Sida 1999. Årsredovisning* (In Swedish). Swedish International Development Cooperation Agency, Stockholm.

Sida, 2002. *Sida 2001. Årsredovisning* (In Swedish). Swedish International Development Cooperation Agency, Stockholm.

Sida, 2003. *Årsredovisning* (In Swedish). Swedish International Development Cooperation Agency, Stockholm.

Sida, 2004. *Evaluation Manual*, Swedish International Development Cooperation Agency, Stockholm.

Sida, 2005. *Årsredovisning* (In Swedish). Swedish International Development Cooperation Agency, Stockholm.

Sinha, S, Beijer, A, Hawkins, J and Teglund, Å, 2001. *Approach and Organization of Sida Support to Private Sector Development*. Swedish International Development Cooperation Agency, Evaluation Report 01/14, October, Stockholm.

Simmonds, N W, 1991. The Earlier Contributions to Tropical Agricultural Research. Paper prepared for IOB/TAA meeting. Tropical Agriculture Association, *Newsletter*, 13(2): 2-7.

SIPRI, 2006. *SIPRI Yearbook 2006*. Armaments, Disarmament and International Security, Stockholm International Peace Research Institute and Oxford University Press, Oxford.

SJFR, 1996. *Ekologisk jordbruks- och trädgårdsproduktion*. Utredning om kunskapsläge, pågående forskning och behov av fortsatt forskning (In Swedish). Skogs- och Jordbrukets Forskningsråd (SJFR), Stockholm.

Sjöberg, C, Bingel, E and Sjöquist, C, 2002. *Från defensiva till pro-aktiva—Drivkrafterna bakom hållbar tillväxt* (In Swedish). Svenskt Näringsliv, Stockholm.

SLU, 2003, *Tillväxt i ett uthålligt samhälle*. Forskningsstrategi för SLU 2003 (In Swedish). Sveriges lantbruksuniversitet (SLU), mimeo.

Smit, J et al., 1996. *Urban Agriculture: Food, Jobs and Sustainable Cities*. United Nations Development Programme, NY Publication Series for Habitat II, Vol. 1, New York.

Smith, J M, 2003. *Seeds of Deception. Exposing Corporate and Government Lies about Safety of Genetically Engineered Food*. Green Books, UK.

Smith, J E, 1990. United States-Mexico Agricultural Trade. *US Davis Law Review*, Spring 1990.

SOU, 2001. *En rättvisare värld utan fattigdom* (In Swedish). Betänkande av den parlamentariska kommittén om Sveriges politik för global utveckling(Globkom), SOU 2001:96, Stockholm.

Spore, 1999. AIDS and agriculture. Red-ribbon farming. *Spore*, 82: 2-3. Technical Centre for Agricultural and Rural Cooperation.

Spore, 2001a. Regional trade. What a way to go. *Spore*, 94: 4-5. Technical Centre for Agricultural and Rural Cooperation..

Spore, 2001b. Protocol jungle hampers organic exports. *Spore*, 94: 9. Technical Centre for Agricultural and Rural Cooperation.

Spore, 2002. Beyond the pail. Dairy development. *Spore*, 102: 4-5. Technical Centre for Agricultural and Rural Cooperation.

Steingraber, S, 1997. *Living Downstream. A Scientist's Personal Investigation of Cancer and the Environment*. Vintage Books, Random House, Inc., New York.

Stent, G, 1978. *The Paradoxes of Progress*. W H Freeman, San Francisco.

Stolt, P, 1998. Rapporter. Svenskt brödvete ratas för sitt kadmiuminnehåll (In Swedish). *Journal of the Swedish Seed Association*, 108(1): 46-47.

Strandberg, S, 1997. *Indien i stort och smått. 1947-1997: Ett halvt sekel av framgång och motgång* (In Swedish). Falun.

Streeten, P, 1974. The limits of development research. *World Development*, 2(10-12).

Swallow, B, Garrity, D and van Noordwijk, M, 2001. The effects of scales, flows, and filters on property rights and collective action in watershed management. *Water Policy* 3: 457-474.

Tabashnik, B E, 1994. Evolution of resistance to *Bacillus thuringensis* toxins. *Annual Review of Entomology*, 39: 47-79.

Taylor, N L et al. (Eds.), 1985. *Clover Science and Technology*. Agronomy: Series of Monographs, 25, American Society of Agronomy, Madison, Wisconsin.

Tweeten, L, 1991. Food for people and profit. Ethics and capitalism. In Bltaz, C V (Ed.), *Ethics and Agriculture. An Anthology on Current Issues in World Context*. University of Idaho Press, Moscow, pp 84-100.

UNCTAD, 2002. *World Investment Report*. United Nations Centre for Trade and Development, Geneva.

UNDP, 2001. *Making New Technologies Work for Human Development*. Human Development Report 2001, United Nations Development Programme, Oxford University Press, New York.

UNDP, 2003. *Human Development Report*. United Nations Development Programme, New York.

UNEP, 2002. *Global Environment Outlook 3 (Geo-3)*. United Nations Environment Programme, Nairobi.

UNEP, 2006. *Building Biosafety Capacity in Developing Countries: Experiences of the UNEP_GEF Project on Development of National Biosafety Frameworks*, www.unep.ch/biosafety/development/devdocuments/UNEPGEFstudyVersion170605.pdf.

UNFCCC, 2001. United Nations Framework Convention on Climate Change. The Marrakesh Declaration, http://www.unfccc.int/.

UN Millennium Project, 2005a. *Investing in Development: A Practical Plan to Achieve the Millennium Development Goals*. Overview, Earthscan, London.

UN Millennium Project, 2005b. *Halving Hunger: It Can Be Done*. Task Force on Hunger, Earthscan, London.

UN Millennium Project, 2005c. *Innovation: Applying Knowledge in Development*. Task Force on Science, Earthscan, London.

USDA, 2001. *Agricultural Statistics 2001*. National Agricultural Statistics Service 2001, US Department of Agriculture, Washington, D.C. US Government Printing Office.

van Emden, H F and Peakall, D P, 1995. *Beyond Silent Spring. Integrated Pest Management and Chemical Safety*. International Centre for Insect Physiology and Ecology and United Nations Environment Programme, Chapman & Hall, London.

van Lautam, E B J and Gerrits, R, 1991. *Bio-Pesticides in Developing Countries. Prospects and Research Priorities*. African Centre for Technology Studies, Nairobi, Kenya and ACTS Biopolicy Institute, Maastricht, the Netherlands.

Virk, P S, Kush, G S and Peng, S, 2004. Breeding to enhance yield potential of rice at IRRI: the ideotype approach. *International Rice Research Notes* (IRRN), IRRI, 29.1/2004, pp 5-9.

Virmani, S S and Kumar, I, 2004. Development and use of hybrid rice technology to increase rice productivity in the tropics. *International Rice Research Notes* (IRRN), IRRI, 29.1/2004, pp 10-19.

Vogel, U, et al., 2005. Turning Policy into Practice: Sida's implementation of the Swedish HIV/AIDS strategy. Sida, EVALUATION 05/21, Stockholm.

vom Saal, F, Montano, M and Wang, M, 1992. Sexual differentiation in mammals. In Colborn, T and Clement, C (Eds.), *Chemically Induced Alterations in Sexual and Functional Development: The Wildlife-Human Connection*. Princeton Scientific Publishing, Princeton, pp 17-83.

von Wright, G H, 1993. *Myten om framsteget*. Tankar 1987-1992 med en intellektuell självbiografi (In Swedish). Albert Bonniers Förlag, Falun.

von Wright, G H, 1994. *Att förstå sin samtid*. Tanke och förkunnelse och andra försök. 1945-1994 (In Swedish). Bonniers, Smedjebacken.

Weatherford, J, 1995. *Das Erbe der Indianer*. Diederichs Verlag, München.

Weidong, G, Fang, J, Zheng, D, Li, Y, Lu, X, Rao, R V, Hodgin, T and Zongwen, Z, 2000. Utilization of germplasm conserved in Chinese national genebanks—a survey. *Plant Genetic Resources Newsletter*, 123: 1-8.

Weiss, R, 2002. War on Disease. *National Geographic*, 201(2): 2-31.

Whitehead, AN, 1926. *Science and the Modern World*. Cambridge University Press, Cambridge. Free Association Books edition 1985 edited by Robert Young, London.

WHO, 2005. Modern food bio-technology, human health and development. World Health Organization, Geneva.

Windels, P et al., 2001. Characterisation of the Round-up Ready soybean insert. *European Food Research Technology* 213: 107-112.

World Bank, 1988. *Rural Development: World Bank Experience, 1965-1986.* Operations Evaluation Department, the World Bank, Washington, D.C.

World Bank, 1998. *Commodity Markets and Developing Countries.* The World Bank. Washington, D.C.

World Bank, 1999. *Greening Industry: New Roles for Communities, Markets and Government.* Oxford University Press, New York.

World Bank, 2001. *Attacking Poverty.* World Development Report 2000/2001, The World Bank, Oxford University Press.

World Bank, 2002. *Building Institutions for Markets.* World Development Report, The World Bank, Oxford University Press.

World Bank, 2005a. *World Development Report 2006: Equity and Development.* The World Bank, Oxford University Press.

World Bank, 2005b. *Agriculture and Achieving the Millennium Development Goals.* Agriculture & Rural Development Department, Report No. 32729-GLB, The World Bank, Washington, D.C.

Worldwatch Institute, 1996. *State of the World 1996.* Worldwatch Institute, Norton, New York.

Worldwatch Institute, 2003. *State of the World 2003.* Special 20th Anniversary Edition, Washington, D.C.

WRI, 1996. World Resources 1996-97. A joint publication by The World Resources Institute, United Nations Environment Programme, United Nations Development Programme and World Bank, Washington, D.C.

WRI, 2003. World Resources 2002-2004: Decisions for the Earth: Balance, Voice, and Power. World Resources Institute, Washington, D.C.

Wuyts, M, Dolny, H and O'Laughlin, B, 2001. *Assumptions and Partnerships in the making of a Country Strategy.* An evaluation of the Swedish-Mozambican experience. Sida Evaluation Report 01/07, August, Stockholm.

Wynne B and Mayer S, 1993. How science fails the environment. *New Scientist,* 138: 33- 35.

Young, A, 2003. *Agroforestry Research, Then and Now.* The evolution of research by the World Agroforestry Centre. Tropical Agriculture Association (TAA), Newsletter, June, pp 3- 6.

Yudelman, M, Ratta, A and Nygaard, D, 1998. *Issues in Pest Management and Food Production. Looking to the Future.* Discussion Paper, International Food Policy Research Institute, Washington, D.C.

Index